Frontiers of Nanoscience

Volume 8

Characterization of Nanomaterials in Complex Environmental and Biological Media

FRONTIERS OF NANOSCIENCE

Series Editor: Richard E. Palmer
The Nanoscale Physics Research Laboratory,
The School of Physics and Astronomy,
The University of Birmingham, UK

Vol. 1 **Nanostructured Materials** edited by
Gerhard Wilde

Vol. 2 **Atomic and Molecular Manipulation** edited by
Andrew J. Mayne and Gérald Dujardin

Vol. 3 **Metal Nanoparticles and Nanoalloys** edited by
Roy L. Johnston and J.P. Wilcoxon

Vol. 4 **Nanobiotechnology** edited by
Jesus M. de la Fuente and V. Grazu

Vol. 5 **Nanomedicine** edited by
Huw Summers

Vol. 6 **Nanomagnetism: Fundamentals and Applications** edited by
Chris Binns

Vol. 7 **Nanoscience and the Environment** edited by
Jamie R. Lead and Eugenia Valsami-Jones

Vol. 8 **Characterization of Nanomaterials in Complex Environmental and Biological Media** edited by
Mohammed Baalousha and Jamie R. Lead

Frontiers of Nanoscience

Volume 8

Characterization of Nanomaterials in Complex Environmental and Biological Media

Edited by

Mohammed Baalousha and Jamie R. Lead
Center for Environmental Nanoscience and Risk,
Arnold School of Public Health
Department of Environmental Health Sciences
University of South Carolina, Columbia,
South Carolina, USA

ELSEVIER

AMSTERDAM • BOSTON • HEIDELBERG • LONDON • NEW YORK • OXFORD
PARIS • SAN DIEGO • SAN FRANCISCO • SINGAPORE • SYDNEY • TOKYO

Elsevier
Radarweg 29, PO Box 211, 1000 AE Amsterdam, Netherlands
The Boulevard, Langford Lane, Kidlington, Oxford OX5 1GB, UK
225 Wyman Street, Waltham, MA 02451, USA

ISBN: 978-0-08-099948-7
ISSN: 1876-2778

For information on all Elsevier publications
visit our website at http://store.elsevier.com/

Contents

Part III
Case Studies

Contributors

Mohammed Baalousha, Center for Environmental Nanoscience and Risk (CENR), Department of Environmental Health Sciences, Arnold School of Public Health, University of South Carolina, Columbia, SC, USA

Xiangyu Bi, School of Sustainable Engineering and the Built Environment, Arizona State University, Tempe, AZ, USA

Alistair Boxall, Environment Department, University of York, Heslington, York, UK

Andy Brown, Institute for Materials Research, School of Chemical and Process Engineering, University of Leeds, Leeds, West Yorkshire, UK

Rik Brydson, Institute for Materials Research, School of Chemical and Process Engineering, University of Leeds, Leeds, West Yorkshire, UK

Agnieszka Dudkiewicz, The Food and Environment Research Agency, Sand Hutton, York, UK

Anne A. Galyean, Bioprocess Measurements Group, Biomolecular Measurement Division, Material Measurement Laboratory, National Institute of Standards and Technology, Gaithersburg, MD, USA; Gillings School of Global Public Health, Department of Environmental Sciences and Engineering, University of North Carolina, Chapel Hill, NC, USA

Rhys Goodhead, Biosciences, University of Exeter, Exeter, UK

Justin M. Gorham, Nano Materials Research Group, Materials Measurement Science Division, Material Measurement Laboratory, National Institute of Standards and Technology, Gaithersburg, MD, USA

Andrew Herzing, Materials Structure and Data, Materials Measurement Science Division, Material Measurement Laboratory, National Institute of Standards and Technology, Gaithersburg, MD, USA

Michael F. Hochella, Jr., Department of Geosciences, Virginia Tech, Blacksburg, VA, USA; Institute for Critical Technology and Applied Science, Environmental Nanoscience and Technology Laboratory, Virginia Tech, Blacksburg, VA, USA

R. David Holbrook, Nano Materials Research Group, Materials Measurement Science Division, Material Measurement Laboratory, National Institute of Standards and Technology, Gaithersburg, MD, USA

Nicole Hondow, Institute for Materials Research, School of Chemical and Process Engineering, University of Leeds, Leeds, West Yorkshire, UK

Victoria Jennings, Biosciences, University of Exeter, Exeter, UK

Bojeong Kim, Department of Earth and Environmental Science, College of Science and Technology, Temple University, Philadelphia, PA, USA

Jamie R. Lead, Center for Environmental Nanoscience and Risk (CENR), Department of Environmental Health Sciences, Arnold School of Public Health, University of South Carolina, Columbia, SC, USA

John Lewis, The Food and Environment Research Agency, Sand Hutton, York, UK

Manuel D. Montano, Department of Chemistry and Geochemistry, Colorado School of Mines, Golden, CO, USA

Soubantika Palchoudhury, Center for Environmental Nanoscience and Risk (CENR), Department of Environmental Health Sciences, Arnold School of Public Health, University of South Carolina, Columbia, SC, USA

John Pettibone, Nano Materials Research Group, Materials Measurement Science Division, Material Measurement Laboratory, National Institute of Standards and Technology, Gaithersburg, MD, USA

James Ranville, Department of Chemistry and Geochemistry, Colorado School of Mines, Golden, CO, USA

Robert Reed, School of Sustainable Engineering and the Built Environment, Arizona State University, Tempe, AZ, USA

Jerome Rose, CNRS-Aix Marseille University, IRD, UM34, UMR 7330, Europole de l'arbois, Aix en Provence, France

Karen Tiede, The Food and Environment Research Agency, Sand Hutton, York, UK

Charles R. Tyler, Biosciences, University of Exeter, Exeter, UK

Paul Westerhoff, School of Sustainable Engineering and the Built Environment, Arizona State University, Tempe, AZ, USA

Preface

In spite of a decade of research in determining the fate, behavior and biological impact of nanomaterials, the precise role of their properties in these processes is still poorly understood. This is due to the multidisciplinary nature of the problem in hand requiring experts in material science, chemistry, physics, biology and toxicology to be brought together. In addition, the complexity of nanomaterial properties, behaviours and dynamic transformations in environmental and biological media makes the problem extremely challenging. The problem is further exacerbated by the lack of nanomaterial characterization methods for complex media, standard sample-preparation methods, technical expertise and understanding and interpretation of the measured parameters. As a result, there are significant discrepancies between measurements performed on different nanomaterial batches by different analytical techniques and among different research groups. Despite the significant progress in the quality of material characterization over the last few years, the majority of research papers describing environmental and toxicological behaviours of nanomaterials do not fully address nanomaterial characterization, and put little emphasis on linking nanomaterial properties to their environmental behaviours. As a result, there are huge discrepancies in the available literature regarding, for instance, the role of size, particles and ions, surface coatings, etc. Therefore, this book was written in part as a survey of the state of the art in nanomaterial characterization for experts as well as for those investigating the fate and effects of nanomaterials. In part, it is also an effort to bridge this knowledge gap and to better understand nanomaterial characterization and the role of nanomaterial properties in controlling their environmental and toxicological behaviour.

The book was organized to accommodate the vision outlined above by summarizing the current state of the art in nanomaterial characterization, with a focus on modern and novel application of techniques that have not been previously examined in detail or techniques that have seen vast methodological improvements in recent years. The book is divided into three parts; **the first part** (Chapters 1 and 2) reviews the current state of the art of nanomaterial toxicity to aquatic organisms and nanomaterial characterization techniques; **the second part** (Chapters 3–7) carefully and critically describes characterization of specific properties of nanomaterials relevant to their environmental behaviours applying a multi-method approach, with specific attention to sample preparation and comparability of measurements performed by different analytical techniques and **the third part** (Chapters 8 and 9) presents two case studies of nanomaterial characterization in consumer products and food stuffs.

More specifically, **Chapter 1** gives a critical review on current nanomaterial toxicity data in aquatic organisms, nanomaterial dose-metric and nanomaterial features that elicit toxicity, the importance of nanomaterial characterization in underpinning nanomaterial health and safety research. **Chapter 2** provides the reader with an overview of the basic concepts of the plethora analytical techniques that can be applied to measure nanomaterial properties of relevance to their environmental behaviour, fate and effects.

Chapter 3 gives an account of nanomaterial size characterization based on microscopy techniques, diffusion coefficient and other advanced approaches such as single particle inductively coupled plasma–mass spectrometry. This chapter presents the general consideration for size analysis, the descriptors of size distributions, the comparison and interconversion between distributions and ends with a discussion of the general considerations for nanomaterial size characterization in environmental systems. **Chapter 4** discusses how to identify and characterize chemical composition and crystal structure (and therefore the exact phase or phases) of nanomaterials (parent forms) as well as their transformed ones (daughter forms) by using analytical transmission electron microscopy and scanning transmission electron microscopy. This chapter gives special attention to data acquisition and analysis methods, as well as sample-preparation procedures in great detail. **Chapter 5** examines the characterization of the three most commonly measured particle concentration metrics, namely mass, surface area and number. The chapter discusses the methods used to measure these metrics, and sample preparation for selected methods, including microscopy and inductively coupled plasma–mass spectrometry. Available analytical techniques for mass, surface area and number concentration measurements are systematically evaluated in terms of sensitivity/detection range for nanomaterial mass, representing nanomaterial surfaces in dispersion, distinguishing nanomaterials in complex medium and minimizing sampling artefacts in number distributions. **Chapter 6** highlights the complex interrelationship between: the form of a nanoparticulate material dispersed in a particular medium; the resultant dissolution or chemical change of the nanomaterials in that delivery medium and the nature and degree of uptake of the nanomaterials by a particular cell or organism. Thus, the chapter reviews the various techniques which are appropriate to determine nanomaterial dispersion, dissolution and dose. **Chapter 7** discusses the importance of nanomaterial surface properties and surface reactions such as surface area, surface atomic arrangements, and adsorption, photocatalytic and redox reactions. This is followed by describing a multi-method approach for the characterization of nanomaterial surface properties and reactions, with special attention given to sample preparation to preserve surface properties. The chapter intentionally distinguishes between surface-specific and non-surface-specific techniques.

Chapter 8 presents a case study on the characterization of common chemical mechanical polishing/planarization nanomaterials such as silica, cerium and aluminium oxides. The chapter investigates nanomaterial removal efficiency using a common industrial on-site treatment strategy (chemical softening and precipitation) and off-site treatment at biological wastewater treatment plants. **Chapter 9** focuses on the characterization of nanomaterials in foodstuffs and food contact materials, and covers approaches for sample preparation prior to analysis; the analytical techniques available for nanomaterial detection and characterization in food matrices and the appropriateness of various QA and QC methodologies. The chapter ends by discussing the specific challenges associated with nanomaterial detection, with analysis in food, beverages and nutraceuticals also discussed.

As the reader will appreciate, due to the breadth of the topic and the skills needed to put this book together, this book would not have been realized without the contributions of the authors of each chapter. Accordingly, we would like to express our gratitude to all authors who have contributed to this work and recognize the efforts that made this publication possible. We would like also to thank various funding bodies supporting the editors in their general research that led the editors to develop this book, in particular, the U.S. National Science Foundation, the Center for Environmental Nanoscience and Risk and the Arnold School of Public Health at the University of South Carolina. Lastly, we would like to thank the publishing team at Elsevier whose help was vital for this book.

Mohammed Baalousha and
Jamie R. Lead

Part I

Nanomaterials and Characterization

Chapter 1

Ecotoxicology of Nanomaterials in Aquatic Systems

Victoria Jennings, Rhys Goodhead and Charles R. Tyler[1]
Biosciences, University of Exeter, Exeter, UK
[1]*Corresponding author: E-mail: C.R.Tyler@exeter.ac.uk*

Chapter Outline

Frontiers of Nanoscience, Vol. 8. http://dx.doi.org/10.1016/B978-0-08-099948-7.00001-4

1. INTRODUCTION

In recent years the development and production of manufactured nanomaterials (NMs), particles with dimensions between 1 and 100 nm, has increased rapidly to supply the expanding nanotechnology industry. Global markets for NMs are expected to exceed $1 trillion by 2015.[1] The small size of NMs results in different and unique properties compared with their larger counterparts and their applications to date are extremely diverse. Indeed, there are currently in excess of 1600 commercially available products containing NMs[2] (Figure 1). Increasing use of NMs inevitably results in increased discharges into the environment, particularly into aquatic environments. The data on environmental levels of NMs, however, are currently scarce, and there has been a strong reliance on modelling for informing on possible exposure scenarios in humans and wildlife. Modelling approaches may not necessarily reflect real-world exposures of ecotoxicological significance, however, because NMs can change fundamentally with the physicochemical conditions of the immediate environment and this can alter their potential for toxicological effects.

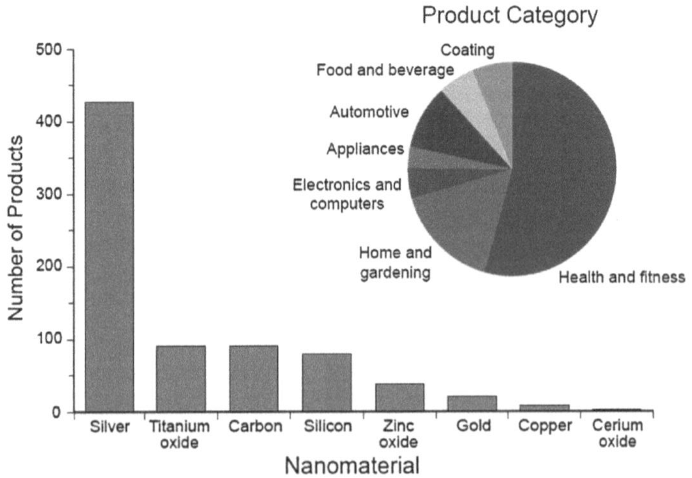

FIGURE 1 Number of products containing commercial NMs (bar chart) and NMs in products as defined by product categories (pie chart) (As listed on the Consumer Products Inventory, The Project on Emerging Nanotechnologies[2]).

Considerable efforts have recently been directed towards research into the toxicological properties of a wide range of NMs to help better understand their potential risks to human health. For the most part these studies have not considered real-world exposure scenarios, but they are nevertheless starting to advance understanding on the toxicology of some NMs. Less attention has been directed at the potential impacts of NMs in wildlife. The aquatic environment acts as a sink for most contaminants discharged, and, as NMs become more prevalent in aquatic systems, it will become increasingly important to assess their potential risks to aquatic wildlife. Current toxicity data for bulk materials cannot be extrapolated in a straightforward manner to nanosized counterparts because both the physicochemical properties associated with nanoscale differ. Furthermore, NMs are often modified using coatings and functional groups that will affect their toxicological properties. Seas, estuaries, streams, rivers, lakes and ponds have very different physicochemical characteristics that will impact the fate and behaviour, and thus the bioavailability and toxicity of NMs (Figure 2).

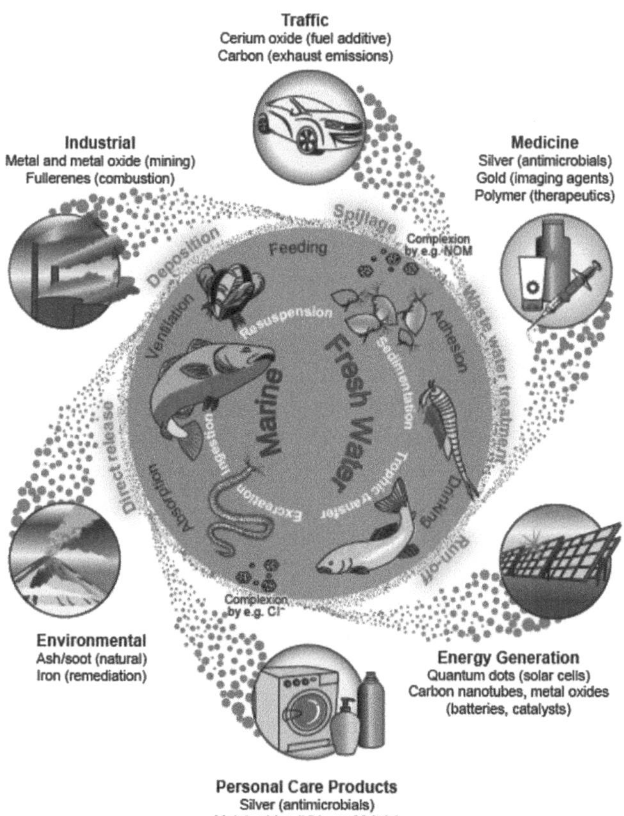

FIGURE 2 Sources of NM to, and their transformations in, aquatic environments.

There are many sources of NMs into aquatic environments (see Section 3) and NMs are likely to be taken up into a range of aquatic organisms across diverse phyla. Bioavailability, and thus potential for toxicity, however, will differ depending on morphologies, habitats and behaviours of these organisms and the physicochemical conditions of their habitat. Developing comprehensive understanding on the risks of NMs and providing appropriate environmental protection for NMs will require understanding of the exposure scenarios for specific wildlife organisms in the different aquatic environments. Understanding on the ecotoxicology of NMs also requires thorough characterisation of the NMs in the exposure medium.

In this chapter, after a brief overview of different NMs, we consider the fate of NMs in aquatic systems and detail the current knowledge on their ecotoxicity for a range of organisms in both freshwater and marine environments. We also identify common effects across these aquatic organisms. To better understand the nature and form of NMs in their natural waters, which will affect their toxicity, thorough characterisation of NMs is essential. We introduce NM characterisation methods to assess physical, chemical and behavioural features, which is covered in subsequent chapters in more detail. We also provide a viewpoint on the challenges we face in linking laboratory-based toxicology for NMs with real-world exposures and effects. Finally, in this chapter we highlight some key emerging issues for NM ecotoxicology research.

2. TYPES, USES AND PROPERTIES OF NMs

The physical and chemical characteristics of materials with dimensions in the nanoscale differ to their bulk counterparts. It is these differences that are exploited in a wide range of applications spanning ultraviolet (UV) protection in sunscreens, fuel additives to enhance combustion efficiency, as antimicrobial agents, and as electrical conductors, to name just a few. These differing properties, however, may also have implications when considering the fate and toxicity of these materials. Smaller particulate size gives NMs a higher surface area to volume ratio which can lead to higher reactivity and for some metal-based materials enhanced rates of dissolution of ions from the NM surface. NM particle size and shape can dictate the route of uptake into cells and reactivity. Furthermore, the small size of NMs means that some can cross biological barriers through routes not normally accessible for their larger counterparts. Some NMs are produced in diverse forms, for example, silver NMs are produced as spherical, rod-shaped and flower-structured materials. Carbon NMs can be in forms of carbon nanotubes (CNT), sheets or spheres, each with the potential for different interactions and behaviour, and thus different toxicities.

There are many forms of NMs entering aquatic environments, including carbon-based, metal/metal oxide, quantum-dots (Q-dots), dendrimers and nanocomposites. The most commonly used NMs currently are metal, metal oxide and carbon-based NMs.[2] Carbon NMs are used for a wide range of applications including gene therapy, in fuel cells utilising their conductive and catalytic properties and in the manufacture of products such as sports racquets and car bodies because of their high tensile strength and light weight.

Metal/metal oxide NMs are similarly used for a wide range of applications, spanning antibacterial agents in medical dressings, brightening agents in paints, fuel catalysts, imaging contrast agents and for groundwater remediation. These materials can be produced with relative ease by reduction or hydrolysis reactions. Nanosized silver (Ag-NMs) is the most commonly used metal NM, predominantly for its antibacterial property. This property is generally associated with ionic silver as the silver NMs (Ag-NMs) undergo dissolution, although toxicity associated with the particle itself is also reported. Antibacterial products that incorporate Ag-NMs include wound dressings, sportswear fabrics, washing machines and air filters. Coating or capping agents are often used to stabilise NMs, and given the nature of some of these capping agents, these can be an additional source of pollutants to aquatic environments. Gold (Au) NMs have size-dependent optical properties, where free electrons follow quantum mechanical rules. Gold NMs are used for a large number of industrial and medical applications, including as nontoxic drug carriers and in both imaging and biodevices for diagnostics. Copper (Cu) NMs too have been used for their antibacterial properties in bioactive antifouling coatings and also in textiles. The optical and semi-conductor properties of copper (I) oxide NMs are being investigated for use in solar energy cells and for catalysis applications (as both catalyst and catalyst support material). Commercially, Cu-NMs have also been used in lubricants as an antifriction agent, in cosmetic products and in batteries.

The photocatalytic properties of metal oxide NMs, such as titanium dioxide (TiO_2), zinc oxide (ZnO) and cerium dioxide (CeO_2) are exploited in solar cells and in water and air purification systems. They also have been exploited for use in personal care products, including sunscreens and cosmetics to provide UV protection in transparent formulations. TiO_2-NMs are also used in paints to act as a whitening pigment. CeO_2-NMs are commonly used as catalysts and in fuels in the automotive industry to increase fuel efficiency and reduce exhaust emissions. This use is likely to contribute to increases in environmental levels of CeO_2-NMs in both the atmosphere and aquatic systems (e.g. via road runoff). Aluminium NMs (usually Al_2O_3) have a wide range of uses including as catalytic agents to improve ignition probability in fuel and to enhance the properties of polymer composites used in electronic circuitry.

Nano-zero-valent iron (nZVI) is used in remediation of waters, sediments and soils to remove organic pollutants. Concentrations of nZVI in slurry used in remediation applications in Europe are in the range of 1 and 30 $g\ L^{-1}$ and a mass in excess of 1 ton can be applied at any one site.[3] This makes this the material of immediate environmental relevance, including for aquatic environments. The magnetic properties of nanosized magnetite (Fe_3O_4) have been utilised for labelling, magnetic separation of biological materials and MRI contrast agents.

Other types of NMs entering aquatic environments include Q-dots, dendrimers and nanocomposites. Q-dots are colloidal semiconductor crystals made from group II and VI elements or group III and V elements. Designed for their optical properties, Q-dots are used in biological imaging, drug delivery, and anticounterfeit pigments. Dendrimers are nano-sized polymers with a branched structure and they can be used to create unique 3D structures for use in biological applications, including drug delivery, sensors and structural processes such as tissue repair. Nanocomposites are materials that contain NMs or nanoscale modifications to enhance specific properties of the material. Currently products classed as related to 'health and fitness' are the most common category containing NMs. Silver NMs are used very widely in sportswear products to provide antibacterial properties.

3. SOURCES, ENTRY AND FATE OF NMs IN AQUATIC ENVIRONMENTS

Materials can occur naturally in the environment at the nanometre-size range and have existed for billions of years. Natural occurrences that derive from sources such as volcanic eruptions and forest fires can generate airborne NMs and nanosized components of natural organic macromolecule (NOM) in water and soil colloids. This has raised debate around the necessity for ecotoxicological testing for nanosized materials because organisms will have inevitably evolved over time to cope with these naturally occurring NMs. In stark contrast with natural NMs, however, engineered NMs have been designed to have specific properties to enable, for example, persistence and in turn are likely to pose different challenges to exposed organisms. Furthermore, for some materials, it is likely that the concentrations of NMs will exceed those found for these naturally occurring materials (e.g. silver).

Aquatic environments act as sinks for anthropogenic discharges and receive NMs from divergent sources, including via wastewater treatment works (WWTWs) and surface runoff into rivers. This will be notable for NMs, such as CeO_2 from fuels and TiO_2 used in exterior paints, for ZnO and TiO_2 entering directly into marine waters from the use of sunscreens, and also accidental releases from industrial plants producing these materials. Aquatic environments also receive NMs indirectly, for example, from soils where NMs are applied through treatments with sewage sludge, via leachate from landfill and/or via precipitation of materials released into the air.

Importantly, no safety guidelines exist currently for NM release into aquatic environments. For metals, the UK and US governments have guidelines for water quality (UK Environmental Quality Standards and US EPA

Aquatic Life Criteria)[4,5] based on metal ions using the biotic ligand model (BLM). Recently the BLM has been applied to metal NMs but this approach is limited due to the high variability in the rate of dissolution for different sized NMs, which is also affected by the nature of the aqueous environment and capping agents present on the NMs. Furthermore, the BLM considers only the dissociating ions from the NMs and does not take into account possible NM-specific biological effects.[6] When considering the ecotoxicity of NMs, we would emphasise that there is not sufficient information available currently to link any adverse effects to individuals or populations at concentrations likely to be found in *most* aquatic environments. Furthermore, information on adverse effects in individuals for any exposures that approach those with (predicted) environmental relevance are still very limited.

3.1 NM Sources and Entry into Aquatic Environments

WWTW discharges are expected to be one of the major sources of NMs into aquatic systems. WWTWs receive significant amounts of NMs from both domestic and industrial sources, and although some are expected to precipitate into the sludge (but may find their way back into aquatic systems via sludge applied to land as fertiliser), the remaining NMs in effluents will enter directly into both freshwater and marine environments. Data on measured levels of NMs in WWTWs influent and effluent are limited and releases of NMs predicted by modelling are highly variable depending on particle type and processes within the specific WWTWs. Measured releases of NMs in WWTW have been reported for C60 and C70 carbon NMs and some metal-based materials. For C60 and C70, carbon NM levels can reach the parts per billion (ppb) range.[7] In a study using a model WWTW, 6% (by weight) of the CeO_2-NMs supplied to the WWTWs were subsequently released in the effluent discharge and addition of associated stabilizing agents increased the amount of CeO_2 passing through the WWTWs into the effluent stream.[8] Predicted effluent concentrations for TiO_2-NMs in WWTW effluents have been reported at between 0.7 and 16 $\mu g\ L^{-1}$. The predicted no effect concentration (PNEC) for TiO_2-NMs is $< 1\ \mu g\ L^{-1}$.[9] One study has reported concentrations of titanium containing NMs ($<0.7\ \mu m$) in WWTW effluents at concentrations in the range $<5-15\ \mu g\ L^{-1}$, exceeding the PNEC value.[10]

The fate and behaviour of silver NMs in WWTWs will depend on many variables including influent concentrations, NM coating, and NM transformation within the WWTW (e.g. sulphidation). Only a relatively small percentage of Ag-NMs pass through WWTWs into the effluent discharged as Ag-NMs, and most are transformed via sulphidation into Ag_2S.[11] The fate of Ag-NMs however is also size-dependent.[12] Sulphidation of Ag-NMs reduces their solubility and toxicity, but the environmental impact of other transformations of Ag-NMs, such as oxidised silver sulphide, has not been established. A modelling approach has predicted effluent concentrations of uncoated Ag-NMs at less than 0.24 μg L^{-1} but higher concentrations for coated Ag-NMs.[13] In a study in Germany, Ag-NMs concentrations in influents were measured at <1.5 μg L^{-1} and at <10 ng L^{-1} in effluents.[14] Predicted environmental concentrations (PECs) for Ag-NMs indicate that at current levels of use they are unlikely to be an environmental problem for most aquatic environments, but their future use (and thus discharge into these environments) is set to increase significantly.[11,13,15]

Other direct sources of NMs into aquatic systems include via industry and hospital discharges, runoff from roads and applications such as agriculture (nanopesticides), exterior paintwork and sunscreens, and via use in soils remediation. There are little data available for the industrial production and release of NMs, and as a consequence studies modelling production levels of NMs have relied on extrapolated data.[16] Hospitals use NMs extensively for health care, for example in imaging (gold and Q-dot NMs), antimicrobial wound dressings (silver and copper NMs) and for drug delivery and gene therapy (polymer/liposome-based NMs). Modelling approaches used to assess gold NM PECs, however, indicate at current usages; even at hot spot discharges from hospitals, they are unlikely to be an environmental problem.[11] In one study the highest predicted water concentration of CeO_2-NMs from road runoff into rivers was 0.02 ng L^{-1}. It is not clear however how representative this is of runoff in waters in close proximity to roads, and after rain storm events.[17] Measured concentrations of TiO_2-NMs in runoff derived from newly painted (experimental) facades, have been measured up to 600 μg L^{-1}.[18] Studies to assess NM release from NM-containing products include the release of up to 650 μg L^{-1} of silver from socks (washed in 500 mL distilled water)[15] and an average concentration of 11 μg L^{-1} of silver in washing machine effluent.[19] For ZnO and TiO_2 in sunscreens up to 25% of sunscreen is washed off on immersion[20] and an estimated 250 tonnes of ZnO and TiO_2-NMs are released into the marine environment from sunscreen use.[21] Analysis of a lake's surface waters over a 12-month period found that residence times for TiO_2-NMs from sunscreens were short, with an increase in TiO_2 concentration only seen during the bathing season.[22] Measured and predicted concentrations of selected NMs in selected effluent discharges and surface and marine waters are shown in Table 1.

Soils can act as a repository for NMs, prior to their entry into aquatic environments, via sources including landfill leachate, atmospheric deposition, direct application of WWTW products to land as fertilisers or direct application of specific NMs for chemical remediation. The nature and fate of NMs in soil will depend on the physicochemical nature of that soil, and this will determine the amount of transfer of NMs into aquatic environments. Soils contain naturally occurring NMs in colloids, including a mixture of NOM, inorganic matter, bacteria and viruses. Interaction between NMs and these colloids affect rates of aggregation, transport, dissolution or deposition and in turn NM release into aquatic systems. In some instances, and as mentioned

TABLE 1 Predicted Environmental Concentrations (PECs) and Modelled and measured Releases of NMs for Selected Sources

Nanomaterial	Concentration	Environment	References
Predicted Environmental Concentrations (PECs)			
Ag	0.088–10,000 ng L^{-1}	Surface water	9,23,24
	16.4–17,000 ng L^{-1}	WWTP effluent	23,24
	1.29–39 mg kg^{-1}	WWTP sludge	23,24
TiO_2	21–10,000 ng L^{-1}	Surface water	9,24–27
	1000–100,000 ng L^{-1}	WWTP effluent	24,28,29
	100–2000 mg kg^{-1}	WWTP sludge	24,27,29
ZnO	1–10, 000 ng L^{-1}	Surface water	24
	220–1420 ng L^{-1}	WWTP effluent	24
	13.6–64.7 mg kg^{-1}	WWTP sludge	24
Carbon-based	0.001–0.8 ng L^{-1}	Surface water	9,24
	3.69–32.66 ng L^{-1}	WWTP effluent	24
	0.0093–0.147 mg kg^{-1}	WWTP effluent	24
Measured Releases of NMs for Selected Single Sources			
All NMs	69 200 metric tons per year	Modelled—global release	30
Ag	650 μg L^{-1}	Measured—release in sock washing	15
	11 μg L^{-1}	Measured—washing machine effluent	19

Continued

TABLE 1 Predicted Environmental Concentrations (PECs) and Modelled and measured Releases of NMs for Selected Sources—cont'd

Nanomaterial	Concentration	Environment	References
TiO_2	600 μg L^{-1}	Measured—runoff from painted facades	18
	Up to 25%	Measured—sunscreen via immersion	20
	250 tonnes	Modelled—sunscreen via immersion	21
ZnO	Up to 25%	Measured—sunscreen via immersion	20
	250 tonnes	Modelled—sunscreen via immersion	21
CeO_2	0.02 ng L^{-1}	Modelled—road runoff	17
	6% By weight	Modelled—WWTW simulation effluent	8
Zero-valent iron	Hundreds of kilograms per square kilometres of contaminated land	Modelled—groundwater remediation applications	3,31

PEC values from Table 1.4 in Ref. 183.

above, contaminated soils are treated directly with NMs, for example, the use of zero-valent iron to clean up heavy metals, pesticides, radionuclides and polychlorinated hydrocarbons[32] and this is now a commercial enterprise using hundreds of kilograms of zero-valent iron NMs per km^2 of contaminated land.[3,31,33] Little is known, however, about the long term effects of these treatments.[34]

Airborne NMs can enter aquatic environments via direct deposition and precipitation and they have been detected in cities in every inhabited continent.[35–40] Major anthropogenic inputs of NMs into the air include cooking fires and vehicle exhaust fumes.[41–43] Fine and ultrafine particles from combustion (and associated chemicals) have long been known to affect respiratory function and induce a range of adverse health effects in humans. Considerable research has been undertaken studying the transport of these airborne particles, their adsorption, subsequent potential to transport other pollutants, and associated health effects.[44] It is wise, therefore, for ecotoxicologists to familiarise themselves with these data, as much can be learned of relevance to the application for understanding NM (potential) effects in other organisms and in other environments (e.g. aquatic environments).

3.2 Environmental Levels of NMs and Their Fate in Aquatic Environments

The ecotoxicology of NMs is hampered by the fact that there are very few data currently on actual environmental levels of NMs or on their characterisation, and PEC levels vary considerably between different reported studies. NMs are likely to be present at ppb or ppt (parts per billion/trillion) levels in the environment and these low concentrations present a set of practical challenge for scientists. The nanosized nature of particles increases the difficulty in sample collection and preparation, and instrument limits of detection, (e.g. ICP-MS, inductively coupled plasma–mass spectrometry) often exceed the quantities of NMs in a sample. Many studies therefore focus on modelling NM release scenarios to make estimations of environmental NM concentrations. The estimates often differ between studies due to the differing parameters used in each model, but through them some basic understanding of likely exposure scenarios is being developed. To generate more realistic PECs comprehensive data on production volume, sources of NM release and fate and behaviour data are required.

Modelled environmental releases and concentrations derived from analyses of a top-down approach have often focused only on NM release from small numbers of NM-containing products.[23,42,45,46] Other models have attempted to consider release into all compartments from all points of discharge. Lack of empirical data for input into these models for marine environments has meant they have relied on estimates of freshwater NM levels (which themselves are extremely limited). Existing models for toxicity used to determine metal toxicity, such as the free ion activity model and biotic ligand model (BLM) may be useful as dissolution is a feature of metal and metal oxide NMs.[47] However, the behaviour of NMs will differ with the physicochemistry of the aquatic environment. Models have predicted the aquatic PECs of silver and titanium dioxide are close to, or above, the PNECs ($<1\ \mu g\ L^{-1}$ for TiO_2, $0.04\ mg\ L^{-1}$ for Ag) for some point source discharges.[9,24] A model for NMs in Swiss rivers, using material flow analysis, found PECs were highly variable between locations (for TiO_2 $11-1623\ ng\ L^{-1}$, Ag $<5\ ng\ L^{-1}$ and ZnO $<100\ ng\ L^{-1}$).[48] A critical analysis of the modelling and field studies conducted assessing the environment levels of NMs has recently been published by Gottschalk and co-authors.[49]

Global release of NMs into water has been estimated at 69,200 metric tons per year.[30] The combinations of NM core material with different surface modifications, coatings, differences in particle size and shape, and the highly dynamic nature of NMs in aquatic systems create a considerable challenge for ecotoxicologists trying to get to grips with the potential risks of NMs in freshwater and marine environments. This is complicated further by the fact that many environmental factors, including water flow rates, sediment composition, vegetation, NOM and wildlife will affect the fate of NMs in aquatic environments and in turn their bioavailability to different organisms (see Figure 2). NMs can interact with other pollutants to affect their toxicities, and trophic transfer of NMs can occur.[50–54] All of these factors add complexity to studies into the toxicological properties of NMs in natural systems and emphasise further the need for comprehensive particle characterisation studies.

Perhaps one of the greatest challenges in the ecotoxicology of NMs is the fact that they are highly dynamic in nature and are constantly undergoing change. In aquatic systems, most NMs are subject to aggregation, dictated by the immediate physicochemical properties of their immediate environment (e.g. free ions in the water column and NOM). Aggregates, 'clusters' or 'clumps' of particles, can form readily and this in turn will confer differing properties on these materials compared with those for individual NMs. Levels of aggregation of NMs depend on features of the NM, such as shape, size and charge, and the media including pH, ionic strength, counter-ion valency, NOM and presence of other NMs or pollutants. Aggregation of NMs will increase sedimentation and thus increase exposure to benthic dwelling organisms, and reduce exposure to organisms living in the water column. Under some environmental conditions, aggregation will occur to the extent that the aggregates are no longer in the nanosize range. Rates of dissolution and fate of the free ions deriving from metal-based NMs will differ markedly between fresh and marine waters. For example, in marine environments Ag^+ derived from Ag-NMs more readily forms AgCl complexes, reducing the levels of dissolved Ag^+ compared within freshwaters.[55] Features such as NM charge and surface chemistry manipulated in the manufacturing processes to improve function e.g. to enhance stability or improve biological compatibility, will inevitably affect the fate, behaviour and toxicity of NMs in aqueous systems, as described in the next section.

4. DOSE-METRICS AND FEATURES OF NMs THAT ELICIT TOXICITY

4.1 Dose-Metrics

When considering the (eco)toxicity of NMs it is necessary to establish a standardised dose-metric. 'Dose-metrics' are measurable properties that correspond to a material's ability to exert an effect, i.e. toxicity. In the case of NMs, applying dose-metrics requires careful consideration. The high surface area to volume ratio increases reactivity compared to bulk counterparts, therefore using material mass per unit volume may not be the most appropriate dose-metric. This was shown in a review of inhalation studies of ultrafine particles (NMs in the range 14–250 nm) where similar adverse effects were seen across a large concentration range of a few $mg\ m^{-3}$ up to $250\ mg\ m^{-3}$.[56] The differing toxicity of the ultrafine particles was not dependent solely on mass per unit volume and particle surface area was better correlated with inflammatory responses and volumetric lung burden for alveolar macrophage clearance of particles. For studies on aquatic organisms there is evidence for a correlation between particle concentration number and effect[57]; reviewed by Wittmaack[58] however particle size rather than number has been linked more with toxicity.[59] Detailed NM characterisation is essential for establishing the appropriate dose-metric data in any given test media in ecotoxicity testing.

4.2 Features of NMs That Elicit Toxicity

Key characteristics of NMs that have been shown to elicit toxicity responses in aquatic species include particle chemistry, size, surface area, morphology (including defects, as observed in CNTs), coatings or functionalisations, material (phase) and solubility (see Figure 3). In some exposure scenarios more than one NM feature may contribute to the toxicity observed. For example solubility, size, shape and composition of seven metal oxide NMs have all been linked with toxicity in exposed rodent fibroblast cell lines.[60]

Furthermore, as stated above, the interactions of the NMs within the exposure environment, especially in aquatic environments, will also play a key role in toxicity. This can include conditions of the immediate environment that affect the physical and chemical characteristics of the NMs such as interaction with polysaccharides, NOM or other toxins present in the environment. Features of selected Ag-NMs reported to be associated with toxicological effects are shown in Table 2.

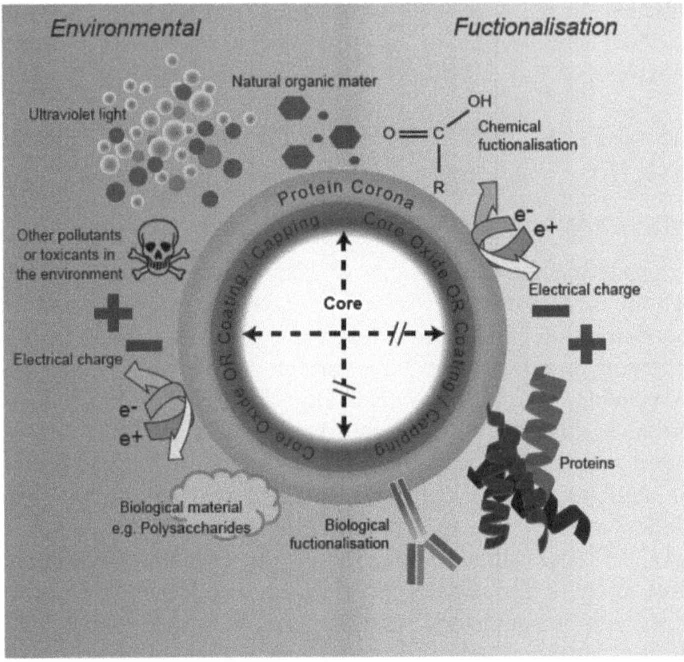

FIGURE 3 Visual representation of features of NMs that can affect their behaviour, transport, interactions and toxicity. This diagram shows a number of modifying factors that are both by design (functionalised) and passively acquired (environmental). These factors can occur independently or act with others, and they include the potential for the core material to be oxidised, coating by a protein corona or capping with a synthetic material, interaction with NOM, changes in surface charge state, transformation after exposure to UV light, other pollutants in the environment, chemical functionalisation and the addition of proteins or antibodies.

4.3 Size, Surface Area and Shape

For a number of different NMs, as size decreases there is an increase in their toxicity (diameter size determined by dynamic light scattering (DLS), transmission electron microscopy (TEM) or nanomaterial tracking analysis (NTA)[70–74]). A decrease in size, compared to bulk materials, increases the percentage of atoms on the surface and can alter the distribution across atomic planes, edges and corners, and thereby these changes drive the reactivity of the NM. For some metal/metal oxide NMs, greater surface area can lead to increased dissolution of ions toxic to organisms.[75] More generally, a smaller size can result in increased reactivity, for example by the generation of reactive oxygen species (ROS).[76] The physical size of NMs will also affect their uptake and distribution in organisms, with smaller NMs more readily able to penetrate into cells to exert intracellular toxicity.[77] Particle size can also determine with which tissues they associate. In fish, for example, smaller sized silver NMs were found to be more readily associated with gill tissues compared with their

TABLE 2 NM Properties Linked with Biological Effects and Specific Examples Given for Silver (Ag) NMs

NM Property	Effect	Comment	Reference
Size	Increased cellular toxicity for exposure to smaller Ag-NMs	Macrophages exposed to 15, 30, 55 nm Ag-NMs	62
	Size-dependent uptake into fish	*Onchorhynchus mykiss* (rainbow trout) exposed to 10–35 nm Ag-NMs	63
Shape	Shape-dependent interactions with gram negative bacteria	*Escherichia coli* exposed to spherical (39 nm) rod (16 × >100 nm) and triangular nano-plate shaped (40 nm edge length) Ag-NMs	64
Charge	Surface charge-dependent toxicity of Ag-NM in bacteria. Negatively charged Ag-Cit*—NMs least toxic, positively charged Ag-BPEI*—NMs most toxic	Gram positive *bacillus* species exposed to (1) uncoated Ag (18 nm), (2) Cit-Ag (10 nm), (3) PVP-Ag (12 nm) and (4) BPEI-Ag (10 nm)—NMs	65
Solubility	Toxicity of Ag-NM attributed to both particle and Ag^+ ions in fish	Perch (*Perca fluviatilis*) and carp (*Carassius carassius*) exposed to 80 nm Ag-NMs	66
	Toxicity Ag^+ > Ag-Cit > Ag-PVP Toxicity predominantly due to Ag^+	Algae *Pseudokirchneriella subcapitata* and *Phaeodactylum tricornutum* and water flea *Ceriodaphnia dubia* exposed to Ag-Cit (14 nm) Ag-PVP (15 nm) -NMs	67
Coating/capping	Toxicity Ag-Cit > Ag-PVP	*P. subcapitata, P. tricornutum and C. dubia* exposed to Ag-Cit (14 nm) Ag-PVP (15 nm)—NMs	67
	Toxicity Ag-GA* > Ag-Cit and Ag-PVP	Medaka *(Oryzias latipes)* fish exposed to Ag-GA (6 nm) Ag-Cit (7 nm), and Ag-PVP (10, 50 nm)—NMs	68
NM preparation/ solvent	Enhanced antibacterial activity for Ag-NMs modified by SDS*, Tween 80* and PVP	Surfactants (SDS, Tween 80) and polymer (PVP), used to stabilise Ag-NM (26 nm) dispersion in exposures to a range of bacteria species	69

Cit, Citrate; BPEI, branched polyethyleneimine; PVP, polyvinylpyrrolidone; GA, Gum Arabic; SDS, sodium dodecyl sulphate; Tween 80, polyoxyethylenesorbitane monooleate.

Table adapted from Table 4 in Ref. 61.

larger counterparts (<600 nm).[63] It should be emphasised, however, that although size has been a driving interest in the ecotoxicity of NMs, there are many published studies that have failed to identify a relationship between particle size, uptake or any associated toxicity. Shape of NMs has also been linked with differences in the toxic effects observed. This could result from either differences in NM shape-dependent characteristics (surface area, exposed surface facets and surface atom oxidation state) and/or differences in the uptake and interaction with cells.[78] An example of this is in zebrafish (*Danio rerio*) where dendritic nickel nanoclusters (60 nm) have been shown to be more toxic than nickel NMs (30, 60 and 100 nm).[79] Needle-shaped CNTs have been shown to induce physical damage of cells.[80]

4.4 Core Chemistry

Core material, reactivity, oxidation state and solubility will all influence the toxicity of the NM to aquatic species. Reactivity, oxidation state and solubility will all also be affected by surface modifications, chemistry and coatings. Differences in the molecular structure of carbon NMs between particle types has been shown to result in different effects for chronic exposures.[81] The differences between crystal structure in rutile and anatase TiO_2-NMs have been shown to be the cause of differing toxicity in cells,[82] and responsible for generating different photocatalytic activity and toxicity in zebrafish.[83] For metal/metal oxides, such as silver and zinc oxide where dissolution occurs, the toxicity observed may simply be an ionic effect (reviewed in Ref. 6). In studies on zebrafish exposed to Ag-NMs, greater uptake of silver ions has been found compared with the bulk silver counterpart and these ions were the major source of material toxicity in those studies.[84,85] There is however, some evidence to suggest that there may be particle related effects for the toxicity of silver NMs as derived from a study on *Chlamydomonas reinhardtii*. Here even when cysteine was used to bind the free Ag ions there was still an enhanced toxicity for Ag-NMs compared with toxicity as a function of silver ion concentration.[86] Furthermore, in a study on zebrafish embryos, gene signatures of toxicity were apparent in the transcriptome for exposure to Ag-NMs that were not accounted for by the signatures for Ag ions alone.[87]

4.5 Functional Groups/Coatings and NM Surface Chemistry

NMs can be modified to improve function, aid dispersal in media, confer specific properties and enhance compatibility with biological systems (e.g. drug delivery). The toxic potency of some metal/metal oxide NMs has been linked with modifications and coatings, including some stabilising agents.[88–90] Silver NMs are often capped with coatings such as citrate, which stabilises the particles by charge, or PEG and PVP, which are steric stabilisers to aid dispersion and reduce dissolution. This capping can markedly affect material stability in aquatic media[91] and affect particle toxicity, as has been shown in exposures to zebrafish embryos.[92] In studies on Q-dots, surface coating has been shown to have a major bearing on their uptake and biological effects in macrophages.[93] For studies on exposures to CNTs, generation of ROS was associated with cell death in exposures to pristine CNTs[94] whereas increased programmed cell death (apoptosis) was shown to occur more for functionalised CNTs.[95]

4.6 Surface Charge

For some NMs, particle charge has been implicated in toxicity.[96,97] Particle charge will relate to the surface chemistry, considered above, and be affected also by the physicochemical properties of the media. Charge plays a critical role in aggregation events and the rate at which they happen. Aggregation is an important factor in determining the bioavailability and uptake of NMs in aquatic environments. As an example, silver ions that are toxic to aquatic species carry a positive charge, but Ag-NMs are slightly negatively charged[98] and this will result in differences in the way in which the particle behaves and interacts with an exposed organism. Sediment dwelling organisms will likely be subject to the greatest exposure levels for most NMs as large aggregates have a tendency to precipitate out of the water column.[99] Even as aggregates, however, some NMs can exert significant toxicity. As an example, micron sized ZnO-NM aggregates flocculating out of the water column caused a greater toxic effects in zebrafish embryos and larvae those resulting from dissolution of the ZnO-NMs into Zn^{2+} ions.[100] Some NMs stabilised with coatings or modifications which change the surface charge are likely to be readily bioavailable to pelagic species though the water column (as well as via the sediments and via the diet).

4.7 Photochemistry

In aquatic media some metal oxide NMs are photocatalytic under exposure to UV light, generating ROS that can be harmful for exposed aquatic wildlife. The effect of this photocatalytic activity has been observed for exposures to TiO_2 and ZnO in the presence of UV light where the enhanced generation of ROS resulted in significant inhibitory effects on growth in cultures of the bacteria, *B. subtilis* and *E. coli*.[33] These findings for the effects of ROS generated by NMs in the presence of UV light have been further supported in a study on zebrafish (23 days, embryo – juvenile life stages) where exposure to two commercially available nano-TiO_2 (12.1 ± 3.7 and 23.3 ± 9.8 nm) under UV illumination (at levels supposed to mimic daylight) resulted in the generation of ROS and in subsequent cellular damage and enhanced mortality (for an exposure to 1 ng TiO_2/mL[83]). Enhanced toxic effects of TiO_2 under UV illumination conditions have also been shown to damage cell structure in human skin fibroblasts[101] and induce DNA damage in cultured rainbow trout *(O. mykiss)* gonad cells.[102]

4.8 Presence of Other Contaminants

Methods for NM production include the use of solvents and catalysts, and these can contribute to the biological effects exerted in exposed organisms. For example, CNTs that do not undergo a purification process have been shown to have residual levels of metal catalysts that can cause toxic effects.[103,104] In a study on carbon NMs, where experimental preparation included the use of tetrahydrofuran (THF), it was shown subsequently that the THF was responsible for the toxic effects originally attributed to the CNTs.[105] Thus, chemicals associated with the production of NMs can have significant bearing on their subsequent toxicity to aquatic organisms. Furthermore, in testing of NMs for toxicity, solvents or other chemicals are often applied to aid their dispersion and maintain them as NMs in aquatic media, and these dispersants can in turn themselves be toxic and or change the toxicity of the NMs. As an example, use of surfactants, such as sodium dodecyl sulphate (SDS) and polyoxyethylenesorbitane monooleate (Tween 80), has been shown to reduce aggregation of Ag-NMs and enhance their toxicity to bacteria.[69] This section illustrates that the solvents or other chemicals associated with the manufactured NMs or for use in aiding their dispersing for testing need to be considered carefully in any assessments on the ecotoxicity of NMs.

4.9 Interactions with Materials in Aquatic Systems

Interactions of NMs with other compounds in the environment can influence their behaviour and therefore bioavailability and toxicity. In aquatic environments NMs will encounter colloidal NOM such as humic and fulvic acids (formed by the degradation of biological matter) which have been shown to stabilise NMs and aid dispersion, potentially enhancing bioavailability. Proteins present in aquatic systems are also able to bind to the surface of NMs to form a corona, altering the surface properties of the NM.[106] Adsorption of other compounds to NMs may cause NMs to act as a 'Trojan horse' and enhance the toxicity of chemicals. Examples of these adsorptions include proteins to CNTs,[107] organic pollutants, such as Napthalene, to C-60 fullerene[108] and cadmium to TiO_2.[109] The observation for a cadmium-enhanced uptake in carp (*Cyprinus carpio*) in the presence of TiO_2 illustrates the need to develop better understanding on how NMs can affect the toxicity of other contaminants in aquatic systems.

5. NM CHARACTERISATION

Understanding ecotoxicity of NMs requires a detailed understanding of the fate and behaviour of the material in the exposure system. Complicating this, the dynamic nature of NMs in aqueous systems means that there is a continuously changing toxicity paradigm. Characterisation requirements in ecotoxicity testing can be divided into three key areas: physical, chemical and behavioural (the latter of which is affected by the former two).

Physical characterisation is used to determine particle size, shape, surface area, surface charge, surface functionalisation, crystal structure, oxidation state and particle distribution and uses various imaging, spectroscopic and separation techniques. The basic principles behind the different characterisation techniques are presented in Chapter 2, and their application to measure specific NM properties, together with critical discussions on sample preparation and data interpretation, is presented in Chapters 3–7. The application of these techniques to characterise NM in complex media is presented in Chapters 8–11.

The most commonly measured physical characteristic of NM is the size which can be determined using a number of different methods including microscopy techniques such as transmission electron microscopy (TEM) and atomic force microscopy (AFM), Field-Flow Fractionation (FFF) that provides average mass or volume-based size distribution of NMs in aqueous, and DLS that gives an intensity-based size distribution. Chemistry of the NM core and surface, as well as dissolution of NMs, is measured via spectrometry and various optical and X-ray-based methods (to identify and quantify elemental composition). Measurement of dissolution for metal/metal oxide-based NMs requires ultra-centrifugation or dialysis coupled with mass spectrometry. The behaviour of NMs in aqueous media will vary greatly as the composition of the media changes over time. Zeta potential is a useful measure of NM behaviour in solution informing on particle stability and surface charge. The more positive or negative (either side of −30 to +30 mV) the potential, the more stable the particles are in suspension.

Characterising the behaviour of NMs, especially in media more representative of 'real-world' exposure scenarios (containing NOM, or mixtures of NM types) can be extremely challenging. The polydispersed nature of the NMs in environmental media results in a bias in DLS measurements towards larger particle sizes or aggregates at the expense of the primary particle size.[110] A further drawback of the DLS technique is the insensitivity of the method for particles at low concentrations (see Ref. 111). This means that DLS is generally restricted to high exposure concentrations that do not reflect real-world exposure conditions.

Using TEM measuring methods for visualising the size and structure of the particles involves actively forcing all particles in suspension onto the TEM grid, and other techniques like AFM are passive, relying on particle diffusion towards mica sheets, which favours smaller particles.[110] Thus, these characterisation techniques provide a biased assessment of NM behaviour in aquatic environments.

Currently there is a lack of any standardised approach for the measurement of NMs in aquatic experimental exposures, although efforts are being made to rectify this. A number of large scale consortiums and EU funded projects are focused on the production of guidelines and standard operating procedures for dispersion protocols and harmonising approaches to the testing of NMs (PROSPECT http://www.nanotechia-prospect.org/, QualityNano http://www.qualitynano.eu/). Laboratory studies need to incorporate more understanding of exposures to NMs in the natural environment. In the ambient environment NM concentrations predicted and relevant for current exposure scenarios in aquatic wildlife are generally low (in the ng L^{-1} range) and occur in a more complex sample matrix compared with that for controlled experimental exposure. Some NMs are sensitive to light, pH and temperature conditions and this means that sample storage and preparation techniques can fundamentally alter the NM. In some cases samples need to be prepared or analysed immediately to avoid physical or chemical transformations (e.g. agglomeration or dissolution). This can be especially difficult when collecting samples in the field or outsourcing characterisation analyses.

In concluding this section, it is critical to the informed interpretation of any NM biological effects study for accurate characterisation of the particles employed. NM characterisation is furthermore essential for meaningful comparison across exposure studies. This characterisation requires a multifaceted approach using a variety of techniques to accurately establish the nature, behaviour and fate of the NMs in the exposure medium.

6. ECOTOXICITY OF NMs IN AQUATIC ENVIRONMENTS

There has been a range of biological effects associated with NMs in aquatic species, but there are many inconsistencies between the studies reported which could be due to differences in the NMs used and/or a lack of characterisation data for the NMs in the exposure systems to allow for effective comparisons. Many studies provide contradictory evidence in terms of NM toxicity and NM toxicity thresholds. Generally the exposures conducted exceed the levels likely to occur in the environment, but some of these studies provide useful information for assessing the toxicological potential for NMs and thus are included in the evidence for toxicological effects in this chapter. A further challenge in making assessments on the toxicity of selected NMs, for example cerium oxide, copper, silver and zinc oxide, is the levels of these metals in waters, sediments and tissues of organisms, which are high naturally. Although acute affects are unlikely to occur for environmentally relevant levels for most NMs, sub-lethal effects have been observed for some species studied and could impact on chronic health. Despite the difficulties in reconciling effects and responses in animals exposed under lab conditions, there are common findings that arise from this work.

6.1 Techniques Used to Detect NMs in Aquatic Organisms

Gaining understanding on the toxicological effects of NMs, or at least directing more targeted effects analyses, requires knowing where these materials partition in the exposed organism. A range of practical techniques have been used to assess the uptake of NMs in organisms, including imaging of fixed and live samples to visualise the NM location, summarized in Table 3. These methods include fluorescence microscopy, coherent anti-stokes raman scattering – CARS, hyperspectral imaging, and electron microscopy as well as a range of methods to measure tissues concentrations (e.g. ICP-MS).

Fluorescence microscopy has been used to establish the ability of NMs to penetrate into body tissues. Use of fluorescent latex and polystyrene particles avoids any effects relating to particle dissolution or labelling of NMs that may alter uptake characteristics.[112,113] Other techniques using fluorescence to detect NMs in tissues include two-photon excitation microscopy, light sheet fluorescence microscopy and variants of confocal microscopy.

CARS has been used for detecting NMs within biological tissues, ranging from isolated cells to whole organisms and including materials spanning metals to polymers. CARS is a label-free imaging technique, providing contrast based on the intrinsic molecular vibrations of a specimen,

TABLE 3 Summary of Microscopy Techniques Used to Assess Uptake of NMs in Organisms

ICP-MS	Excellent Quantification	Poor Localisation Ability
Electron microscopy	Very high imaging resolution.	Terminal sampling of the organism and extensive sample preparation are required. Quantification is tedious and analysis practically confined to part of the tissue or cell.
Dark-field single optical microscopy	Real-time imaging with nanometre resolution. Ability to count individual particles.	High intensity illumination can lead to photodamage of tissue. Minimal tissue penetrance.
Confocal microscopy	Thin optical sectioning capabilities. Quantification is possible, but limited to certain probes (e.g. fluorescence, which is an indirect measurement.	Sample needs modification by contrast agent. Low spatial resolution.
CARS microscopy	Label free, IR wavelengths reduce tissue photodamage and allow deep tissue penetration. Provides complementary information on chemical makeup.	~200 nm resolution. Information derived is not easily quantified.

CARS, coherent anti-stokes Raman scattering.

circumventing the need for chemical perturbation by exogenous labels. Key advantages of CARS include chemically selective live imaging and substantial depth penetration. CARS has been used to show NMs internalised in both cell cultures and whole organisms[114] and examples for exposed aquatic life are shown in Figure 4. A comprehensive review on CARS and how it compares with other imaging techniques mentioned for tracing NMs in cells and body tissues of exposed animals is provided in Ref. 116. More recent advances in the application of hyperspectral imaging allow for distinguishing between different NMs and for studies into mixture effects of NMs. However, there is a

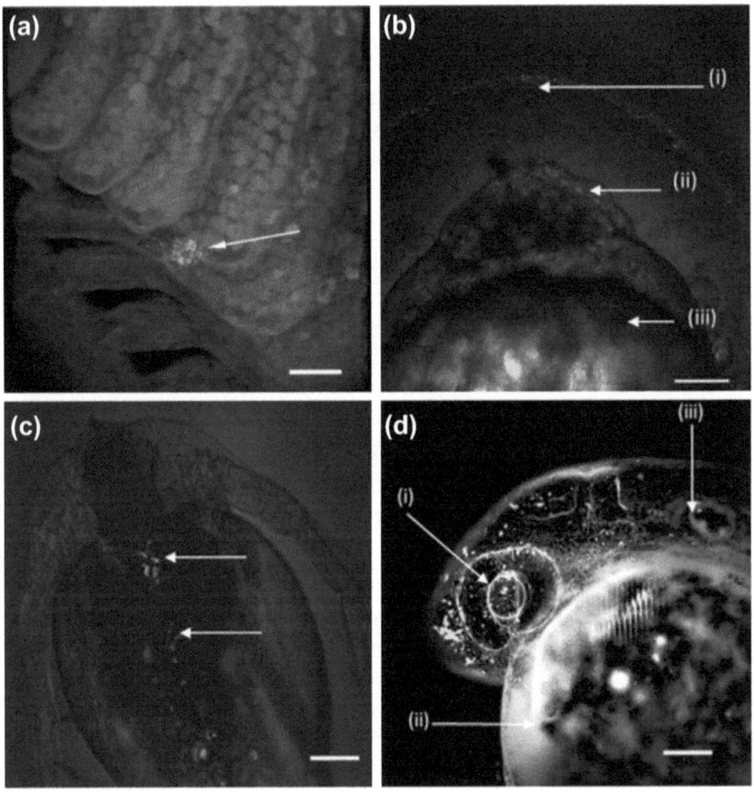

FIGURE 4 CARS images depicting NM contained in body tissues for aquatic animals exposed to metal/metal oxide NMs. (a) Gill arch of a rainbow trout (*O. mykiss*), showing the presence of TiO_2-NMs (arrowed) taken up from an aqueous exposure - reproduced from Ref. 115 (b) zebrafish (*D. rerio*) embryos microinjected at the one cell stage with CeO_2-NMs (i, contrasted blue), imaged 3.5 h post fertilisation, (ii) indicates the mass of dividing cells and (iii) the yolk sac. (c) ZnO particles depicted in red (arrowed) shown to be retained in the gut of a daphnia after an oral exposure. (d) A zebrafish embryo microinjected with Ag-NMs at the one cell stage and imaged at 24 hpf, the yellow components of the image are the Ag-NMs dispersed throughout the organism; the (i) eye, (ii) ear and (iii) yolk sac are marked for orientation. Figures (b) and (c) reproduced from Ref. 116. Scale bars represent 100 μm.

relatively high detection limit for this method and spectral libraries to date are limited. There is nevertheless significant future potential for this imaging method for establishing what features of NMs confer biochemical changes in exposed tissues.

7. EVIDENCE FOR TOXICOLOGICAL EFFECTS OF NMs IN AQUATIC ORGANISMS

There is a growing body of evidence for effects of NMs in aquatic wildlife across different phyla in both marine and freshwater environments, with a focus on silver and carbon NMs. Most studies conducted into the effects of NMs in aquatic organisms have been undertaken in model species and often adopting standardised toxicity testing approaches (e.g. OECD Test Guidelines). To what extent these species and standardised test are suitable for assessing the ecotoxicological effects and impacts of NMs in the natural environment, however, is not clear and has not yet received significant critical debate. We would argue for the need to consider what species are most likely to be targets for exposures to NMs than is perhaps considered currently. Nevertheless, the data derived from the reported exposure studies have started to address the issues of the bioavailability of NMs to aquatic plants, invertebrates and vertebrates, their routes of uptake (e.g. via ingestion or absorption across respiratory skins surfaces) from the water/sediment and their potential routes of entry into cells (via transporters or endocytosis). Approaches for studying the toxicological effects of NMs reported in the literature extend from in vitro studies directed at understanding mechanism of action and higher throughput of screening for possible effects, in vivo studies focused more on NM bioavailability and whole organism responses, to more complex mesocosm studies simulating conditions of the natural environment and extending to investigating transfer of NMs through the aquatic food web. Here we analyse effects reported for aquatic organisms spanning bacteria to fish and attempt to highlight common effect measures and findings across these diverse phyla. A summary of toxicity data for TiO_2-NMs can be found in Table 4, highlighting the variability in data across studies using common laboratory models.

7.1 Evidence for NM Toxicity in Aquatic Bacteria

Bacteria are abundant in most aquatic environments and they include primary producers and aid in the turnover of nutrients. NMs such as silver, copper and zinc oxide have been proven to have antibacterial properties and are used commercially for that purpose. The anti-bacterial activity of silver NMs has been associated principally with the release of Ag^{+} ions and generation of ROS.[76,128] This may also be the case for ZnO-NMs.[129,130] Antibacterial effects of these metal oxide NMs however have also been attributed directly to the

TABLE 4 Concentrations for TiO_2-NMs Eliciting Toxicity Responses in Common Test Species

NM	Organism	Particle Size (nm)	EC_{50} (mg L^{-1})	References
TiO_2	*Pseudokirchneriella Subcapitata* Algae	<10	241 (95.6–609)	118
		25–70	5.83 (3.75–7.58)	119
		30	71.1 (59.4–85.1)	118
		<100	>100[a]	120
TiO_2	*Daphnia magna* Water flea	≤20	35.306 (25.627–48.928)	121
		20–30	>100	122
		21	>100	123
		50 × 10	>100	122
		10–20	5.5	105

NM	Organism	Particle Size (nm)	Effect	Comment	References
TiO_2	*Danio rerio* zebrafish	20.5 ± 6.7	$LC_{50} > 10$ mg L^{-1}	48 h, adults/juveniles	124
		≤20	No toxicity	96 h, embryos/larvae, 0–500 mg L^{-1}	125
		21 ± 1	LC_{50} 290 mg L^{-1}	5 days, embryos plus humic acid (HA)	126
			LC_{50} 156 mg L^{-1}		126
		20–70	LC_{50} 124.5 mg L^{-1}	96 h, adult	127

[a]*IC_{50} value.*
Reviewed in Ref. 117.

ability of the particles, rather than the free ions, to cause oxidative stress through generation of ROS in bacteria.[33,131] Fullerene (C60) suspensions in water have also been shown exert antibacterial effects on *B. subtilis* (0.1 and 1 mg L^{-1})[132] and *E. coli* (140 μM).[133] In soil, a link has been drawn between NOM and a reduction in antibacterial activity of fullerenes, and this may also be the case for fullerenes in water.[134] Reduction in growth in the bacteria *Pseudomonas fluorescens* has been shown for an exposure to citrate-capped Ag-NMs at 2000 μg L^{-1}, in freshwater.[57] There is also evidence that Ag-NMs can have detrimental effects on communities of bacteria, in biofilms. Biofilms are multi-species communities of bacteria which provide important functions in ecosystems, including cycling of nutrients and as a food source for other organisms. In an exposure to Ag-NMs (20 nm, polymer coated) biofilm growth was reduced at environmentally relevant exposure concentrations (5 μg L^{-1}).[135] Recovery from these effects has been shown to occur, possibly resulting from selection of silver tolerant species. In a study on marine biofilms, citrate-capped Ag-NMs (65 ± 30 nm; 200 mg L^{-1}) similarly induced a significant decrease in the biofilm volume and biomass[136] however no such effects were seen for exposure to another Ag-NMs (uncapped, 58.6 ± 18.6 nm; 1000 mg L^{-1}) on marine sediment biofilms.[137] Thus the collective evidence for NM toxicity to bacteria is far from consistent. As silver is a known biocide and has been synthesised as such for decades, an increase in levels in the environment is a cause for concern. Measurements of environmental NMs with photocatalytic activity or known antimicrobial properties, in strategic hotspots, are warranted given the potential far reaching consequences on bacterial populations fundamental for maintaining ecosystem services.

7.2 Evidence for NM Toxicity in Aquatic Algae

Algae are vital in aquatic food webs as primary producers and in generating oxygen as a by-product of photosynthesis. Any disruption to algal populations could therefore profoundly affect aquatic ecosystems. Dose-dependent toxicity for TiO_2 and Ag-NMs has been shown in various species of algae, which appears to differ widely in their sensitivities. As an example, TiO_2 toxicity in the algae *P. subcapitata* has been shown between 1 and 5 mg L^{-1}[119,138] but in *Desmodesmus subspicatus* a considerably higher concentration (44 mg L^{-1}) of the same TiO_2 was required to exert any toxicity.[139] Contrasting with these findings, no effects were observed for TiO_2 on *P. subcapitata*,[124,140] further illustrating the difficulties in aligning effects comparisons across different studies without comprehensive reporting on the NM exposure conditions and fate. In marine algal species, decreased growth rates (50% growth inhibition, IC_{50}) have been observed in marine diatom *P. tricornutum* exposed to citrate-capped Ag-NMs (14 nm; 2380 ± 1880 μg L^{-1}) and PVP-capped Ag-NMs (15 nm, 3690 ± 2380 μg L^{-1}).[67] In an exposure to Ag-NMs (58 ± 27 nm) in sea lettuce (*Ulva lactuca,* a type of macro-algae), there was a reduced yield of photosystem II, but only at concentrations >55 μg L^{-1}.[141] In further studies using marine algal species, decreased growth rates have been reported for exposure to ZnO-NMs, but not to TiO_2-NMs.[142] ZnO-NMs have also been shown to inhibit growth of *Chaetoceros gracilis* and *Thalassiosira pseudonana* at concentrations >10 mg L^{-1}.[143] ZnO-NMs have been shown to be toxic to *P. tricornutum*[143] and *T. pseudonana* and *Skeletonema costatum.*[21] Mechanisms of toxic mode action for effects in algae are not well documented. For some of the exposures described, very high NM concentrations will physically shield algae from the light, reducing their photosynthetic capacity, and thus growth (and potentially survival) which is not a nano-specific effect.

7.3 Evidence for NM Toxicity in Aquatic Invertebrate Species

Aquatic invertebrates span animals of great diversity with a range of morphologies and feeding behaviours and inhabiting different compartments in aquatic environments. Exposure scenarios therefore for NMs may differ widely. Invertebrates that live in the water column are likely to have lower exposure to NMs compared with species that inhabit the benthos, where many NMs will accumulate due to aggregation and precipitation out from water column. Filter feeding invertebrates that extract food from the water may selectively remove NMs forming as aggregates in the water column for ingestion due to their similarity in size to natural food sources. Most of the data available for exposure and effects of NMs in aquatic invertebrates derive from studies in daphnia, as an OECD test guideline species, which reside in the water column, and the benthic dwelling organisms – mussels, polychaete worms and fairy shrimp, which are common test organisms.

Planktonic crustaceans and rotifers that live in the water column are likely to be exposed to freely dispersed NMs. More than 150 studies into the effects of NMs in the freshwater crustacean *Daphnia magna* have been reported (results for 'research articles' obtained when searching 'ISI Web of Science' in 2014 using the topic keywords 'daphnia', 'NM' and 'toxicity', Year Published, 2000–2014). Exposures of daphnia to NMs have been shown to cause abnormal swimming,[144] inhibit growth and reproduction[145] and induce DNA damage (exposure to 15 and 30 nm CeO_2-NMs).[80] Reported toxicity values from different studies on *D. magna* exposed to TiO_2-NMs (for 48 h) range widely between 5.5 and >100 mg L^{-1}, with no link between NM size or surface area (reviewed in Ref. 117). C60 fullerenes (prepared by stirring, 0.5–5 mg L^{-1}) have been shown to increase mortality, delay moulting and cause a decrease in offspring production.[146]

In acute lethal tests, the median lethal concentrations (LC_{50}) for the marine copepod *Tigriopus japonicus* and the amphipod *Elasmopus rapax,* exposed to ZnO-NMs (26.2 ± 5.1 nm, at 4 and 40 mg L^{-1}) for 96 h were 0.68 and 0.95 mg L¯1, respectively. Similar results were seen for exposures to ionic Zn

(14 mg L^{-1}) with LC_{50} values of 1.14 mg L^{-1} (*T. japonicus*) and 0.80 mg L^{-1} (*E. rapax*), indicating ZnO-NM toxicity is due primarily to dissolution.[21] Exposure of freshwater amphipods (*Hyalella azteca*) and midges (*Chironomus dilutus*) to CNTs found very low toxicity (at 1.00 g L^{-1}, dry wt). The preparation methods for CNT were shown to contribute the CNT toxicity; sonicated CNTs further decreased amphipod survival and growth of the amphipods, and midges exposed to nitric acid-cleaned CNTs were significantly reduced.[147]

Benthic species exhibit a wide range of feeding methods, including as filter feeders, deposit feeders, scavengers and predators and therefore exposure scenarios and associated risk for NM exposure will differ widely between these species. Benthic polychaete worms ingest large volumes of sediment where most NMs are likely to accumulate and a range of toxic effects of NMs have been reported, including DNA damage and oxidative stress. DNA damage has been reported in lugworm *Arenicola marina* exposed to TiO_2-NMs (32 nm) at concentrations $\geq$2 g/g.[148] Effects of TiO_2-NMs on behaviour in lugworms, as measured by the mass of casts produced, were also reported. In their studies, Galloway et al. reported no significant effects in the lugworm for single-walled carbon nanotubes (SWCNT). In the ragworm, *Hediste diversicolor,* oxidative stress responses have been observed for exposure to CuO-NMs (197 nm, 10 μg L^{-1}), characterised by increases in glutathione-s-transferase (GST) and catalase (CAT) activity, and due to generation of ROS.[149] In an exposure of *Nereis diversicolor* to citrate-capped Ag-NMs (30 $\pm$ 5 nm, 250 ng/g sediment) for 10 days via the diet, silver particles were found associated with villi in the gut lumen and in endocytotic pits, identifying a possible route for internalisation.[150] In the estuarine sediment dwelling amphipod *Leptocheirus plumulosus*, LC_{50} values for ZnO-NMs (20–30 nm) and CuO-NMs (200–1000 nm) have been reported at 763 μg/g sediment and 868 μg/g sediment, respectively.[151] No abnormalities were observed in *L. plumulosus* for exposures to insoluble CeO_2 and TiO_2-NMs up to 10 mg L^{-1}.[152]

Echinoderms, that include sea urchins and starfish, utilise a water vascular system for movement and to prevent the body from collapsing. To supply these vascular systems they draw water into their bodies and thus they are potentially vulnerable to taking up NMs (including aggregates) directly from the water. Exposure of the sea urchin *Paracentrotus lividus* to SnO_2- (61 nm), CeO_2 (50–105 nm) and Fe_3O_4 (20–30 nm)-NMs showed their entry into the digestive, immune and reproductive systems, and they induced damage to the endoplasmic reticulum and golgi apparatus, but only at high exposures (0.1–10 mg L^{-1}).[153] In white sea urchin embryos (*Lytechinus pictus*) exposed to ZnO-NMs (18 nm) the LC_{50} (median effective, mortality) was 99.5 μg L^{-1} and differed from that for bulk and ionic zinc exposures, and the toxicity was attributed to dissolution.[152]

Bivalves are benthic animals that will actively take up NMs from the water column though their gills that are a specialised surface aiding in the transfer of food to the mouth. NMs drawn into bivalves first encounter the gills and can then pass into the digestive system via the mouth if the particles are selected for digestion.[154] A number of studies have reported accumulation of NMs in the digestive glands indicating transport from the gills to the digestive system. High levels of accumulation in the digestive gland in *Mytilus edulis* and *Mytilus galloprovincialis* have been shown for exposures to Au-NMs (5.3 ± 1 nm, 750 mg L^{-1})[155] and polyethylene glycol (PEG)-capped Fe_3O_4-NMs (50 nm, 370 ug L^{-1}),[156] respectively, for exposure in aqueous. In a study on *M. edulis* exposed to radio-labelled Ag-NMs (<40 nm, 0.7 mg L^{-1}), 7% of the material up taken was subsequently found in the extrapallial fluid (used to keep the mussel moist within the shell).[157] Both PEG-capped Fe_3O_4 and Q-dot-NMs (elliptical in shape − 12 × 6 nm) have been shown to penetrate into the haemolymph in mussels for exposures via the water.

TiO_2-NMs (22 nm) have been shown to cause a significant reduction in lysosomal stability in the digestive gland (at ≥0.2 mg L^{-1}) and in haemocytes (≥1 mg L^{-1}) of *M. galloprovincialis.*[158] Exposures of oysters *Crassostrea virginica* to C60 (1−500 ppb) found accumulation in liver cells, effects on embryo development and significant lysosomal destabilisation at concentrations >100 ppb.[159] Although not acutely toxic to the Abalone, *Haliotis diversicolor supertexta,* uncapped TiO_2-NMs (≤10 nm; 96 h) were shown to cause lipid peroxidation and increased nitric oxide production, linked with oxidative stress, for exposure at concentrations ≥1 mg L^{-1}. These effects, however, may not necessarily be a 'nano-' related, and in this work there was a lack of any comparison with bulk and soluble TiO_2.[160]

7.4 Evidence for NM Toxicity in Aquatic Vertebrates (Fish)

Considerable research effort has been directed towards NM toxicity to fish but there is an absence of data for aquatic mammals and birds. Data in fish are available for a range of species, both marine and freshwater, but most studies have been centred on the zebrafish (*D. rerio*), and predominantly zebrafish embryos. Effects observed in zebrafish embryos exposed to NMs include mortality, hatching delays and various developmental malformations (metal NMs,[161] metal oxide NMs,[125] and carbon NMs[162]). Photoactivation has been shown to increase the toxicity of TiO_2-NM in developing zebrafish embryos through increased ROS generation.[83,126] Zebrafish embryos, however, are surrounded by a chorion with pores that are in the nanometre scale, and this is likely to highly be restrictive for uptake of many NMs.

Adult zebrafish have been exposed to a range of NMs to assess lethality, tissue loading, cellular responses such as oxidative stress, and changes in gene expression. Exposures of adult zebrafish to Ag-NMs (PVP coated, 81 nm) recorded an LC_{50} of 84 μg L^{-1} and 25 μg L^{-1} for Ag ions (48 h[163]). Another study, exposing zebrafish (24 h) to 5−20 nm Ag-NMs recorded an LC_{50} of 250 mg L^{-1}, and both oxidative stress and apoptosis were observed in liver tissue.[164] Gene expression studies in fish for exposures to high concentrations of Ag-NMs[84,87] and Cu-NMs[85] have indicated that the toxicological effects seen were primarily related to dissolution and metal ions for these NMs. No uptake or biological effects were observed for semi-static aqueous exposures of adult zebrafish to 10 nm CeO_2-NMs or 34 nm TiO_2-NMs (0.05–5 mg L^{-1}).[165]

In other fish species, exposures have been undertaken for silver, zinc, and titanium dioxide. In rainbow trout (*O. mykiss*) evidence of uptake and oxidative stress has been reported for a 14-day aqueous exposure to 21 nm TiO_2-NMs (0.1, 0.5 or 1.0 mg L^{-1}).[166] These findings however contrast with another aqueous exposure to 34 nm TiO_2-NMs (0.05, 0.5, 5 mg L^{-1}) in rainbow trout where no uptake or effects were seen.[165] In juvenile estuarine sheepshead minnow (*Cyprinodon variegatus*) exposed to 10 μg L^{-1} 35 nm citrate-capped Ag-NMs for 28 days, bioaccumulation of silver occurred in tissues, not seen for the ionic Ag exposure (1 μg L^{-1}).[167] Bioaccumulation of Ag in the tissues of adult sheepshead minnows was also observed in this study and a thickening of the gill epithelia was seen after 4 days, that was significantly greater in fish exposed to Ag-NMs compared with for exposure to Ag^+. By 28 days the thickening had reduced, but it was still significantly greater in the Ag-NMs exposed fish. In a study on rainbow trout exposed to Ag-NMs (16.6 nm) when the fish were acclimating to an increase in salinity (which induces drinking to overcome the water loss via osmosis), silver was found to increase in body tissues with increasing dose of Ag-NMs (0.1–32 mg L^{-1}) and salinity (0.4–12 psu).[168] This finding would indicate that stressed fish are more susceptible to NM uptake and to biological effects on exposure to NMs. Disruption seen in olfaction (smell) in carp (*Carassius carassius*) and perch (*Perca fluviatilis*) for exposures to Ag-NMs has been attributed to both particles and silver ions (~80 nm, 0.45–45 μg L^{-1}).[66] TiO_2-NMs have been reported to cause gill pathologies such as thickening of the lamellae and oxidative stress in juvenile rainbow trout (in a 14-day semi-static exposure.[166]

CNTs, at high exposure concentrations (between 0.1 and 0.5 mg L^{-1}), have been shown to cause cellular damage in the gill, and to induce lipid peroxidation and oxidative stress in juvenile rainbow trout.[169] In further study from the same laboratory, SWCNTs were subsequently reported to have minimal effects on tissue histopathology, haematocrit and movement behaviours and the author suggests that dispersal agents, in this case SDS, may have been responsible for the toxicity of CNTs.[170]

7.5 Toxicity of NMs in 'Real World' Aquatic Environments

Through laboratory-based experiments studying trophic transfer and larger scale mesocosm studies, researchers are starting to develop approaches for NM exposures that are more representative of 'real-world' environments. Furthermore there have also been some studies investigating ecosystem-level effects of NM exposure. Trophic transfer of NMs has been shown in laboratory experiments in both freshwater and marine species (reviewed in Ref. 52). Examples of this include for uptake of Ag-NMs by *D. magna* via a diet of Ag-NM dosed algae[171] and into the marine medaka fish *(Oryzias melastigma)* via Ag-NM dosed brine shrimp.[53] Transfer of TiO_2-NMs into zebrafish via dosed daphnia has also been observed[54]; however, there was no evidence of biomagnification of the TiO_2. In the marine mussel, *M. galloprovincialis*, CeO_2-NMs taken in sorbed to phytoplankton was shown to be cleared via the pseudofaeces (i.e. the mussels actively rejected rather than digested the dosed phytoplankton).[50] These studies on tropic transfer, however, have often employed dosing levels that far exceeds those likely to occur in the environment, to help in subsequent particle detection. Future work of this nature would benefit from exposures that are more environmentally relevant.

Mesocosms provide mini ecosystems for both investigating trophic transfer of NMs and allowing for transformations of NMs that may occur in natural waters. Unlike in the natural world, they allow for control on the physicochemistry of the environment and the organisms included. In mono- and poly-cultures of algae in a mesocosm representing a stream, 27 nm TiO_2-NMs (dosed at 0.1 or 1.0 ppm) were up taken from the surface waters and underwent trophic transfer. The algal polyculture accrued greater amounts of Ti than the mono-cultures for three different algal species and Ti accumulated in the tissues of the herbivorous snail *Physa acuta* that consumed the algae.[172] In outdoor mesocosms dosed with 10 $\mu g\ L^{-1}$ CuO-NMs (10–100 nm, average 29.5 nm)[173] over 21 days, bioaccumulation occurred in the bivalve mollusc *Scrobicularia plana* and polychaete worm *H. diversicolor* and in both species toxicity was attributed to the CuO-NMs rather than dissolved Cu. In a second study, using the same experimental set up, exposure to both CuO-NMs and 40 nm Ag-NMs resulted in bioaccumulation in the same study species and a nano-specific toxic effect was reported.[174] Transfer of gold nanorods has also been observed in a mesocom designed to replicate an estuarine environment containing microbes, biofilms, snails (*Ilyanassa obsoleta*), clams (*Mercenaria mercenaria*), shrimp (*Palaemonetes pugio*) and fish (*C. variegatus*).[175] The mesocoms received a single dose of gold nanorods (65 nm length $\times$ 15 nm diameter) to produce a theoretical water concentration of 7.08×10^8 particles mL^{-1} and over a period of 12 days the gold partitioned into all the organisms, with the clams and biofilms containing the greater quantities by mass. In a pond mesocosm experiment dosed with CeO_2-NMs via the water (8 nm, 1 $mg\ L^{-1}$) the NMs underwent aggregation and settlement where they became available to benthic species and were subsequently taken up and induced oxidative stress in the snail, *Planorbarius corneus*.[176]

7.6 Common Effects of NMs in Aquatic Species

NM uptake and toxicity has been reported across diverse phyla in aquatic environments. Tissue burdens of NMs are generally at the initial point of uptake in the organism, the gills or digestive tract, and/or in organs that mediate toxicant excretion, including the liver and kidney. The most commonly reported sub-lethal effects of NMs include oxidative stress responses and DNA damage, but these studies are biased to exposures that far exceed environmentally relevant concentrations. It is the case that where exposures bear more environmental relevance, the biological findings are more variable.

A reoccurring theme for studies on the toxicity of metal/metal oxide NMs to aquatic organism studies is the impact of NM dissolution, and the debate on whether ions are the sole cause of the toxicity seen. This is difficult to resolve for many reasons, not least because the rate of dissolution will vary between exposure media and for different types of (and preparations of the same) particles. The general consensus is however that, for silver NMs, it is the ions that are the primary factor driving toxicity on exposed aquatic organisms, spanning bacteria,[119,124,177] algae,[67,86] invertebrates,[21,151] and fish (zebrafish).[92,178] This is also likely to be the case for Cu-NMs[119] and ZnO-NMs.[21,129,143] However, responses at the level of the transcriptome in fish do suggest that for Ag-NMs[84,87] and Cu-NMs[85] there are some toxicological effects also related to the particle itself.

Another common feature relating to the toxicity of NMs across different species is that for some reported effects it is the material in which the NMs have prepared that is toxic, rather than the NM itself. Examples of this include for some of the early studies on fish where the effects of fullerenes, observed in juvenile largemouth bass (*Micropterus salmoides*)[179] have subsequently been attributed to the solvent (TNF) used in the preparation process for fullerenes. Studies on TNF in zebrafish have shown it reduces larval hatching rates, induces developmental delays and decreases survival rates[162,180] and studies on solvent free preparations of fullerene have found no enzyme effects in the livers of fathead minnows (*Pimephales promelas*),[146] and a lack of any uptake of fullerenes into the brains of carp (*C. carpio*),[181] that were reported for fullerenes in TNF in the original studies of Ref. 179 on largemouth bass. Toxic effects of carbon NMs observed in *D. magna* have also been linked to preparation solvent.[105] and THF has been shown to increase the toxicity of TiO_2-NMs exposed to *D. Magna*.[105]

8. GAPS IN OUR KNOWLEDGE AND FUTURE CHALLENGES IN NM AQUATIC ECOTOXICOLOGY

8.1 Linking Laboratory-Based Studies with the 'Real World'

Gaining better understanding on the fate, transport and uptake of NMs into exposed organisms is a priority for advancing understanding on their bioavailability for hazard identification and risk assessment. Furthermore, establishing links between specific features of NM properties and biological effects observed in relevant test scenarios is fundamental for advancing NM ecotoxicology. This in turn requires extensive NM characterisation and this is a major reason for the synthesis in this book. Replicating 'real-world' environments for toxicity testing in the laboratory, however, is extremely difficult due to the dynamic nature of the NMs even under relatively stable chemico-physical environmental conditions. Introduction of test organisms into the tank of water containing the NM will likely alter the form of the NMs and thus affect their bioavailability. In the natural environment NMs will encounter complex media with differing properties that will affect the fate and behaviour of the NMs. NOM present in natural waters for example has been shown to impact the fate and behaviour of NMs and different types and mixtures of NMs may potentially cause additive or protective effects on toxicity. Understanding the toxicity of NMs to wildlife in natural systems will require replicating the exposure conditions that are most relevant for the NM of concern.

The relevance of current standardised toxicity tests for assessing the environmental risk of NMs is therefore brought into question. Furthermore, current preparation techniques used for laboratory exposure experiments with NMs include the use of solvents or capping agents to aid dispersal, and these potentially contribute to toxicity. These 'additives' require testing separately to distinguish their effects from those of the NMs. It is important to recognise also, that NMs available commercially often behave differently to NMs synthesised in laboratories for research purposes, or from those that are released from NM containing products (and which have undergone ageing).

Model species are useful for toxicity studies as they are well characterised and easy to maintain however, as identified above, they may not be the more relevant for exposures to NMs in natural systems. Model species can be used to determine fundamental principles such as the mechanisms of action but their response and sensitivity could differ to species found in aquatic environments that are most at risk to NM exposure. Understanding of effects in species found in the water polluted with NMs will aid assessment of the risk to those populations and ecosystems.

8.2 Future Challenges in the Ecotoxicology of NMs

Nanotechnology continues to develop rapidly and we need to ensure new NMs entering aquatic environments do not pose a significant health concern and to effectively assess their safety. The rate of development of new types of NMs with even wider ranges of materials and modifications, properties and functions, however, make this a considerable challenge in the absence of any agreed **standardised** testing protocols. NM in medicines potentially poses one of the greater challenges as they are being designed specifically to be bioactive to improve bioavailability and uptake of therapeutic agents. In turn these properties will likely enhance their bioavailability to non-target aquatic species (reviewed in Ref. 182).

Setting appropriate context for laboratory-based work is vital for better evaluating the risks that NMs may pose to aquatic organism. This is challenged by the practical limitations for characterising NMs in the natural environment, due to lack of accurate assessments on environmental releases, and more so due to instrument (in)sensitivity. Further development of instrumentation for detection on NMs in the environment and in exposed organisms will aid significantly in developing our understanding on to what extent (or not) NMs truly pose a significant threat to aquatic wildlife and ecosystem functioning.

ACKNOWLEDGEMENTS

This work was supported by the Natural Environmental Research Council (NE/H013172/1) and NANOMILE EU FP7. Engineered nanomaterial mechanisms of interactions with living systems and the environment: 2013-2016 (www.nanomile.eu/)

REFERENCES

1. Roco MC, Bainbridge WS. *Societal implications of nanoscience and nanotechnology - Introduction.* 2001 (Page 3 "2. Nanotechnology Goals").
2. PEN. http://www.nanotechproject.org/, http://www.nanotechproject.org/cpi/about/analysis/; 2014.
3. Mueller NC, Braun J, Bruns J, Cernik M, Rissing P, Rickerby D, et al. Application of nanoscale zero valent iron (NZVI) for groundwater remediation in Europe. *Environ Sci Pollut Res* 2012;**19**(2):550–8.
4. SEPA. *Hydrogeological risk assessment for landfills and the derivation of control and trigger levels.* Version 2.12. SEPA Technical Guidance Note. 2005. Table 1, Pages 29–46.
5. EPA U. *National recommended water quality criteria.* 2009.
6. Shaw BJ, Handy RD. Physiological effects of nanoparticles on fish: a comparison of nanometals versus metal ions. *Environ Int* 2011;**37**(6):1083–97.
7. Farre M, Perez S, Gajda-Schrantz K, Osorio V, Kantiani L, Ginebreda A, et al. First determination of C-60 and C(70) fullerenes and N-methylfulleropyrrolidine C-60 on the suspended material of wastewater effluents by liquid chromatography hybrid quadrupole linear ion trap tandem mass spectrometry. *J Hydrol* 2010;**383**(1–2):44–51.
8. Limbach LK, Bereiter R, Mueller E, Krebs R, Gaelli R, Stark WJ. Removal of oxide nanoparticles in a model wastewater treatment plant: influence of agglomeration and surfactants on clearing efficiency. *Environ Sci Technol* 2008;**42**(15):5828–33.
9. Mueller NC, Nowack B. Exposure modeling of engineered nanoparticles in the environment. *Environ Sci Technol* 2008;**42**(12):4447–53.
10. Kiser MA, Ladner DA, Hristovski KD, Westerhoff PK. Nanomaterial transformation and association with fresh and freeze-dried wastewater activated sludge: implications for testing protocol and environmental fate. *Environ Sci Technol* 2012;**46**(13):7046–53.
11. Kaegi R, Voegelin A, Sinnet B, Zuleeg S, Hagendorfer H, Burkhardt M, et al. Behavior of metallic silver nanoparticles in a pilot wastewater treatment plant. *Environ Sci Technol* 2011;**45**(9):3902–8.

12. Kaegi R, Voegelin A, Ort C, Sinnet B, Thalmann B, Krismer J, et al. Fate and transformation of silver nanoparticles in urban wastewater systems. *Water Res* 2013;**47**(12):3866–77.
13. Hendren CO, Badireddy AR, Casman E, Wiesner MR. Modeling nanomaterial fate in wastewater treatment: Monte Carlo simulation of silver nanoparticles (nano-Ag). *Sci Total Environ* 2013;**449**:418–25.
14. Li LXY, Hartmann G, Doblinger M, Schuster M. Quantification of nanoscale silver particles removal and release from municipal wastewater treatment plants in Germany. *Environ Sci Technol* 2013;**47**(13):7317–23.
15. Benn TM, Westerhoff P. Nanoparticle silver released into water from commercially available sock fabrics. *Environ Sci Technol* 2008;**42**(11):4133–9.
16. Hendren CO, Mesnard X, Droege J, Wiesner MR. Estimating production data for five engineered nanomaterials as a basis for exposure assessment. *Environ Sci Technol* 2011;**45**(7):2562–9.
17. Johnson AC, Park B. Predicting contamination by the fuel additive cerium oxide engineered nanoparticles within the United Kingdom and the associated risks. *Environ Toxicol Chem* 2012;**31**(11):2582–7.
18. Kaegi R, Ulrich A, Sinnet B, Vonbank R, Wichser A, Zuleeg S, et al. Synthetic TiO_2 nanoparticle emission from exterior facades into the aquatic environment. *Environ Pollut* 2008;**156**(2):233–9.
19. Farkas J, Peter H, Christian P, Urrea JAG, Hassellov M, Tuoriniemi J, Gustafsson S, et al. Characterization of the effluent from a nanosilver producing washing machine. *Environ Int* 2011;**37**(6):1057–62.
20. Danovaro R, Bongiorni L, Corinaldesi C, Giovannelli D, Damiani E, Astolfi P, et al. Sunscreens cause coral bleaching by promoting viral infections. *Environ Health Perspect* 2008;**116**(4):441–7.
21. Wong SWY, Leung PTY, Djurisic AB, Leung KMY. Toxicities of nano zinc oxide to five marine organisms: influences of aggregate size and ion solubility. *Anal Bioanal Chem* 2010;**396**(2):609–18.
22. Gondikas AP, von der Kammer F, Reed RB, Wagner S, Ranville JF, Hofmann T. Release of TiO_2 nanoparticles from sunscreens into surface waters: a one-year survey at the old Danube Recreational Lake. *Environ Sci Technol* 2014;**48**(10):5415–22.
23. Blaser SA, Scheringer M, MacLeod M, Hungerbuehler K. Estimation of cumulative aquatic exposure and risk due to silver: contribution of nano-functionalized plastics and textiles. *Sci Total Environ* 2008;**390**(2–3):396–409.
24. Gottschalk F, Sonderer T, Scholz RW, Nowack B. Modeled environmental concentrations of engineered nanomaterials (TiO(2), ZnO, Ag, CNT, fullerenes) for different regions. *Environ Sci Technol* 2009;**43**(24):9216–22.
25. Arvidsson R, Molander S, Sanden BA, Hassellov M. Challenges in exposure modeling of nanoparticles in aquatic environments. *Hum Ecol Risk Assess* 2011;**17**(1):245–62.
26. Gottschalk F, Nowack B. The release of engineered nanomaterials to the environment. *J Environ Monit* 2011;**13**(5):1145–55.
27. Praetorius A, Scheringer M, Hungerbuehler K. Development of environmental fate models for engineered nanoparticles-a case study of TiO_2 nanoparticles in the Rhine river. *Environ Sci Technol* 2012;**46**(12):6705–13.
28. Kiser MA, Westerhoff P, Benn T, Wang Y, Perez-Rivera J, Hristovski K. Titanium nanomaterial removal and release from wastewater treatment plants. *Environ Sci Technol* 2009;**43**(17):6757–63.

29. Westerhoff P, Song G, Hristovski K, Kiser MA. Occurrence and removal of titanium at full scale wastewater treatment plants: implications for TiO(2) nanomaterials. *J Environ Monit* 2011;**13**(5):1195–203.
30. Keller AA, McFerran S, Lazareva A, Suh S. Global life cycle releases of engineered nanomaterials. *J Nanopart Res* 2013;**15**(6).
31. Tratnyek PG, Johnson RL. Nanotechnologies for environmental cleanup. *Nano Today* 2006;**1**(2):44–8.
32. Zhang WX. Nanoscale iron particles for environmental remediation: an overview. *J Nanopart Res* 2003;**5**(3–4):323–32.
33. Adams LK, Lyon DY, Alvarez PJJ. Comparative eco-toxicity of nanoscale TiO_2, SiO_2, and ZnO water suspensions. *Water Res* 2006;**40**(19):3527–32.
34. Grieger KD, Fjordboge A, Hartmann NB, Eriksson E, Bjerg PL, Baun A. Environmental benefits and risks of zero-valent iron nanoparticles (nZVI) for in situ remediation: risk mitigation or trade-off? *J Contam Hydrol* 2010;**118**(3–4):165–83.
35. Buzorius G, Hameri K, Pekkanen J, Kulmala M. Spatial variation of aerosol number concentration in Helsinki city. *Atmos Environ* 1999;**33**(4):553–65.
36. Hughes LS, Cass GR, Gone J, Ames M, Olmez I. Physical and chemical characterization of atmospheric ultrafine particles in the Los Angeles area. *Environ Sci Technol* 1998;**32**(9): 1153–61.
37. Li CS, Lin WH, Jenq FT. Characterization of outdoor submicron particles and selected combustion sources of indoor particles. *Atmos Environ Part B-Urban Atmos* 1993;**27**(4): 413–24.
38. Maenhaut W, Schwarz J, Cafmeyer J, Annegarn HJ. Study of elemental mass size distributions at Skukuza, South Africa, during the SAFARI 2000 dry season campaign. *Nucl Instrum Methods Phys Res Sect B* 2002;**189**:254–8.
39. Morawska L, Thomas S, Bofinger N, Wainwright D, Neale D. Comprehensive characterization of aerosols in a subtropical urban atmosphere: particle size distribution and correlation with gaseous pollutants. *Atmos Environ* 1998;**32**(14–15):2467–78.
40. Trier A. Submicron particles in an urban atmosphere: a study of optical size distributions .1. *Atmos Environ* 1997;**31**(6):909–14.
41. Buzea C, Pacheco II, Robbie K. Nanomaterials and nanoparticles: sources and toxicity. *Biointerphases* 2007;**2**(4):MR17–71.
42. Park B, Donaldson K, Duffin R, Tran L, Kelly F, Mudway I, et al. Hazard and risk assessment of a nanoparticulate cerium oxide-based diesel fuel additive - a case study. *Inhal Toxicol* 2008;**20**(6):547–66.
43. Zhiqiang Q, Siegmann K, Keller A, Matter U, Scherrer L, Siegmann HC. Nanoparticle air pollution in major cities and its origin. *Atmos Environ* 2000;**34**(3):443–51.
44. Oberdorster G. Pulmonary effects of inhaled ultrafine particles. *Int Arch Occup Environ Health* 2001;**74**(1):1–8.
45. Boxall ABA, Chaudhry Q, Sinclair C, Jones A, Aitken R, Jefferson B, et al. *Current and future predicted environmental exposure to engineered nanoparticles*. York: Central Science Laboratory; 2007.
46. O'Brien N, Cummins E. Ranking initial environmental and human health risk resulting from environmentally relevant nanomaterials. *J Environ Sci Health Part A* 2010;**45**(8): 992–1007.
47. Choi O, Cleuenger TE, Deng B, Surampalli RY, Ross Jr L, Hu Z. Role of sulfide and ligand strength in controlling nanosilver toxicity. *Water Res* 2009;**43**(7):1879–86.

48. Gottschalk F, Ort C, Scholz RW, Nowack B. Engineered nanomaterials in rivers - exposure scenarios for Switzerland at high spatial and temporal resolution. *Environ Pollut* 2011;**159**(12).
49. Gottschalk F, Sun T, Nowack B. Environmental concentrations of engineered nanomaterials: review of modeling and analytical studies. *Environ Pollut* 2013;**181**:287–300.
50. Conway JR, Hanna SK, Lenihan HS, Keller AA. Effects and implications of trophic transfer and accumulation of CeO_2 nanoparticles in a marine mussel. *Environ Sci Technol* 2014; **48**(3):1517–24.
51. Holbrook RD, Murphy KE, Morrow JB, Cole KD. Trophic transfer of nanoparticles in a simplified invertebrate food web. *Nat Nanotechnol* 2008;**3**(6):352–5.
52. Hou WC, Westerhoff P, Posner JD. Biological accumulation of engineered nanomaterials: a review of current knowledge. *Environ Sci Processes Impacts* 2013;**15**(1):103–22.
53. Wang J, Wang W-X. Low bioavailability of silver nanoparticles presents trophic toxicity to Marine medaka (*Oryzias melastigma*). *Environ Sci Technol* 2014;**48**(14):8152–61.
54. Zhu X, Wang J, Zhang X, Chang Y, Chen Y. Trophic transfer of TiO_2 nanoparticles from daphnia to zebrafish in a simplified freshwater food chain. *Chemosphere* 2010;**79**(9): 928–33.
55. Tappin AD, Barriada JL, Braungardt CB, Evans EH, Patey MD, Achterberg EP. Dissolved silver in European estuarine and coastal waters. *Water Res* 2010;**44**(14):4204–16.
56. Oberdorster G. Significance of particle parameters in the evaluation of exposure-dose-response relationships of inhaled particles. *Part Sci Technol* 1996;**14**(2):135–51.
57. Fabrega J, Fawcett SR, Renshaw JC, Lead JR. Silver nanoparticle impact on bacterial growth: effect of pH, concentration, and organic matter. *Environ Sci Technol* 2009;**43**(19): 7285–90.
58. Wittmaack K. In search of the most relevant parameter for quantifying lung inflammatory response to nanoparticle exposure: Particle number, surface area, or what? *Environmental Health Perspectives* 2007;**115**(2):187–94.
59. Limbach LK, Li YC, Grass RN, Brunner TJ, Hintermann MA, Muller M, Gunther D, Stark WJ. Oxide nanoparticle uptake in human lung fibroblasts: Effects of particle size, agglomeration, and diffusion at low concentrations. *Environmental Science & Technology* 2005;**39**(23):9370–6.
60. Brunner TJ, Wick P, Manser P, Spohn P, Grass RN, Limbach LK, et al. In vitro cytotoxicity of oxide nanoparticles: comparison to asbestos, silica, and the effect of particle solubility. *Environ Sci Technol* 2006;**40**(14):4374–81.
61. Fabrega J, Luoma SN, Tyler CR, Galloway TS, Lead JR. Silver nanoparticles: behaviour and effects in the aquatic environment. *Environ Int* 2011a;**37**(2):517–31.
62. Carlson C, Hussain SM, Schrand AM, Braydich-Stolle LK, Hess KL, Jones RL, et al. Unique cellular interaction of silver nanoparticles: size-dependent generation of reactive oxygen species. *J Phys Chem B* 2008;**112**(43):13608–19.
63. Scown TM, Santos EM, Johnston BD, Gaiser B, Baalousha M, Mitov S, et al. Effects of aqueous exposure to silver nanoparticles of different sizes in rainbow Trout. *Toxicol Sci* 2010;**115**(2):521–34.
64. Pal S, Tak YK, Song JM. Does the antibacterial activity of silver nanoparticles depend on the shape of the nanoparticle? A study of the gram-negative bacterium *Escherichia coli*. *Appl Environ Microbiol* 2007;**73**(6):1712–20.
65. El Badawy AM, Silva RG, Morris B, Scheckel KG, Suidan MT, Tolaymat TM. Surface charge-dependent toxicity of silver nanoparticles. *Environ Sci Technol* 2011;**45**(1): 283–7.

66. Bilberg K, Doving KB, Beedholm K, Baatrup E. Silver nanoparticles disrupt olfaction in Crucian carp (*Carassius carassius*) and Eurasian perch (*Perca fluviatilis*). *Aquat Toxicol* 2011;**104**(1–2).
67. Angel BM, Batley GE, Jarolimek CV, Rogers NJ. The impact of size on the fate and toxicity of nanoparticulate silver in aquatic systems. *Chemosphere* 2013;**93**(2):359–65.
68. Kwok KWH, Auffan M, Badireddy AR, Nelson CM, Wiesner MR, Chilkoti A, et al. Uptake of silver nanoparticles and toxicity to early life stages of Japanese medaka (*Oryzias latipes*): effect of coating materials. *Aquat Toxicol* 2012;**120**:59–66.
69. Kvitek L, Panacek A, Soukupova J, Kolar M, Vecerova R, Prucek R, et al. Effect of surfactants and polymers on stability and antibacterial activity of silver nanoparticles (NPs). *J Phys Chem C* 2008;**112**(15):5825–34.
70. Karlsson HL, Gustafsson J, Cronholm P, Moller L. Size-dependent toxicity of metal oxide particles-A comparison between nano- and micrometer size. *Toxicol Lett* 2009;**188**(2): 112–8.
71. Merrifield RC, Cole P, Lead JR. The size related toxicity of cerium oxide nanoparticles. *Geochim Cosmochim Acta* 2010;**74**(12):A700-A700.
72. Pelletier DA, Suresh AK, Holton GA, McKeown CK, Wang W, Gu B, et al. Effects of engineered cerium oxide nanoparticles on bacterial growth and viability. *Appl Environ Microbiol* 2010;**76**(24):7981–9.
73. Renwick LC, Brown D, Clouter A, Donaldson K. Increased inflammation and altered macrophage chemotactic responses caused by two ultrafine particle types. *Occup Environ Med* 2004;**61**(5):442–7.
74. Van Hoecke K, Quik JTK, Mankiewicz-Boczek J, De Schamphelaere KAC, Elsaesser A, Van der Meeren P, et al. Fate and effects of CeO_2 nanoparticles in aquatic ecotoxicity tests. *Environ Sci Technol* 2009;**43**(12):4537–46.
75. Oberdorster G. Toxicology of ultrafine particles: in vivo studies. *Philos Trans R Soc Lond Ser A* 2000;**358**(1775):2719–39.
76. Choi O, Hu Z. Size dependent and reactive oxygen species related nanosilver toxicity to nitrifying bacteria. *Environ Sci Technol* 2008;**42**(12):4583–8.
77. Nativo P, Prior IA, Brust M. Uptake and intracellular fate of surface-modified gold nanoparticles. *Acs Nano* 2008;**2**(8):1639–44.
78. Albanese A, Sykes EA, Chan WCW. Rough around the edges: the inflammatory response of microglial cells to spiky nanoparticles. *Acs Nano* 2010;**4**(5):2490–3.
79. Ispas C, Andreescu D, Patel A, Goia DV, Andreescu S, Wallace KN. Toxicity and developmental defects of different sizes and shape nickel nanoparticles in zebrafish. *Environ Sci Technol* 2009;**43**(16):6349–56.
80. Liu S, Wei L, Hao L, Fang N, Chang MW, Xu R, et al. Sharper and faster "Nano darts" kill more bacteria: a study of antibacterial activity of individually dispersed pristine single-walled carbon nanotube. *Acs Nano* 2009;**3**(12):3891–902.
81. Arndt DA, Moua M, Chen J, Klaper RD. Core structure and surface functionalization of carbon nanomaterials alter impacts to daphnid mortality, reproduction, and growth: acute assays do not predict chronic exposure impacts. *Environ Sci Technol* 2013;**47**(16):9444–52.
82. Sayes CM, Wahi R, Kurian PA, Liu YP, West JL, Ausman KD, et al. Correlating nanoscale titania structure with toxicity: a cytotoxicity and inflammatory response study with human dermal fibroblasts and human lung epithelial cells. *Toxicol Sci* 2006;**92**(1):174–85.
83. Bar-Ilan O, Louis KM, Yang SP, Pedersen JA, Hamers RJ, Peterson RE, et al. Titanium dioxide nanoparticles produce phototoxicity in the developing zebrafish. *Nanotoxicology* 2012;**6**(6):670–9.

84. Griffitt RJ, Hyndman K, Denslow ND, Barber DS. Comparison of molecular and histological changes in zebrafish gills exposed to metallic nanoparticles. *Toxicol Sci* 2009; **107**(2):404–15.
85. Griffitt RJ, Weil R, Hyndman KA, Denslow ND, Powers K, Taylor D, et al. Exposure to copper nanoparticles causes gill injury and acute lethality in zebrafish (*Danio rerio*). *Environ Sci Technol* 2007;**41**(23):8178–86.
86. Navarro E, Baun A, Behra R, Hartmann NB, Filser J, Miao A-J, et al. Environmental behavior and ecotoxicity of engineered nanoparticles to algae, plants, and fungi. *Ecotoxicology* 2008;**17**(5):372–86.
87. van Aerle R, Lange A, Moorhouse A, Paszkiewicz K, Ball K, Johnston BD, de-Bastos E, Booth T, Tyler CR, Santos EM. Molecular Mechanisms of Toxicity of Silver Nanoparticles in Zebrafish Embryos. *Environ Sci Technol* 2013;**47**(14):8005–14.
88. Gu Y-J, Cheng J, Lin C-C, Lam YW, Cheng SH, Wong W-T. Nuclear penetration of surface functionalized gold nanoparticles. *Toxicol Appl Pharmacol* 2009;**237**(2):196–204.
89. Pan Y, Neuss S, Leifert A, Fischler M, Wen F, Simon U, et al. Size-dependent cytotoxicity of gold nanoparticles. *Small* 2007;**3**(11):1941–9.
90. Singh S, Shi T, Hohr D, Martra G, Fubini B, Borm P, et al. Interleukin-8 expression by titanium dioxide in A549 cells: effects of particle surface area and surface coating. *Toxicol Appl Pharmacol* 2004;**197**(3):137–375. (Abstract on page 307).
91. Tejamaya M, Roemer I, Merrifield RC, Lead JR. Stability of citrate, PVP, and PEG coated silver nanoparticles in ecotoxicology Media. *Environ Sci Technol* 2012;**46**(13):7011–7.
92. Osborne OJ, Johnston BD, Moger J, Balousha M, Lead JR, Kudoh T, et al. Effects of particle size and coating on nanoscale Ag and TiO_2 exposure in zebrafish (*Danio rerio*) embryos. *Nanotoxicology* 2013;**7**(8):1315–24.
93. Clift MJD, Rothen-Rutishauser B, Brown DM, Duffin R, Donaldson K, Proudfoot L, et al. The impact of different nanoparticle surface chemistry and size on uptake and toxicity in a murine macrophage cell line. *Toxicol Appl Pharmacol* 2008;**232**(3):418–27.
94. Isakovic A, Markovic Z, Todorovic-Markovic B, Nikolic N, Vranjes-Djuric S, Mirkovic M, et al. Distinct cytotoxic mechanisms of pristine versus hydroxylated fullerene. *Toxicol Sci* 2006;**91**(1):173–83.
95. Magrez A, Kasas S, Salicio V, Pasquier N, Seo JW, Celio M, et al. Cellular toxicity of carbon-based nanomaterials. *Nano Lett* 2006;**6**(6):1121–5.
96. Lockman PR, Koziara JM, Mumper RJ, Allen DD. Nanoparticle surface charges alter blood–brain barrier integrity and permeability. *J Drug Target* 2004;**12**(9–10):635–41.
97. Mayer A, Vadon M, Rinner B, Novak A, Wintersteiger R, Froehlich E. The role of nanoparticle size in hemocompatibility. *Toxicology* 2009;**258**(2–3):139–47.
98. Liu J, Hurt RH. Ion release kinetics and particle persistence in aqueous nano-silver colloids. *Environ Sci Technol* 2010;**44**(6):2169–75.
99. Romer I, White TA, Baalousha M, Chipman K, Viant MR, Lead JR. Aggregation and dispersion of silver nanoparticles in exposure media for aquatic toxicity tests. *J Chromatogr A* 2011;**1218**(27):4226–33.
100. Zhu X, Wang J, Zhang X, Chang Y, Chen Y. The impact of ZnO nanoparticle aggregates on the embryonic development of zebrafish (*Danio rerio*). *Nanotechnology* 2009;**20**(19).
101. Vileno B, Lekka M, Sienkiewicz A, Jeney S, Stoessel G, Lekki J, et al. Stiffness alterations of single cells induced by UV in the presence of NanoTiO(2). *Environ Sci Technol* 2007; **41**(14):5149–53.
102. Vevers WF, Jha AN. Genotoxic and cytotoxic potential of titanium dioxide (TiO_2) nanoparticles on fish cells in vitro. *Ecotoxicology* 2008;**17**(5):410–20.

103. Cheng JP, Flahaut E, Cheng SH. Effect of carbon nanotubes on developing zebrafish (*Danio rerio*) embryos. *Environ Toxicol Chem* 2007;**26**(4):708–16.
104. Murray AR, Kisin E, Leonard SS, Young SH, Kommineni C, Kagan VE, et al. Oxidative stress and inflammatory response in dermal toxicity of single-walled carbon nanotubes. *Toxicology* 2009;**257**(3):161–71.
105. Lovern SB, Klaper R. *Daphnia magna* mortality when exposed to titanium dioxide and fullerene (C-60) nanoparticles. *Environ Toxicol Chem* 2006;**25**(4):1132–7.
106. Tenzer S, Docter D, Kuharev J, Musyanovych A, Fetz V, Hecht R, et al. Rapid formation of plasma protein corona critically affects nanoparticle pathophysiology. *Nat Nanotechnol* 2013;**8**(10):772–U1000.
107. Dutta D, Sundaram SK, Teeguarden JG, Riley BJ, Fifield LS, Jacobs JM, et al. Adsorbed proteins influence the biological activity and molecular targeting of nanomaterials. *Toxicol Sci* 2007;**100**(1):303–15.
108. Cheng XK, Kan AT, Tomson MB. Naphthalene adsorption and desorption from aqueous C-60 fullerene. *J Chem Eng Data* 2004;**49**(3):675–83.
109. Zhang X, Sun H, Zhang Z, Niu Q, Chen Y, Crittenden JC. Enhanced bioaccumulation of cadmium in carp in the presence of titanium dioxide nanoparticles. *Chemosphere* 2007;**67**(1):160–6.
110. Baalousha M, Ju-Nam Y, Cole PA, Gaiser B, Fernandes TF, Hriljac JA, et al. Characterization of cerium oxide nanoparticles-Part 1: size measurements. *Environ Toxicol Chem* 2012;**31**(5):983–93.
111. Domingos RF, Baalousha MA, Ju-Nam Y, Reid MM, Tufenkji N, Lead JR, et al. Characterizing manufactured nanoparticles in the environment: multimethod determination of particle sizes. *Environ Sci Technol* 2009;**43**(19):7277–84.
112. Kashiwada S. Distribution of nanoparticles in the see-through medaka (*Oryzias latipes*). *Environ Health Perspect* 2006;**114**(11):1697–702.
113. Snell TW, Hicks DG. Assessing toxicity of nanoparticles using *Brachionus manjavacas* (Rotifera). *Environ Toxicol* 2011;**26**(2):146–52.
114. Scown TM, Goodhead RM, Johnston BD, Moger J, Baalousha M, Lead JR, et al. Assessment of cultured fish hepatocytes for studying cellular uptake and (eco)toxicity of nanoparticles. *Environ Chem* 2009;**7**(1):36–49.
115. Moger J, Johnston BD, Tyler CR. Imaging metal oxide nanoparticles in biological structures with CARS microscopy. *Opt Express* 2008;**16**(5):3408–19.
116. Goodhead RM, Moger J, Galloway TS, Tyler CR. Tracing engineered nanomaterials in biological tissues using coherent anti-stokes Raman scattering (CARS) microscopy – a critical review. *Nanotoxicology*, in press.
117. Menard A, Drobne D, Jemec A. Ecotoxicity of nanosized TiO_2. Review of in vivo data. *Environ Pollut* 2011;**159**(3):677–84.
118. Hartmann NB, Von der Kammer F, Hofmann T, Baalousha M, Ottofuelling S, Baun A. Algal testing of titanium dioxide nanoparticles-Testing considerations, inhibitory effects and modification of cadmium bioavailability. *Toxicology* 2010;**269**(2–3):190–7.
119. Aruoja V, Dubourguier H-C, Kasemets K, Kahru A. Toxicity of nanoparticles of CuO, ZnO and TiO_2 to microalgae *Pseudokirchneriella subcapitata*. *Sci Total Environ* 2009;**407**(4):1461–8.
120. Blaise C, Gagne F, Ferard JF, Eullaffroy P. Ecotoxicity of selected nano-materials to aquatic organisms. *Environ Toxicol* 2008;**23**(5):591–8.
121. Zhu X, Zhu L, Chen Y, Tian S. Acute toxicities of six manufactured nanomaterial suspensions to *Daphnia magna*. *J Nanopart Res* 2009;**11**(1):67–75.

122. Wiench K, Wohlleben W, Hisgen V, Radke K, Salinas E, Zok S, et al. Acute and chronic effects of nano- and non-nano-scale TiO_2 and ZnO particles on mobility and reproduction of the freshwater invertebrate *Daphnia magna*. *Chemosphere* 2009;**76**(10):1356–65.
123. Zhu X, Chang Y, Chen Y. Toxicity and bioaccumulation of TiO_2 nanoparticle aggregates in *Daphnia magna*. *Chemosphere* 2010;**78**(3):209–15.
124. Griffitt RJ, Luo J, Gao J, Bonzongo JC, Barber DS. Effects of particle composition and species on toxicity of metallic nanomaterials in aquatic organisms. *Environ Toxicol Chem* 2008;**27**(9):1972–8.
125. Zhu XS, Zhu L, Duan ZH, Qi RQ, Li Y, Lang YP. Comparative toxicity of several metal oxide nanoparticle aqueous suspensions to zebrafish (*Danio rerio*) early developmental stage. *J Environ Sci Health Part A Tox Hazard Subst Environ Eng* 2008;**43**(3):278–84.
126. Yang SP, Bar-Ilan O, Peterson RE, Heideman W, Hamers RJ, Pedersen JA. Influence of humic acid on titanium dioxide nanoparticle toxicity to developing zebrafish. *Environ Sci Technol* 2013;**47**(9):4718–25.
127. Xiong D, Fang T, Yu L, Sima X, Zhu W. Effects of nano-scale TiO_2, ZnO and their bulk counterparts on zebrafish: acute toxicity, oxidative stress and oxidative damage. *Sci Total Environ* 2011;**409**(8):1444–52.
128. Lok C-N, Ho C-M, Chen R, He Q-Y, Yu W-Y, Sun H, et al. Silver nanoparticles: partial oxidation and antibacterial activities. *J Biol Inorg Chem* 2007;**12**(4):527–34.
129. Jiang W, Mashayekhi H, Xing B. Bacterial toxicity comparison between nano- and micro-scaled oxide particles. *Environ Pollut* 2009;**157**(5):1619–25.
130. Kang S, Mauter MS, Elimelech M. Microbial cytotoxicity of carbon-based nanomaterials: implications for river water and wastewater effluent. *Environ Sci Technol* 2009;**43**(7):2648–53.
131. Thill A, Zeyons O, Spalla O, Chauvat F, Rose J, Auffan M, et al. Cytotoxicity of CeO_2 nanoparticles for *Escherichia coli*. Physico-chemical insight of the cytotoxicity mechanism. *Environ Sci Technol* 2006;**40**(19):6151–6.
132. Lyon DY, Adams LK, Falkner JC, Alvarez PJJ. Antibacterial activity of fullerene water suspensions: effects of preparation method and particle size. *Environ Sci Technol* 2006; **40**(14):4360–6.
133. Brunet L, Lyon DY, Hotze EM, Alvarez PJJ, Wiesner MR. Comparative photoactivity and antibacterial properties of C-60 fullerenes and titanium dioxide nanoparticles. *Environ Sci Technol* 2009;**43**(12):4355–60.
134. Li Q, Mahendra S, Lyon DY, Brunet L, Liga MV, Li D, et al. Antimicrobial nanomaterials for water disinfection and microbial control: potential applications and implications. *Water Res* 2008;**42**(18):4591–602.
135. Doiron K, Pelletier E, Lemarchand K. Impact of polymer-coated silver nanoparticles on marine microbial communities: a microcosm study. *Aquat Toxicol* 2012;**124**:22–7.
136. Fabrega J, Zhang R, Renshaw JC, Liu W-T, Lead JR. Impact of silver nanoparticles on natural marine biofilm bacteria. *Chemosphere* 2011b;**85**(6):961–6.
137. Bradford A, Handy RD, Readman JW, Atfield A, Muehling M. Impact of silver nanoparticle contamination on the genetic diversity of natural bacterial assemblages in estuarine sediments. *Environ Sci Technol* 2009;**43**(12):4530–6.
138. Hall S, Bradley T, Moore JT, Kuykindall T, Minella L. Acute and chronic toxicity of nano-scale TiO_2 particles to freshwater fish, cladocerans, and green algae, and effects of organic and inorganic substrate on TiO_2 toxicity. *Nanotoxicology* 2009;**3**(2):91–7.
139. Hund-Rinke K, Simon M. Ecotoxic effect of photocatalytic active nanoparticles TiO_2 on algae and daphnids. *Environ Sci Pollut Res* 2006;**13**(4):225–32.

140. Velzeboer I, Hendriks AJ, Ragas AMJ, Van de Meent D. Aquatic ecotoxicity tests of some nanomaterials. *Environmental Toxicology and Chemistry* 2008;**27**(9):1942–7.
141. Turner A, Brice D, Brown MT. Interactions of silver nanoparticles with the marine macroalga, *Ulva lactuca*. *Ecotoxicology* 2012;**21**(1):148–54.
142. Miller RJ, Lenihan HS, Muller EB, Tseng N, Hanna SK, Keller AA. Impacts of metal oxide nanoparticles on marine phytoplankton. *Environ Sci Technol* 2010;**44**(19):7329–34.
143. Peng X, Palma S, Fisher NS, Wong SS. Effect of morphology of ZnO nanostructures on their toxicity to marine algae. *Aquat Toxicol* 2011;**102**(3–4):186–96.
144. Asghari S, Johari SA, Lee JH, Kim YS, Jeon YB, Choi HJ, et al. Toxicity of various silver nanoparticles compared to silver ions in *Daphnia magna*. *J Nanobiotechnol* 2012;**10**.
145. Zhao C-M, Wang W-X. Comparison of acute and chronic toxicity of silver nanoparticles and silver nitrate to *Daphnia magna*. *Environ Toxicol Chem* 2011;**30**(4):885–92.
146. Oberdorster E, Zhu SQ, Blickley TM, McClellan-Green P, Haasch ML. Ecotoxicology of carbon-based engineered nanoparticles: effects of fullerene (C-60) on aquatic organisms. *Carbon* 2006;**44**(6):1112–20.
147. Mwangi JN, Wang N, Ingersoll CG, Hardesty DK, Brunson EL, Li H, et al. Toxicity of carbon nanotubes to freshwater aquatic invertebrates. *Environ Toxicol Chem* 2012;**31**(8): 1823–30.
148. Galloway T, Lewis C, Dolciotti I, Johnston BD, Moger J, Regoli F. Sublethal toxicity of nano-titanium dioxide and carbon nanotubes in a sediment dwelling marine polychaete. *Environ Pollut* 2010;**158**(5):1748–55.
149. Buffet P-E, Tankoua OF, Pan J-F, Berhanu D, Herrenknecht C, Poirier L, et al. Behavioural and biochemical responses of two marine invertebrates *Scrobicularia plana* and *Hediste diversicolor* to copper oxide nanoparticles. *Chemosphere* 2011;**84**(1):166–74.
150. Garcia-Aonso J, Khan FR, Misra SK, Turmaine M, Smith BD, Rainbow PS, et al. Cellular internalization of silver nanoparticles in gut epithelia of the estuarine polychaete nereis diversicolor. *Environ Sci Technol* 2011;**45**(10):4630–6.
151. Hanna SK, Miller RJ, Zhou D, Keller AA, Lenihan HS. Accumulation and toxicity of metal oxide nanoparticles in a soft-sediment estuarine amphipod. *Aquat Toxicol* 2013;**142**:441–6.
152. Fairbairn EA, Keller AA, Maedler L, Zhou D, Pokhrel S, Cherr GN. Metal oxide nanomaterials in seawater: linking physicochemical characteristics with biological response in sea urchin development. *J Hazard Mater* 2011;**192**(3):1565–71.
153. Falugi C, Aluigi MG, Chiantore MC, Privitera D, Ramoino P, Gatti MA, et al. Toxicity of metal oxide nanoparticles in immune cells of the sea urchin. *Mar Environ Res* 2012;**76**:114–21.
154. Canesi L, Ciacci C, Fabbri R, Marcomini A, Pojana G, Gallo G. Bivalve molluscs as a unique target group for nanoparticle toxicity. *Mar Environ Res* 2012;**76**:16–21.
155. Tedesco S, Doyle H, Blasco J, Redmond G, Sheehan D. Oxidative stress and toxicity of gold nanoparticles in *Mytilus edulis*. *Aquat Toxicol* 2010;**100**(2):178–86.
156. Hull MS, Vikesland PJ, Schultz IR. Uptake and retention of metallic nanoparticles in the Mediterranean mussel (*Mytilus galloprovincialis*). *Aquat Toxicol (Amsterdam, Netherlands)* 2013;**140–141**:89–97.
157. Zuykov M, Pelletier E, Demers S. Colloidal complexed silver and silver nanoparticles in extrapallial fluid of *Mytilus edulis*. *Mar Environ Res* 2011;**71**(1):17–21.
158. Canesi L, Fabbri R, Gallo G, Vallotto D, Marcomini A, Pojana G. Biomarkers in *Mytilus galloprovincialis* exposed to suspensions of selected nanoparticles (nano carbon black, C60 fullerene, Nano-TiO_2, Nano-SiO_2). *Aquat Toxicol* 2010;**100**(2):168–77.
159. Ringwood AH, Levi-Polyachenko N, Carroll DL. Fullerene exposures with oysters: embryonic, adult, and cellular responses. *Environ Sci Technol* 2009;**43**(18):7136–41.

160. Zhu X, Zhou J, Cai Z. TiO(2) nanoparticles in the marine environment: impact on the toxicity of tributyltin to abalone (*Haliotis diversicolor supertexta*) embryos. *Environ Sci Technol* 2011;**45**(8):3753–8.

161. Asharani PV, Yi L, Gong Z, Valiyaveettil S. Comparison of the toxicity of silver, gold and platinum nanoparticles in developing zebrafish embryos. *Nanotoxicology* 2011;**5**(1):43–54.

162. Zhu XS, Zhu L, Li Y, Duan ZH, Chen W, Alvarez PJJ. Developmental toxicity in zebrafish (*Danio rerio*) embryos after exposure to manufactured nanomaterials: buckminsterfullerene aggregates (nC(60)) and fullerol. *Environ Toxicol Chem* 2007;**26**(5):976–9.

163. Bilberg K, Hovgaard MB, Besenbacher F, Baatrup E. In vivo toxicity of silver nanoparticles and silver ions in zebrafish (*Danio rerio*). *J Toxicol* 2012;**2012**.

164. Choi JE, Kim S, Ahn JH, Youn P, Kang JS, Park K, et al. Induction of oxidative stress and apoptosis by silver nanoparticles in the liver of adult zebrafish. *Aquat Toxicol* 2010;**100**(2): 151–9.

165. Johnston BD, Scown TM, Moger J, Cumberland SA, Baalousha M, Linge K, et al. Bioavailability of nanoscale metal oxides TiO_2, CeO_2, and ZnO to fish. *Environ Sci Technol* 2010;**44**(3):1144–51.

166. Federici G, Shaw BJ, Handy RD. Toxicity of titanium dioxide nanoparticles to rainbow trout (*Oncorhynchus mykiss*): gill injury, oxidative stress, and other physiological effects. *Aquat Toxicol* 2007;**84**(4):415–30.

167. Griffitt RJ, Brown-Peterson NJ, Savin DA, Manning CS, Boube I, Ryan RA, et al. Effects of chronic nanoparticulate silver exposure to adult and juvenile sheepshead minnows (*Cyprinodon variegatus*). *Environ Toxicol Chem* 2012;**31**(1):160–7.

168. Salari Joo H, Kalbassi MR, Yu IJ, Lee JH, Johari SA. Bioaccumulation of silver nanoparticles in rainbow trout (*Oncorhynchus mykiss*): influence of concentration and salinity. *Aquat Toxicol (Amsterdam, Netherlands)* 2013;**140–141**:398–406.

169. Smith CJ, Shaw BJ, Handy RD. Toxicity of single walled carbon nanotubes to rainbow trout, (*Oncorhynchus mykiss*): respiratory toxicity, organ pathologies, and other physiological effects. *Aquat Toxicol* 2007;**82**(2):94–109.

170. Boyle D, Fox JE, Akerman JM, Sloman KA, Henry TB, Handy RD. Minimal effects of waterborne exposure to single-walled carbon nanotubes on behaviour and physiology of juvenile rainbow trout (*Oncorhynchus mykiss*). *Aquat Toxicol* 2014;**146**:154–64.

171. McTeer J, Dean AP, White KN, Pittman JK. Bioaccumulation of silver nanoparticles into *Daphnia magna* from a freshwater algal diet and the impact of phosphate availability. *Nanotoxicology* 2014;**8**(3):305–16.

172. Kulacki KJ, Cardinale BJ, Keller AA, Bier R, Dickson H. How do stream organisms respond to, and influence, the concentration of titanium dioxide nanoparticles? A mesocosm study with algae and herbivores. *Environ Toxicol Chem* 2012;**31**(10):2414–22.

173. Buffet P-E, Richard M, Caupos F, Vergnoux A, Perrein-Ettajani H, Luna-Acosta A, et al. A mesocosm study of fate and effects of CuO nanoparticles on endobenthic species (*Scrobicularia plana, Hediste diversicolor*). *Environ Sci Technol* 2013;**47**(3):1620–8.

174. Buffet P-E, Zalouk-Vergnoux A, Chatel A, Berthet B, Metais I, Perrein-Ettajani H, et al. A marine mesocosm study on the environmental fate of silver nanoparticles and toxicity effects on two endobenthic species: the ragworm *Hediste diversicolor* and the bivalve mollusc *Scrobicularia plana*. *Sci Total Environ* 2014;**470**:1151–9.

175. Ferry JL, Craig P, Hexel C, Sisco P, Frey R, Pennington PL, et al. Transfer of gold nanoparticles from the water column to the estuarine food web. *Nat Nanotechnol* 2009;**4**(7): 441–4.

176. Tella M, Auffan M, Brousset L, Issartel J, Kieffer I, Pailles C, et al. Transfer, transformation, and impacts of Ceria nanomaterials in aquatic mesocosms simulating a pond ecosystem. *Environ Sci Technol* 2014;**48**(16):9004–13.
177. Franklin NM, Rogers NJ, Apte SC, Batley GE, Gadd GE, Casey PS. Comparative toxicity of nanoparticulate ZnO, bulk ZnO, and $ZnCl_2$ to a freshwater microalga (*Pseudokirchneriella subcapitata*): the importance of particle solubility. *Environ Sci Technol* 2007;**41**(24):8484–90.
178. Yeo M-K, Kang M. Effects of nanometer sized silver materials on biological toxicity during zebrafish embryogenesis. *Bull Korean Chem Soc* 2008;**29**(6):1179–84.
179. Oberdorster E. Manufactured nanomaterials (Fullerenes, C-60) induce oxidative stress in the brain of juvenile largemouth bass. *Environ Health Perspect* 2004;**112**(10):1058–62.
180. Henry TB, Menn F-M, Fleming JT, Wilgus J, Compton RN, Sayler GS. Attributing effects of aqueous C-60 nano-aggregates to tetrahydrofuran decomposition products in larval zebrafish by assessment of gene expression. *Environ Health Perspect* 2007;**115**(7): 1059–65.
181. Shinohara N, Matsumoto T, Gamo M, Miyauchi A, Endo S, Yonezawa Y, et al. Is lipid peroxidation induced by the aqueous suspension of fullerene C-60 nanoparticles in the brains of *Cyprinus carpio*? *Environ Sci Technol* 2009;**43**(3):948–53.
182. Mahapatra I, Clark J, Dobson PJ, Owen R, Lead JR. Potential environmental implications of nano-enabled medical applications: critical review. *Environ Sci Process Impacts* 2013; **15**(1):123–44.
183. Baalousha M, How W, Valsami-Jones E, Lead J. In: *Overview of Environmental Nanoscience In: Frontiers of Nanoscience*, vol. 7. Oxford: Elsevier; 2014, ISBN 978-0-08-099408-6. p. 1–54.

Chapter 2

Overview of Nanomaterial Characterization and Metrology

R. David Holbrook,*[,1] Anne A. Galyean,[§,‡] Justin M. Gorham,* Andrew Herzing[†] and John Pettibone*
*Nano Materials Research Group, Materials Measurement Science Division, Material Measurement Laboratory, National Institute of Standards and Technology, Gaithersburg, MD, USA; [§]Bioprocess Measurements Group, Biomolecular Measurement Division, Material Measurement Laboratory, National Institute of Standards and Technology, Gaithersburg, MD, USA; [‡]Gillings School of Global Public Health, Department of Environmental Sciences and Engineering, University of North Carolina, Chapel Hill, NC, USA; [†]Materials Structure and Data, Materials Measurement Science Division, Material Measurement Laboratory, National Institute of Standards and Technology, Gaithersburg, MD, USA
[1]Corresponding author: E-mail: dave.holbrook@nist.gov

Chapter Outline

Frontiers of Nanoscience, Vol. 8. http://dx.doi.org/10.1016/B978-0-08-099948-7.00002-6
2015 Published by Elsevier Ltd.

1. INTRODUCTION

Nanotechnology is based on the recognition that materials possessing nanometer-scale dimensions (i.e., nanomaterials) have unique, exploitable properties when compared with their bulk-scale counterparts.[1–3] These nanomaterials can be engineered using different elements and in many diverse forms, including but not limited to, nanomaterials, nanotubes, nanocrystals, nanoropes, nanofibers, nanocages, and nanorods.[4] The diversity of shapes and composition, the various methods of fabrication, the use of assorted surface coatings, and the subsequent behavior of these nanomaterials all speak to the complexity of using engineered nanomaterials effectively and safely. An important, if not key, parameter in reducing this inherent complexity is in conducting proper nanomaterial characterization. Characterization is important for at least three main reasons. Quantifying critical properties of a specific nanomaterial is necessary (1) to provide a comprehensive understanding of the starting material; (2) to understand how these same properties are altered in situ by a specific environment; and (3) to correlate these measured nanomaterial properties with a specific measured response. Yet, a major knowledge gap exists regarding the development of property–response relationships for a broad range of nanomaterials, and thus, identifying the key nanomaterial property (or properties) for a given behavior remains elusive.

The importance of proper nanomaterial characterization cannot be understated. Stefaniak et al.[5] argues that incomplete nanomaterial characterization limits the reliability (and perhaps validity) of conclusions drawn from scientific investigations, which may impede both future research in and commercialization of nanotechnology. By way of example, a review by Hansen et al.[6] of over 400 published nanotoxicology studies concluded that nanomaterial characterization was either poorly described or not at all conducted. Similarly, a review of the in vitro and in vivo genotoxicity literature indicated that comparing published data from disparate studies is difficult since detailed nanomaterial characterization is often missing.[7] Consequently, building a general consensus framework from different investigations without proper characterization is nearly impossible, even when similar nanomaterials are used.[8] Such uncertainty from the scientific community will also not favor timely environmental, health and safety guidelines from regulatory agencies.

Given the importance of nanomaterial characterization to the nanotechnology sector, it should come as no surprise that much emphasis has been placed on the following related areas: adaptation and optimization of existing characterization techniques for nanomaterials; development of new, robust characterization techniques; and creation of guidance documents and reference materials to assess method performance and laboratory proficiency. Yet a central question remains for nanomaterial characterization—what physiochemical parameter(s) should be measured for proper characterization?

This question has been widely debated (e.g., Ref. 5 and references therein), and although the answer is somewhat dependent upon the application or purpose, there are several measurable parameters that are considered critical in providing an overall nanomaterial characterization description. Broadly speaking, analytical characterization techniques can be divided into three main areas: physical, chemical, and behavioral. Physical characterization pertains to the nanomaterials' architecture and includes the size, size distribution, crystal structure, morphology, and/or surface area of a specific nanomaterial. Chemical and elemental characterization is focused on atomic compositions (including purity), bonding states, and oxidation states which could include bulk and surface measurements that provide information regarding atomic concentrations, specimen purity, and/or reactivity. Behavioral characterization, where the measured property indicates how a nanomaterial responds to external factors, is strongly dependent on the ambient environment. Surface charge, zeta potential, and particle stability (both colloidal and dissolution potential) will all depend on the specific matrix. As such, behavioral characterization descriptors all **require** context (quantified descriptors of the ambient environment) for relevance. While the above list is by no means exhaustive, these parameters are most likely to be deemed important for a wide range of applications including human and environmental risk assessments. Tiede et al.[9] provides a comprehensive discussion of the importance of characterization parameters for developing a better understanding of nanomaterial behavior.

Although the difficulty in developing nanomaterial property−response relationships has been mentioned, nanomaterial characterization also poses a significant challenge. In theory, nanomaterial characterization appears to be very straightforward—obtain a sample, measure various parameters for that sample, analyze the measured data, and finally report analyzed data. Practically, however, there can be many technical and methodological nuances that can inadvertently introduce experimental bias during nanomaterial characterization measurements. A well-known example of such bias occurs when quantifying the average diameter of a heterogeneous particle mixture using dynamic light scattering (DLS)[10]; the size distribution is shifted to the most efficient scatters (larger particles) thereby underrepresenting (or completely hiding) the population of smaller particles, which can also be an artifact of the parameters used in the software. This example serves to highlight that a thorough understanding of all issues that could influence a specific measurement for a given nanomaterial and instrument is a vital necessity for the successful analyst. In subsequent sections, some issues are highlighted for different characterization techniques, if for no other reason than to draw the analyst's attention away from simply obtaining data (e.g., blindly using an instrument) and more toward performing *quality and reliable measurements.*

1.1 Instrument Calibration, Traceability, and Validation

In an ideal world, all analytical measurements made in one location would be consistent with those made elsewhere. Some steps to consider in realizing this goal include the following: ensuring that measurement results from a specific instrument are accurate compared with a known entity (e.g., physical standard) and is operating within proper instrumentation guidelines (calibration), having an unbroken chain of calibrations using proper materials (such as reference materials and/or certified reference materials) that have been characterized by a metrologically valid procedure with associated uncertainty (traceability), and performing periodic assessments to ensure neither instrument nor samples have drifted outside the original, intended purpose (metrological validation). These steps, along with ensuring that staff making the actual measurements are both qualified and competent, will assist in creating robust, reliable, and reproducible measurements.

1.2 Reference Materials and Documentary Standards

The use of available physical standards, documentary standards, and standardized protocols can be quite helpful in achieving measurement confidence. Nanomaterial physical standards, such as the different types of reference materials,[11,12] are typically small quantities of identical materials that have been well-characterized for one or more characterization parameters. These standards can be used to calibrate instruments, assess measurement procedures, assignment of values to other materials (e.g., reference material equivalents) or used in a quality control procedure. Documentary standards are established by consensus, approved by a recognized body, and provide detailed guidelines, rules, or characteristics in an effort to achieve the most consistent and reproducible results. Recognized standard bodies for nanomaterials include the International Organization for Standardization (ISO), ASTM International, and International Electrotechnical Commission (ICE). ISO/TR 13014:2012, for example, provides guidance for the physicochemical characterization of manufactured nanomaterials for toxicological testing purposes.[13] Standardized protocols typically contain precise instructions for performing a specific method.[14,15] Such protocols often contain a detailed evaluation of important operational parameters to better assist the reader in achieving experimental reproduction.

1.3 Sample Preparation

Meticulous sample preparation can be an important step in minimizing sample artifacts. Nanomaterials are notorious for physical and chemical changes in ambient or controlled environments[16] and, compared with their bulk counterparts, can offer a new set of procedural challenges to the analyst before any measurements are even conducted. Aside from analyst safety concerns,[17] sample composition, concentration, type, and matrix will all need to be considered, as well as the analysis environment of the required instrumentation.[18,19] For example, does sample preparation or analysis require ultrahigh vacuum (UHV) levels (where liquids would quickly evaporate, thereby potentially induce agglomeration) or can samples remain fully hydrated during analysis? Each sample and measurement requirement will need to be evaluated on a case-by-case basis. Indeed, sample preparation for measurement of specific properties and in specific environments is discussed more thoroughly in Chapters 3 through 11 of this book.

1.4 Sample Size

Microscopy techniques are often used to quantify particle size, particle size distribution, and morphology of a given sample. Such techniques rely on the collection of representative images at sufficient magnification to allow resolution of individual particles. Statistical errors in the particle size and particle size distribution measurements may be introduced if an insufficient number of particles are counted.[20,21] Comprehensive procedures to minimize these errors, including guidelines for estimating the number of particles that need to be analyzed for a given level of confidence, are available.[22,23] Assuming a lognormal distribution, the number of particles required for counting is dependent on the evaluation basis (i.e., mass-median diameter vs mean volume diameter), the acceptable error, and the geometric standard deviation of the population. Depending on requirement of accuracy, the number of particles to be counted may range from several hundreds to several thousands.[23]

In the following discussion, techniques are grouped according to their primary purpose (physical (Section 1), chemical (Section 2), behavioral (Section 3), and combined (Section 4) characterization methods), but there can be substantial overlap in these characterization areas for some techniques.

Since a detailed discussion of each technique is beyond the scope of this chapter, the reader is directed to the cited references, which should provide a more thorough technical discussion. The reader is also directed to several review papers that can provide additional detail (e.g., Refs. 24,25 and references therein).

2. PHYSICAL CHARACTERIZATION

Physical characterization measurements are intended to represent the structure and/or architecture of a specific nanomaterial. Descriptive parameters such as particle size, particle size distribution, shape, morphology, and surface area would be considered physical in nature. Many of these techniques require proper sample-handling protocols in order to reduce artifacts. For example, a number of physical characterization techniques are carried out in UHV conditions, and sample preparation for introduction into UHV systems may promote nanomaterial agglomeration/aggregation.

In the physical characterization section, specific imaging (subsection 1.1), fractionation (1.2), scattering (1.3), and surface area (1.4) measurement techniques are described. Further discussion on size/shape and size/shape distribution can be found in Chapter 3, composition (phase and crystallinity) in Chapter 4, concentration in Chapter 5, and physical surface properties in Chapter 7.

2.1 Electron and Scanning Probe Microscopy

The signals resulting from the scattering of energetic electrons by the atoms within a specimen contain a wealth of information related to its structure and composition, and these signals underlie the family of characterization techniques known as electron microscopy and microanalysis. These techniques are capable of much higher spatial resolutions than those achievable by optical microscopy due to the much shorter wavelength of the incident radiation. For example, 30 keV electrons exhibit a wavelength of $\approx$ 41 pm compared to 390–700 nm for visible light. In fact, modern instrumentation has enabled the acquisition of atomic-scale structural and chemical information to be nearly routine, making electron microscopy an indispensable tool for nanomaterial characterization. In this section, an overview will be given of the two most widely used classes of electron microscope (EM) instruments: the scanning electron microscope (SEM) and the transmission electron microscope (TEM). While the physics underlying these instruments are the same, their practical use differs enough to warrant a separate discussion of each. The reader is also directed to the following references for discussion on sample preparation considerations for electron microscopy, including surface contamination, surface charging, and outgassing.[18,20,26,27]

For nanomaterial characterization, the spatial resolution and chemical sensitivity of electron beam techniques can be especially powerful. However, care must be taken since the incident electron beam can significantly alter or entirely destroy the specimen being analyzed via beam-induced ionization and sputtering.[28] In addition, except in specialized equipment, these techniques must be carried out under conditions of high vacuum (HV). Not only does this limit the types of specimens that can be analyzed, but can also produce physical and chemical alterations in samples which are sensitive to the gaseous environment in which they are contained.

2.1.1 Scanning Electron Microscopy

In the SEM, an electron gun assembly produces a beam of electrons with a primary energy that typically falls in the range of 500–40 keV depending upon the application.[29,30] This beam is focused to a small spot by electromagnetic lenses and is then scanned over the surface of the specimen in a raster using scan coils to control its position. Serially collecting one or more of the signals resulting from the interaction between the incident electrons and the atoms within the specimen produces a spatially resolved image. A wide variety of signals related to the structural and chemical makeup of the specimen is accessible in the SEM, so it is important for the analyst to consider the problem at hand and to properly choose the experimental parameters to ensure that the data collected will be most useful.

Micro- and nanoscale morphology and topography are typically assessed using the secondary electron (SE) signal, consisting of low-energy (< 50 eV) electrons ejected from the sample by inelastic scattering of the primary beam. Due to the low energy which they possess, only those SE's generated close to the specimen surface and within a short distance of the beam position can escape the specimen and contribute to the collected image signal. Because of this, both the lateral and depth resolutions of SE images can be very high (< 1 nm).

Chemical contrast can be gained by collecting the backscattered electron (BSE) signal. In this case, primary beam electrons are elastically reflected back toward the direction of incidence without significant energy loss. Since more massive atoms backscatter electrons more strongly than lower mass atoms, the intensity at any given pixel in a BSE image is proportional to the average atomic number present at that point in the specimen. Topographic contrast and edge effects are reduced in this image mode relative to SE imaging, and thus interpretation of the contrast can be much more quantitative. For nanomaterial characterization, this imaging mode is extremely useful for enhancing the contrast between heavy metal nanomaterials from a lower Z matrix such as that encountered in supported metal catalysts or particles embedded in organic matter.

One of the chief advantages of SEM imaging is the extremely large depth of field which the technique is capable of. This can be several tens of microns in magnitude, meaning that all areas of a complex, topographic specimen can be imaged simultaneously. For example, supported metal catalysts consist of nanomaterials dispersed over larger substrate particles. The latter will often agglomerate to form larger, 3-D networks which are quite complex and exhibit a significant degree of topography. These networks can be directly imaged in the SEM owing to the large depth of field, and the 3-D structure of the agglomerates, degree of agglomeration, etc., can be readily determined.

2.1.2 Focused Ion Beam-SEM

The SEM is often outfitted with a second column which produces a focused beam of heavy ions in order to locally remove material and reveal interior features.[31] The focused ion beam (FIB) source can consist of any low-melting-temperature metal or gaseous plasma. In the former case, gallium is the most widely used source material, while xenon is usually employed for FIBs with a plasma source. For nanomaterial analysis, the main application of the FIB is the characterization of nanomaterial distributions embedded within a matrix. This is done using a "slice-and-view" type technique, wherein a small amount of material is removed by the ion beam and the freshly exposed, underlying face is then analyzed using any of the SEM signals previously discussed. By repeating this removal/analysis sequence many times, the three-dimensional distribution of nanomaterials and their elemental composition can be ascertained.

2.1.3 Transmission Electron Microscopy

In the TEM, a thin, electron transparent specimen and the various signals generated by electron scattering are collected after the incident beam traverses the specimen thickness.[32–34] Much higher beam energies are utilized in this case than those employed in the SEM, with most instrumentation operating in the 80–300 keV range. These higher beam energies improve the electron transmission behavior, allowing specimens with larger thicknesses to be reliably analyzed. More importantly, the higher energy beam also improves the spatial resolution of the instrument, both through the reduced wavelength of the high-energy electrons as well as the limited spatial extent of the beam–specimen interaction volume. With recent advances in electron optics (e.g., aberration correctors, high-stability sources, etc.) and sample preparation, instrumentation capable of carrying out quality imaging and analysis at spatial resolutions in the order of 100 pm is now widely accessible.

The TEM can be utilized in two distinct modes of operation. In the first of these, conventional TEM (CTEM) modes, the electron optics in the instrument are configured so as to produce a parallel and static beam which evenly illuminates the specimen. The transmitted electron signal can then be collected in a variety of ways enabling a host of imaging methods. Most commonly, a bright-field (BF) image is obtained by utilizing a small objective aperture to collect only the electrons scattered to very low angles. In this way, all regions of the specimen that scatter electrons to higher angles than those permitted by the objective aperture will be dark in the image, while those scattering to lower angles will be bright. Depending on the desired information, this single imaging mode is often sufficient for nanomaterial characterization; allowing the analyst to reliably characterize the size and shape of individual particles as well as the distribution of both features present within a larger population.

Further information about the specimen can be collected via other imaging signals in the TEM. For instance, if the particles analyzed are crystalline, Bragg diffraction of the electron beam can be utilized for structural determination. The signal collected in CTEM is typically integrated over a large area of the specimen (i.e., > 500 nm), and thus, for nanomaterial characterization, will represent the diffraction characteristics of several tens or hundreds of particles depending on the density of particle coverage and the size of the illuminated area. While this negates the possibility of performing single crystal diffraction experiments, polycrystalline patterns can still be used to characterize the structures present in the sampled population. Alternatively, the objective aperture can be positioned around one of the reflections in the diffraction pattern. In this case, a dark-field (DF) image is formed using only these diffracted electrons, and this provides an indication of the spatial distribution of specific structural features.

Finally, a particularly powerful technique for nanomaterial characterization can be utilized if the TEM is equipped with a coherent electron source such as a field-emission electron gun: the so-called high-resolution TEM (HR-TEM) or lattice-imaging mode.[35] In HR-TEM, a large objective aperture is employed so as to allow the unscattered beam and one or more coherently diffracted beams to contribute to the image. The interaction of these coherent beams produces what amounts to an interference pattern in the image, which, under certain conditions, can be interpreted as a representation of the crystalline lattice present in the specimen. An example of this is given in Figure 1

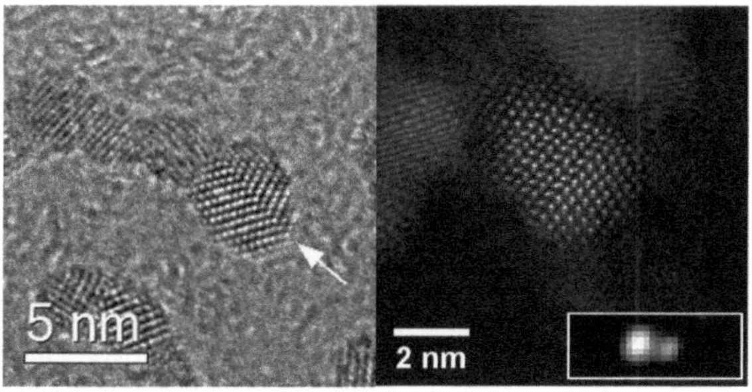

FIGURE 1 Left figure: High-resolution transmission electron microscope lattice image of CdSe quantum dots showing lattice planes of the Wurtzite structure along with a twin boundary (arrow) in one of the larger particles. Right figure: Scanning transmission electron microscope-high-angle annular dark field image from the same specimen showing atomic number contrast between the Cd and Se atoms within the structure (see inset).

(left panel), which shows an HR-TEM image of a CdSe quantum dot. The lattice planes are clearly resolved in this case, showing that the particle exhibits the hexagonal Wurtzite-type structure. A twin boundary is also evident in the largest particle present in the image. Extreme care must be taken in the interpretation of such images, since phase shifts due to specimen thickness and microscope defocus can result in an erroneous assignment of atomic positions. Because of this, a series of images is often acquired over a range of microscope defocus conditions, and these experimental data are then compared to computer simulations in order to ensure proper interpretation.[36]

Aside from CTEM mode, most instruments can also be operated in **scanning TEM (STEM)** mode,[37,38] which can be useful for nanomaterial characterization and is commonly performed in dedicated TEM/STEM instruments due to sample constraints. In STEM, the beam is focused to a small spot and scanned over the specimen surface. BF and DF images can be collected similar to those previously described for CTEM, and, due to the nature of the detector configuration in STEM, multiple signals can be acquired simultaneously. One particularly useful mode not available in CTEM is known as high-angle annular dark field (HAADF) imaging. In this case, an annular detector is used to integrate the signal scattered to very high angles (e.g., 70–150 mrads), which is dominated by Rutherford-type, elastic scattering. This signal carries two important characteristics. The first is that it is largely incoherent so the thickness/defocus considerations previously described for HR-TEM imaging are not as significant in this case. Secondly, the probability of scattering to these high angles carries an atomic number (Z) dependence, enabling the so-called Z-contrast imaging where the intensity at any pixel in the image is proportional to the mean atomic number at that point in the specimen. An example of this is shown in Figure 1 (right panel), which presents a STEM-HAADF image of the same CdSe quantum dot specimen that generated the HR-TEM image discussed previously. In this case, the alternating "zig-zag" pattern typical of the Wurtzite structure is detected, with each alternating unit consisting of a pair of Cd and Se atoms. If the image is zoomed to the level of a single atom pair (see inset), there is clear differential contrast between the two atoms due to the difference in atomic number ($Z_{Cd} = 48$, $Z_{Se} = 34$).

2.1.4 Environmental SEM/Environmental TEM/Wet Cell

The previous discussion has been limited to traditional electron microscopy techniques, all of which are carried out under HV or UHV conditions to limit electron scattering by ambient gas molecules. However, introducing the specimen to such an environment can have unpredictable effects on its structure and chemistry, especially if it is in a more reactive nanoparticulate form. For this reason, techniques have been developed whereby electron microscopy and microanalysis can be carried out in gaseous or liquid environments.[39] This is accomplished by isolating the region of the instrument where the specimen is located from the rest of the instrument. In this way, the majority of the beam flight path takes place in an HV (i.e., low scattering) environment and, no less importantly, the sensitive components of the instrument such as the electron gun are protected. One way that this is currently done is by employing a multistage differential pumping aperture, and this design underlies the environmental SEM and TEM (commonly referred to as ESEM and ETEM, respectively). The development of these techniques has provided an enhanced understanding of the active mechanisms for carbon nanotube (CNT) growth[40] as well as the changes undergone by oxide nanomaterials during catalytic reactions.[41]

An alternative method of isolation is to enclose the specimen within a small vessel, which is then introduced into the SEM chamber or TEM column. The vessel, commonly referred to as a "wet cell," is fabricated to allow the incident probe and desired signal to pass through a thin yet water impermeable membrane.[42] This so-called "wet-cell" approach is most commonly used for imaging particles in liquid, whereas the ESEM/ETEM is typically used for controlling the gaseous environment surrounding the particles. The "wet-cell" can be fully isolated, with no liquid exchange occurring once it has been sealed and inserted into the instrument. Alternatively, the cell can be outfitted with fluid flow channels connecting the cell interior to the exterior of the instrument. A pump is then used to induce liquid exchange from an external reservoir, allowing the specimen to be analyzed under dynamic flow conditions. While the technique is relatively new, already several reports of liquid-born nanomaterial characterization have been published.[42–46]

2.1.5 Atomic Force Microscopy

Atomic force microscopy (AFM) is a powerful scanning probe microscopy technique that can be used to analyze any solid surface at very high spatial resolutions and provide information on morphology and surface properties. A cantilever is mounted onto a stage where the tip's location is noted by a laser detector. The tip is brought into close proximity with a surface to the point where the attractive and repulsive forces associated with the sample can be "detected" by the probe. Piezoelectric scanners sweep the probe laterally in two-dimensions. When the probe is scanned over the specimen, the potential energy difference between the cantilever tip and the sample surface causes a vertical displacement of the probe. Depending on the mode of operation, the AFM uses this tip-sample force as the imaging signal and a feedback loop is used to maintain a constant force on the tip by adjusting the sample height. The adjustments are recorded and translated into a three-dimensional map of the sample surface. AFM images consistently achieve Angstrom-level resolution in the z-dimension (height), while resolution in the lateral dimensions are limited by the curvature of the tip (typically 10 nm or more). A direct link between the experimental signal and underlying physical forces has been established, where measured forces can be tractably interpreted to yield information about the strength and geometry of chemical bonds between single atoms.[47,48] Furthermore, the analysis can be carried out in ambient conditions.

AFM can be used to characterize height, size, shape, aspect ratio, and surface morphology of individual particles.[49] In AFM experiments, particle size is defined as the maximum height of the particle and particle size distributions may be calculated using three-dimensional analysis software. AFM images have three dimensions of data, which allow for nanomaterial diameter, volume, and surface area to be calculated. These measurements are affected by the probe geometry and the cantilever is often calibrated in the z-dimension for particle size measurements. Other parameters, such as the particle volume and circumference, critically depend on the cantilever geometry.[47]

Particles deposited on rough or curved substrates generally make characterization difficult. AFM imaging requires that particles be rigidly adhered to a substrate, that particles are well-dispersed on the substrate, and that the substrate roughness is less than the size of the nanomaterials. There are numerous combinations of nanomaterials, substrates, and adhesives, however, that have been demonstrated to be suitable for AFM and each requires adequate surface treatments. Various sample and surface treatments for nanomaterial characterization have been discussed in the literature.[50]

2.2 Fractionation

Matrix variability and sample polydispersity in nanomaterial samples can often pose measurement challenges for many direct characterization techniques. Nondestructive fractionation methods can be interfaced with characterization techniques to reduce sample complexity and typically obtain indirect particle size measurements by calibration with known size standards relative to retention time. These indirect measurements, however, are heavily dependent on the reproducibility of operating conditions and therefore care must be taken.

2.2.1 Field Flow Fractionation

Field flow fractionation (FFF) is a class of analytical methods suitable for the separation and characterization of nanomaterials and shares the most common likeness with liquid chromatography (LC). The mechanism for separation, however, departs from the classical enthalpic interactions with a stationary phase used in LC methods. Instead, a field is applied normal to a laminar flow through a narrow channel, which results in a parabolic flow profile effectively separating different analytes into distinct regions of the velocity profile and results in detection out of the channel at different times.

Commercially available types of FFF instruments include flow FFF (FlFFF) and sedimentation FFF (SdFFF). In SdFFF, a centrifugal force is applied and the separation occurs due to the difference between centrifugal and buoyant forces (i.e., nanomaterials can be separated based on size and density), which can be used to calculate nanomaterial diameters of known density. The lower size limit for nanomaterials capable of being fractionated with SdFFF is related to the instrument's capability to generate the field (Force = *buoyant mass* × (*angular frequency*)^2 × *radius of curvature*) and the density of the nanomaterials, and is near 10–20 nm for more dense metal and metal oxide nanomaterials and larger for less dense materials (e.g., polystyrene).

In FlFFF, the applied field is created by a pressure differential at the bottom, permeable wall (accumulation wall) of the channel. In normal mode elution, the analytes that are retained in equilibrium with the applied force perpendicular to the channel wall and the diffusion of each analyte. Thus, smaller nanomaterials with the largest diffusion coefficients (D) will elute before larger nanomaterials and sizes ranging 1 nm to micrometer sizes can be separated. The retention time, t_R, of the nanomaterials can be used to determine D, and a hydrodynamic diameter can be calculated based on the Stokes–Einstein equation. Calibrated retention can be used to measure the size of monodisperse particles or the size distribution of polydisperse mixtures when band-broadening effects are known.[51] However, fractionation is not limited to only spherical nanomaterials. Separation of different shapes (e.g., rods, plates, etc.) can also be accomplished with FFF, which can have distinct mechanisms for elution compared to spheres, and is ongoing research in the field.

An inherent advantage in FlFFF is the flow field interacts with and affects all solute species without applying selectivity.[52] This also results in the ability to remove small species, including dissolved ions, small molecules, and some small nanomaterials, during the equilibration (focusing step), which likely would be present in increasing complex matrices. The focusing step can greatly minimize sample preparation and allows examination of the nanomaterials in a less-perturbed environment than many other instruments, providing a distinct advantage for samples that may change physically (e.g., agglomerate) or chemically (e.g., oxidize) during more extensive sample preparation (e.g., UHV environments). Furthermore, FFF can be hyphenated with online detectors, such as inductively coupled plasma–mass spectrometry (ICP-MS), UV-Vis, and light scattering, to multiplex nanomaterial characterization data of distinct populations in situ, or samples can be fractionated and collected to be characterized off-line.

Although a large size range of nanomaterials can be fractionated with FlFFF, instrument and separation optimization can be quite complex due to the numerous parameters affecting retention of the nanomaterials and can result particle–particle interactions within the channel, particle–membrane interactions, lower recovery, or poor retention reproducibility.[53] Thus, the use of appropriate calibrants is necessary for accurate determination of D, which is commonly overlooked and misreported due to the lack of calibration materials and validated methods. For example, polystyrene and Au NMs of the same nominal sizes did not elute simultaneously and demonstrate a commonly occurring error. Further demonstration and discussion on the evaluation of calibrants, as well as strategies for optimizing separation conditions, can be found elsewhere.[53–55]

2.2.2 Size Exclusion Chromatography

Size exclusion chromatography (SEC) is also referred to as gel chromatography, gel filtration, and gel permeation chromatography, is a technique in which particles or molecules in solution are separated for analysis via size exclusion. The solution to be analyzed is introduced to the SEC column, which is packed with a stationary phase consisting of small porous beads. As the solution travels down the column, the beads retain particles depending on their size and shape, with larger species passing through more quickly due to weaker interaction with the pores and the beads themselves. Peak elution times can thus be quantitatively correlated to the hydrodynamic volume of the analytes. This relationship comes at the cost of chromatographic resolution, as SEC is a technique that requires column calibration in order to determine the statistical average molecular weight or size. Absolute molecular weight or size measurements are possible, however, when coupled to online detectors that analyze, for example, light-scattering characteristics.[56]

The observed correlation based on the hydrodynamic volume became accepted as the basis of universal SEC calibration, but the interpretation of SEC data is not fully understood.[57] Both theory and computer simulations assume that the separation process is determined by the partitioning of solute particles between two phases: a dilute bulk solution phase located in the interstitial space and confined solution phases within the pores of the column-packing material. Based on this theory, the relevant size parameter to the partitioning of polymers in pores is the mean maximal projection onto a line.

Major disadvantages of SEC include possible interactions of the solute with the solid phase or the limited size separation range of the columns, which may not include the size range of all nanomaterials and their aggregates. Nanomaterials stabilized by electrostatic or steric means typically render traditional SEC stationary phases inadequate for effective separation.[58] In order to accommodate the wide range of nanomaterial functionalization, specific methods have been employed to overcome the problem of solid-phase interactions include the addition of capping agents to the mobile phase and analyte recycling.[59] Samples should be concentrated to a small volume and proper buffer selection should be practiced for effective resolution and minimal solid-phase interactions.

2.2.3 Analytical Ultracentrifugation

Analytical ultracentrifugation (AUC) is a technique in which a sample is subjected to a high centrifugal field and is fractionated via the mass-dependent sedimentation behavior of the species present. In contrast to many methods, samples are characterized in their native state in free solution, and therefore AUC is not limited by complications due to interactions with matrices or surfaces. AUC is nondestructive and samples may be recovered for further analysis from both on- and off-line detectors, which measures optical absorption, scattering, or interference detectors.[60,61] AUC allows the hydrodynamic and thermodynamic characterization of nanomaterials in situ through two primary experiment types: sedimentation velocity (SV) and sedimentation–diffusion equilibrium (SDE) ultracentrifugation.

SV utilizes hydrodynamic theory and the motion of solutes in high centrifugal fields to define the size, shape, and interactions of macromolecules. The conversion of sedimentation coefficient distributions to a particle size distribution highly relies heavily on knowledge of the density of the particle. In a sufficiently concentrated dispersion of nanomaterials, the sedimentation velocity is sensitive to the nature and strength of interactions between particles.[61] SE is a thermodynamic method where equilibrium concentration gradients at lower centrifugal fields are analyzed to define molecular mass, assembly stoichiometry, association constants, and solution nonideality.[60] Additionally, SV and SE experiments provide a means for studying equilibrium constants or aggregation of particles. In contrast to SE data, SV experimental data may even yield information about reaction rate constants.[61]

Nanomaterial dispersions may vary in shape, density, and size, and result in wide sedimentation coefficient distributions. As such, large aggregates or small impurities present in low concentrations may not be detected.[61] Pure nonionic analytes, or charged species that can be prepared in an isoionic form, may be analyzed directly in an appropriate solvent. Difficulties arise with ionic species, such as charge-stabilized nanomaterials, resulting in nonideal sedimentation behavior.[61] Adequate compensation with buffer and the addition of supporting electrolytes can be used if needed.

2.3 Optical Scattering

The interaction between nanomaterials and light can be an effective characterization tool. The energy source produces a wave vector that can be modified relative to the physical characteristics of the particle, such as size and shape, while the frequency remains unchanged or experiences absorption. The intensity of scattered light is proportional to the particle size ($\propto$ radius^6) and inversely proportional to the wavelength of light ($1/\lambda^4$), therefore, larger particles scatter more light than smaller particles and shorter wavelengths are scattered more strongly than longer wavelengths.[10,62] The resulting signal can then be measured by several techniques applicable to nanomaterial characterization.

2.3.1 Dynamic Light Scattering

DLS is also known as photon correlation spectroscopy or quasi-elastic light scattering, observes the time-dependent fluctuation in the light-scattering intensity of particles measured by a digital autocorrelator. The fluctuations are due to random Brownian motion and are directly related to the translational rate of diffusion of the particle through the solvent. In turn, this diffusion coefficient is related to the hydrodynamic radius of the particle through the Stokes–Einstein equation, which closely reflects the apparent size adopted by a solvated, tumbling molecule.[10] DLS is one of the few nondestructive techniques that can estimate the average hydrodynamic diameter with minimal sample processing.

In DLS, the observed scattering intensity is a function of the scattering angle, particle size and shape, the instrument optics, and the duration of the measurement. In addition, for irregularly shaped particles larger than approximately half the wavelength of the laser, scattering intensity is a function of their form factor. Because the relationship between particle size and scattering intensity is highly nonlinear, obtaining the true particle size distribution remains challenging.[10]

For example, the presence of very large particles, like dust and nanomaterial aggregates, causes extraneous light scattering. These larger particles not only significantly limit the reproducibility of the measurements, but they also artificially broaden the apparent size distribution and, in some cases, could completely obscure a numerically dominant smaller size fraction. Therefore, in highly polydisperse systems, samples must be fractionated or filtered prior to DLS analysis. Another consideration is analyte concentration: the weak scattering intensity of metallic nanomaterials < 50 nm makes their detection difficult and less reproducible at very low concentrations.[63]

2.3.2 Static Light Scattering

Static light scattering (SLS) or laser diffraction, is a technique that measures the scattering angle and the scattering intensity of incident light and can be used to determine the size of the nanomaterials based on Mie theory. Commonly, the Rayleigh–Gans–Debye (RGD) approximation is used for dilute solutions, where the refractive index of the nanomaterial and solvent is similar, but the approximation has been shown to correspond well with orthogonal measurements of size strictly outside this condition. In general, to verify appropriate use of the RGD approximation or other formalisms to derive size, orthogonal measurements are always recommended. From an instrument perspective, a well-collimated and highly coherent beam of polarized light is used to analyze a solution containing a suspension of the analyte of interest in SLS. The electric field of the incident light is oriented perpendicular to the plane in which the intensity and angular dependence of the subsequently scattered light is to be measured. SLS is one of the few absolute methods available for molecular mass and structure determination over a broad range of molecular weights.

If the angular dependence of the scattered light is measured, it is possible to determine the size of the analyte. For SLS, the size measurement is known as the root mean square (RMS) radius and is a measure of particle size weighted by the mass distribution about its center of mass. The RMS radius can be related to the geometrical dimensions of the particle if the conformation is known (e.g., random coil, sphere, or rod), and depends on the internal mass distribution of the particle. The intensity of the light scattered by a particle is directly proportional to the molar mass and concentration of particles in solution, making SLS a powerful technique for monitoring the presence and formation of aggregates.[10,62]

Sample preparation considerations are similar to those for DLS, where the presence of large particles may interfere with the analysis of smaller particles and very low concentrations make scattering intensities difficult to reproduce. Additionally, the RGD approximation applies to dilute solutions, with negligible interactions between sample particles, solvent molecules, or each other. Further, the calculations assume that the refractive index of the particle is essentially indistinguishable from the solvent.[10,62] Data analysis,[64] instrument modifications,[65] and/or use of this technique as a detector following a fractionation step can be done in order to compensate for samples that cannot be prepared in a way that meets these criteria.

2.3.3 Small-Angle X-ray Scattering

Small-angle X-ray scattering (SAXS) is a technique where the elastic scattering of X-rays to very low angles (0.1°−10°) is used to measure the electronic density fluctuations within a specimen. Because SAXS measurements use X-rays at high scattering angles, it can examine length scales from Angstroms to micrometers and allows for the analysis of the interior structure of disordered systems and for the direct determination of structural information for systems with large-scale random density fluctuations (changes in electron density) in real time.

SAXS measurements typically produce two-dimensional scattering patterns that are circularly averaged into one-dimensional average intensities. The scattering patterns provide information about the size, shape, and spatial arrangements of the nanomaterials. The methods used for data analysis will depend on the parameter(s) of interest and any a priori knowledge of the samples. Through careful analysis of a high-quality scattering profile, the zero-angle scattered intensity and the RMS radius, both relating to the size and shape of the scattering particle, can be tractably calculated with high precision using Guinier's approximation. Multiple populations of sufficiently distinct size can result in distinct features in the scattering curve, but broad distributions of species make identification of distinct populations difficult. To interpret SAXS data in terms of structure, it is useful to perform a Fourier transform of the scattering profile to obtain the interatomic distance distribution function of the scattering particle. The determination of physical characteristics, such as molecular weight, particle volume, and specific surface area (see following section), is feasible only if scattering intensity is measured on an absolute scale. The concentration and molecular weight of the analyte is proportional to the scattering signal, therefore, accurate sample concentration must be available to calculate the molecular weight from the scattering data.[66,67] However, assumptions about shape (spheres) are commonly used for in situ monitoring of nanomaterial formation,[68] but form factors can also be determined from SAXS experiments.

The requirements for sample preparation for elucidating parameters of interest will change similar to the methods applied for analysis. For quality SAXS measurements of precise nanostructures, sample preparation is incredibly important, since the scattering signal must be measured from monodisperse particles of identical composition in order to obtain accurate structural parameters. The sample to be measured must be highly pure and sample aggregation should be avoided.[66] However, as mentioned above, an advantage of SAXS is the ability to better statistically measure distributions of products with minimal sample preparation compared to EM methods. A disadvantage to the SAXS measurements for determining nanomaterial distributions is the necessity of known shape factors, which would need to be determined with coupled EM measurements, but the use of orthogonal techniques for validation is a general requirement for validating many nanomaterial measurements as outlined in this chapter.

2.4 Specific Surface Area Measurements

A large fraction of the atoms in a nanomaterial is present at its surface and it is these atoms that often determine the properties of the particle. Therefore, a measure of the specific surface area (SSA), or the total surface area per unit of mass, is often essential to understanding the behavior of a nanomaterial system. A surface area measurement is usually the accessible surface and determined using gas adsorption methods. By collecting physio-sorption isotherm data, which is relationship between the amount of gas adsorbed and the equilibrium pressure, methods can be applied to derive characteristics of the nanomaterials. The most common method applied to derive an SSA is the Brunauer–Emmett–Teller (BET) method.[69] A narrow range of relative pressures are used from the isotherm to derive the monolayer capacity of the gas on the sample from the BET equation, which is subsequently used to calculate a BET SSA, notated SA_{BET} to differentiate the value from other methods. Sample preparation is important in gas adsorption experiments to remove adsorbed gases, water, or other surface "impurities" that could affect the gas adsorption. Also, the BET method is a simplified approach with assumptions that do not hold in all cases for nanomaterial characterization, thus, further verification may be necessary, and methods for extending confidence are found elsewhere.[70] Although there are limitations to the BET method, it is a fast, inexpensive method for producing a relative measure of the SSA of a large number of nanomaterials when strict attention to detail of sample preparation and implementation of calibrants, when possible, are employed.

The biggest disadvantage of this measurement technique is that it can be used only for dried samples. Consequently, aggregation/agglomeration may lead to lower reported SSA values for tightly bound nanomaterials,[70] which may necessitate a disaggregation or fractionation process on postdried particles prior to analysis that increases the accessible surface area.

3. CHEMICAL AND ELEMENTAL CHARACTERIZATION

Elemental and chemical characterization of nanomaterials are extremely important component of any research study. There are many techniques that are capable of providing elemental information and some of those also provide information on the chemistry of the element probed, yet they vary as to what is being probed and how the sample is impacted.

In the following section, the analytical techniques have been broken down into two sections: methods which traditionally analyze deep into or through a material are referred to as **bulk** sensitive (subsection 2.1), and those which focus on the top-most layer of a material are referred to as **surface** sensitive (2.2). For the sake of this section, a 10 nm sampling depth or less will be considered *surface sensitive*, while all deeper sampling depths (including those without a true sampling depth) will be defined as *bulk sensitive*. A more in-depth discussion of the techniques introduced below can be found in Chapters 4 (Compositions, Phase, and Crystallinity), 5 (Concentration), and 7 (Surface Properties).

Furthermore, from the perspective of how the sample is being measured, most techniques can be classified as either destructive or nondestructive techniques. Destructive techniques will be considered those that irreversibly alter the physicochemical properties of a sample in a quantifiable manner. Nondestructive techniques are those which probe elemental and chemical compositions without altering them in any permanent way. It is important to note that these classifications for a given technique are somewhat generic and can be sample dependent. In the following sections, we will provide a brief overview of several techniques from the standpoint of what each technique is measuring, benefits and challenges associated with performing the measurement, and how the data are analyzed and interpreted. Additionally, examples of the application of these techniques to nanomaterial-related studies will be provided. As a general guide, it is the authors' opinion that the best approach for characterizing the elemental or chemical characteristics of nanomaterials or any other sample is to employ more than one technique.

3.1 Bulk Elemental and Chemical Characterization

3.1.1 Inductively Coupled Plasma-Mass Spectrometry

ICP-MS is a bulk sensitive technique that is used to determine the elemental composition of a sample with very high precision.[71,72] The sample is injected into an "analytical zone" of high temperature plasma (T = 6000−7000° K) effectively vaporizing, atomizing, and finally ionizing all material present,[71] such that ICP-MS is classified as a destructive characterization technique. The charged ions are directed through and focused by an ion lens and are guided into a mass spectrometer (e.g., with a quadrupole mass analyzer), where the ions are separated based on their mass to charge (*m/z*) ratio and detector quantified.[71] With the appropriate controls and calibrations for a given material, ICP-MS is capable of providing quantitative information in regards to the elemental composition and mass. Depending on the element of interest, the limits of detection typically range from nanomoles per mol to picomoles per mol.

ICP-MS has been widely used for characterizing the elemental concentrations of nanomaterial suspensions and powders, which are often solubilized prior to analysis via acid digestion.[73,74] Due to the ability to measure trace levels of elements in a solution, ICP-MS has been widely used in environmental studies in attempts to measure, for example, the amount of silver nanomaterials released in environmentally relevant conditions.[75] It has also been used as a quality control verification for the purity of CNTs after removal of any residual catalysts.[76]

3.1.1.1 Single Particle ICP-MS

Single particle ICP-MS (SP-ICP-MS) is a burgeoning technique used to monitor the changes in nanomaterial distributions and dissolved ion concentrations in a single measurement of well-characterized nanomaterial samples (composition, phase, shape, size, etc.) with the sensitivity of ICP-MS. The dissolved ion concentration is related to the baseline intensity using the measured calibration curve. The size distribution is derived from the response of single nanomaterial events ionized in the plasma that result in signal intensities that are converted to a diameter using a geometric (spherical) model and bulk density of the analyte. Sample preparation requires dilute solutions to ensure single nanomaterial ionization events occur more slowly than the acquisition time step (usually $\leq 10\ \mu s$) employed (i.e., statistical probability of a single particle reaching the plasma per integration time step). Calculating the transport efficiency and examination of size-dependent response factors for the analyte of interest are also important factors in more accurate calibration. Multiple methods are available for calculating transport efficiency and summarized elsewhere.[77] The limit of detection for reliably measuring Ag and Au NMs is reported to be near 20 nm[78]; therefore, nanomaterials with diameters less than this size will not be reliably detected nor will they contribute to the dissolved fraction and can result in a less accurate assessment of the size distribution. The development of methods for unknown samples is part of an ongoing research area.[79] Full nanomaterial detection, characterization, and (semi)quantification will likely always require multiple measurement techniques with well-defined protocols and calibration methods.

An example of the benefits associated with examination of nanomaterials in complex media using hyphenated systems from a polydisperse nanomaterial sample is demonstrated with FFF-SLS.[80] A broad distribution of nanomaterials that results in an inaccurate size distribution using batch light scattering methods can be more accurately assessed using an optimized fractionation FFF method,[54,81] where the fractionated populations of species can be characterized with the online light scattering detectors (DLS and SLS) to collect shape and size information and minimize bias from the largest species present.

3.1.2 X-ray Energy Dispersive Spectroscopy

X-ray energy dispersive spectroscopy (XEDS, EDS, or EDX) is the principal means for carrying out compositional analysis in the SEM and TEM. The technique relies on the ionization of atoms in the specimen by the ejection of an inner shell electron by an incident electron. The atom can then relax in a number of ways, one of which is through the loss of energy by an outer shell electron, which fills the hole left behind by the ejected electron (Figure 3, left and top right). This energy loss results in the emission of an X-ray photon (also known as X-ray fluorescence), and the energy of this X-ray is the difference in the energy levels between the ejected and outer shell electrons involved. Thus, the energy of the emitted X-ray is element specific, and can be used to perform chemical analysis with an energy dispersive detector. Elements present in concentrations $> 1\%$ by mass are usually detectable, although mitigating factors such as spectral overlap can complicate such detection. For a thorough treatment of these processes, refer to Goldstein et al.[29]

Quantification of the XEDS spectral signal can be accomplished by properly correcting the raw signal through background subtraction, correction for absorption, and fluorescence effects, and comparison to spectra acquired under similar conditions from reference materials of known composition. Given sufficient spectral counts and a reliable quantification approach, XEDS can be used to determine composition within $\pm 1\%$ for a 95% confidence interval. Meeting these criteria can be challenging for nanomaterials due to the inherently small volume of material being analyzed. However, by carefully quantifying the signal acquired from individual nanomaterials, a reliable description of the chemical heterogeneity in the population can be produced. Alternatively, the analyst can also acquire a full XEDS spectrum at each pixel in a given image raster to form a hyperspectral image that can reveal the spatial distribution of elemental species within an individual nanomaterial.[38,82]

STEM mode in the TEM is particularly useful for carrying out very high spatial resolution spectroscopic analysis via XEDS in an analogous fashion to that described previously for the SEM. The difference in this case is that the interaction volume between the beam and the specimen is much smaller due to the thin nature of the specimen itself, the smaller probe size, and the high energy of the electron beam. An example of this for nanomaterial characterization is given in Figure 2 (left panel), which shows an ADF STEM image of

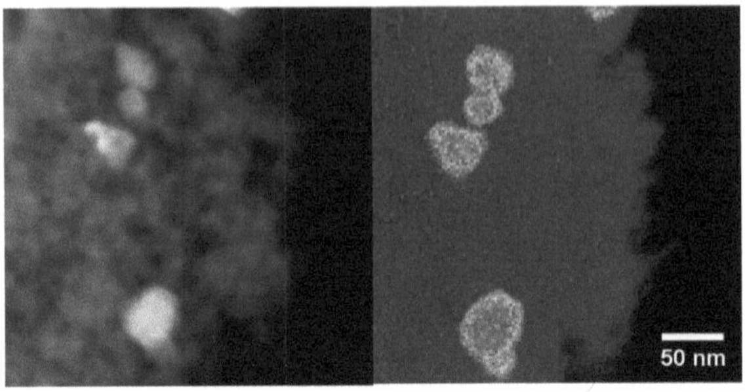

FIGURE 2 Left figure: Annular dark-field scanning transmission electron microscope image of a supported metal catalyst consisting of nanoscale Au-Pd particles supported on an Al_2O_3 substrate. Right figure: Color overlay of elemental X-ray energy dispersive spectroscopy maps showing the spatial extent of the Au (red (light gray in print versions)) and Pd (green (white in print versions)).

a heterogeneous catalyst consisting of Au-Pd particles on an Al_2O_3 support. In this case, the metal particles appear brighter in the ADF image due to their increased scattering. By collecting a full XEDS spectrum at each pixel in this image space, the elemental distribution in this region can be fully mapped. The results of this type of hyperspectral imaging can be seen in Figure 2 (right panel), which presents a color overlay of the individual elemental maps extracted for Au (red), Pd (green), and Al (blue). Due to the processing history of this catalyst, a core–shell morphology has developed, whereby the surface of most particles has been significantly enriched in Pd with an Au-rich core being left behind. Such spectroscopic analysis of nanomaterials can be extremely powerful, and has seen application for nanomaterial characterization. These results will only improve further as new detector designs continue to increase performance, which allow for increased analytical sensitivity and even three-dimensional characterization of nanoscale particles.

3.1.3 Raman Spectroscopy

Raman spectroscopy is a bulk sensitive technique in which the vibrational states present in a material are determined by the inelastic scattering of light.[82] Typically, a laser is employed as the light source, and, depending on the laser power, Raman can vary in its destructiveness. Specifically, an incident photon with a given energy excites a vibrational state with a characteristic frequency. When the system relaxes, a second photon is emitted. Occasionally, the system relaxes to a higher vibrational level than it was in originally. In this case, commonly referred to as Stokes scattering, the reemitted photon exhibits a lower energy than the incident light and this difference is indicative of the vibrational states present.[83,84] Alternatively, if the specimen is already excited to a vibrational state above its ground level prior to further excitation by the incoming photon, then it may subsequently relax all the way to its ground state. This is known as anti-Stokes scattering and results in an emitted photon with higher energy than that of the incoming light.

The spatial resolution of traditional Raman spectroscopy is on the micrometer scale, and is therefore of limited utility for nanomaterial characterization. Several variations of this technique, however, do exist and are more commonly employed for nanomaterial characterization. The most notable of these variations include **Surface Enhanced Raman Spectroscopy (SERS)** and **Near-Field Scanning Optical Microscopy (NSOM) Raman**, or nano-Raman.[83,85] Both techniques improve the spatial resolution of Raman to a few hundred nanometers. SERS, for example, can be employed to look at molecules absorbed to a nanomaterial surface, while NSOM uses a small aperture to focus the scattering.[86,87]

3.1.4 X-ray Absorption Spectroscopy

X-ray absorption spectroscopy (XAS) is a technique that provides a bulk measurement of the elemental and chemical composition in a specimen by virtue of its X-ray absorption characteristics. XAS requires a high intensity, coherent X-ray beam which is tunable over a wide energy range and therefore must be carried out using a synchrotron radiation source.[88] XAS spectra contain three classes of features which are typically inspected. The first energy range is the X-ray absorption edge itself and is due to the absorption of incident photons by core level electron ejection. The features within a range of 50−100 eV above the edge are typically referred to as the **X-ray Absorption Near Edge Structure**, while the more subtle spectral features present at still higher energies (up to 500 eV or greater above the edge) are referred to as the **Extended X-ray Absorption Fine Structure (EXAFS)**.[88,89] By analyzing these features, the elemental composition, oxidation state, and electronic structure of the specimen can be ascertained[88−90] by referencing to known structures. XAS has been employed in the characterization of many nanomaterial-containing systems. For example, EXAFS have been employed as a means to analyze the sulfidation processes associated with silver nanomaterials by calculating silver sulfide concentrations and relating the findings to *Escherichia coli* growth rates and dissolution processes.[91,92] Other studies have employed XAS to help chemically identify zero-valent gold nanomaterials within synthesized from a precursor in biological systems such as plant systems such as *Sesbania drummondii*.[93]

3.1.5 X-ray Diffraction

X-ray radiation exhibits a wavelength which is on the order of the typical interatomic spacings present in crystalline solids, and is therefore of critical importance in structural characterization. Of particular utility is X-ray diffraction (XRD), which is broadly defined as the elastic scattering of a coherent X-ray source by the periodic lattice inside a specimen.[94−96] Due to coherent interference effects, strong intensity maxima will exist at well-defined scattering angles given by the Bragg equation.[97] Through careful analysis of these peak positions, structural determination of unknown specimens can be achieved. In the case of nanomaterials, where the probe is incident upon many particles during the analysis, the collected spectrum represents a summation of the scattering characteristics of all particles in the analytical volume. Thus characterization of the individual particles is usually not possible. However, the aggregated data are very useful for determining the structures present in a larger group.

XRD is also commonly used to measure the mean particle size in a population of nanomaterials via the Scherrer equation.[94] It is important to account for instrumental sources of peak broadening, and this is usually done by prior measurement of a reference material.[94] Since peak broadening is insignificant for larger particles, this technique is generally useful only for nanoscale crystallites with diameters below 100–200 nm. Conversely, broadening from very small particles (i.e., $< 2-3$ nm) becomes so severe that the peak width cannot be adequately measured or even detected. Finally, it is important to note that the Scherrer equation does not account for internal particle strain and defects, which can also contribute to beam broadening. For this reason, the particle size as measured by XRD should be considered a lower bound rather than an absolute measure of the mean crystallite size. Other approaches are available for the analyst to employ, and are described more fully in standard texts.[94,96] When the particle shape is anisotropic and/or the particle size distribution is nonuniform, this technique is less informative and should be interpreted with additional analyses such as transmission electron microscopy.

3.2 Surface Elemental and Chemical Characterization

By design, engineered nanomaterials have a very high surface area to volume ratio. Depending on size and confirmation, the majority of atoms that comprise the nanomaterial may be found on the surface when compared to a bulk particle. Consequently, a thorough understanding of the surface composition is a critical parameter when understanding how nanomaterials will behave in a given environment.

In most surface-sensitive techniques, incident radiation (e.g., X-rays, electrons, ions) interacts with the surface ejecting interfacial species which can be analyzed in various ways and related to the elemental, chemical, or molecular structure from which the species originated. The techniques achieve their surface sensitivity by virtue of the lower mean free path of scattering associated with electrons and ions within the specimen, which is far lower than that of a photon with similar energy. Those electrons and ions generated far from the specimen surface will either be fully absorbed or be further scattered prior to escaping the specimen (inelastic scattering).

The quality of the specimen surface is of paramount importance for all of these techniques, and therefore, sample preparation is a primary concern. First, the nanomaterials being analyzed must be introduced into an HV to UHV environment, with HV operating between 10^{-3} Pa and 10^{-6} Pa and UHV at 10^{-6} Pa to 10^{-9} Pa.[98] These low pressures are required to minimize additional inelastic scattering of the incident probe and ejected surface species from gaseous molecules as well as to prevent the development of an adsorbed surface contamination layer, which would mask and/or interfere with some of the measured intensity of the underlying elemental species.[99] All samples therefore must be fully dried and have low vapor pressures in order to maintain the UHV environment.[100] Second, when employing the following techniques for characterization of nonconducting surfaces, surface charging can alter the results. This can sometimes be alleviated by surface coatings and/or by using low-energy electron sources for charge compensation.[98,101,102]

3.2.1 Auger Electron Spectroscopy

Auger electron spectroscopy (AES) is a surface-sensitive technique performed in an electron beam instrument (i.e., SEM, TEM, etc.) wherein the energy of electrons emitted during relaxation of ionized atoms is used to determine the specimen composition.[103] Typical AES instrumentation employs a focused electron beam ($E_i = 2-5$ kV) to eject a core level electron (E_1) from an atom in the specimen as displayed in Figure 3. As in the previously described XEDS process, another electron from a higher energy level (E_2) relaxes to occupy the

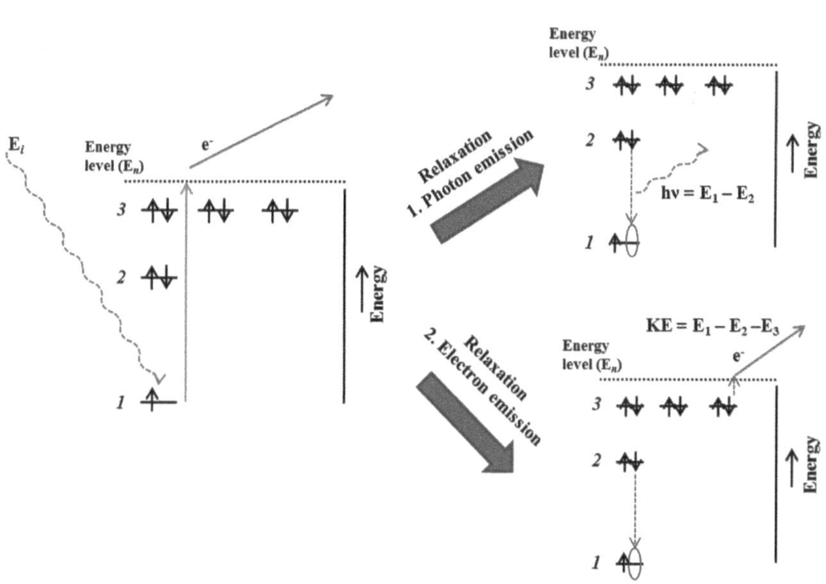

FIGURE 3 (Left) Emission of a core level electron (E_1) by an energetic incident source (i.e., electron or X-ray). (Top right) Relaxation by filling of the vacancy with an electron from a higher energy level. In this case, an X-ray is how the excess energy is released. (Bottom right) Relaxation process where the excess energy is imparted into another ejected electron. *Adapted from Ref. 83.*

level vacated by the ejected electron. As a result of employing an electron beam, AES is generally considered a minimally destructive technique as it can chemically alter the surface. In contrast to the XEDS, the excess energy generated through this relaxation step is released as kinetic energy imparted to a third electron, or the Auger electron, which is ejected from the energy level it occupies (E_3). The Auger electron emission process is independent of E_i and can be calculated by the following equation[103]:

$$KE = E_1 - E_2 - E_3 - WF \quad (1)$$

where WF is the work function of the specimen and the analyzer.[83,98,102] As in the case of XEDS, the energy of the emitted Auger electrons is indicative of the elemental identity of the ionized atom. The probability of Auger emission is generally higher for light elements, with the notable exceptions of hydrogen and helium which do not contain sufficient numbers of electrons for Auger emission. Heavier elements preferentially undergo X-ray fluorescence, so AES is often used in concert with XEDS for analyzing specimens which contain both low- and high-atomic number elements. Finally, due to the limited escape depth of low-energy Auger electrons, AES exhibits excellent surface sensitivity, which is dictated by the kinetic energy imparted.

Typically, raw AES spectra consist of broad peaks due to Auger emission on a broad background continuum due mostly to SEs and multiple scattering of Auger electrons. The signal-to-background ratio for the Auger peaks is quite low, even under ideal conditions, so spectra are often analyzed by calculating the derivative of the electron count signal with respect to energy (dN/dE). With the appropriate controls and calibration factors, quantitative and semiquantitative information can be gleaned regarding the elemental concentration of all elements in a given sample.[98,104] The collection of Auger spectra from individual particles is also possible. Alternatively, elemental mapping can be carried out by slowly scanning the beam over the specimen surface and collecting the spatially resolved signal, often referred to as Auger electron microscopy.[100] For example, Rades et al. utilized SE imaging to locate the positions of silica nanomaterials on a substrate, and subsequently collected AES spectra at these positions to determine their composition.[105]

3.2.2 Electron Energy-Loss Spectroscopy

Electron energy-loss spxectroscopy (EELS) is another method for carrying out nanoscale spectroscopy in the STEM[106,107] and, while not specifically a surface-sensitive technique, is used on samples that are usually less than 100 nm thick. In this technique, the energy lost by beam electrons due to inelastic scattering by the specimen is analyzed in order to determine the elemental identity of atoms present or even their valence and bonding states. The low energy-loss range of EELS spectra typically contains a very intense peak at zero loss, corresponding to electrons that have lost little or no energy, followed by a broader peak positioned in the range of 5–30 eV loss. This peak is due to plasmon scattering, which arises from the collective oscillation of valence electrons in the specimen, and it can be used to locally measure the dielectric and optical properties of a specimen. Recently this has been utilized to map the plasmonic modes present on the surface of metallic nanowires,[108,109] and to observe the shift in optical frequency due to quantum confinement in one-dimensional nanostructures.[110]

At higher losses, characteristic edges are present which are located at the ionization threshold for the inner shell electrons of atoms present in the specimen. Therefore, this inelastic signal can be used to ascertain the identity and concentration of elements within the specimen in a complimentary fashion to that of XEDS. In particular, due to the underlying physics and detector limitations in XEDS, EELS is much better suited to light element characterization.

The sensitivity of core-loss EELS is strongly dependent upon the particular element in question as well as the thickness of the specimen. In fact, EELS has been employed to determine the elemental identity of isolated individual atoms[111] as well as single atoms in a matrix.[112,113] Generally speaking, in favorable conditions the sensitivity of EELS is very high, and concentrations less than 1 atomic % can be readily detected.[107] To achieve such results, the edge intensity must first be extracted from the background signal due to the tail of the plasmon peak, multiple scattering events, and intensity from other core-loss edges in the spectrum present at lower energy losses than the edge in question. Ensuring that the specimen thickness does not exceed the mean free path for inelastic scattering, which depends on the beam energy as well as specimen composition and density, can minimize this background signal. This is generally not a problem for dispersed nanomaterials, as their thickness is inherently small, however, it is an area of concern for particles embedded in a matrix.

Core-loss EELS analysis has been widely used in nanomaterial characterization, and advances in instrumentation have accelerated this trend. Several examples include the use of the sulfur L-edge at $\approx$ 165 eV to determine the location and thickness of the passivating ZnS shell which often surrounds CdSe quantum dot structures[114] and characterization of particle evolution in Pt-Co catalysts.[115]

3.2.3 Time-of-Flight Secondary Ion Mass Spectrometry

Time-of-flight secondary ion mass spectrometry (ToF-SIMS) is one of the most surface-sensitive chemical characterization techniques available and is considered a destructive technique since it must remove the topmost surface atoms and molecules before subsequent ionization and detection. ToF-SIMS operates by bombarding a specimen surface with a focused ion beam (e.g., Ar, C_{60}, SF_6, etc.) with a high kinetic energy (1−25 kV) in UHV.[100,101] The incident ion beam sputters material from the top few atomic layers of a target surface in the form of single atoms, molecules, or molecular fragments, all of which can be in either ionic or neutral states. The ions are directed via an electric field toward a time-of-flight tube that separates them by mass based on the time required for them to reach a detector positioned at the end of the tube.

Researchers have recently employed ToF-SIMS to better understand the chemical and physical properties of nanomaterials and nanomaterials in suspensions. For example, SIMS has been employed in conjunction with inkjet printing to better understand the chemical characteristics of titanium dioxide nanomaterials deposited on a surface.[116] In addition, it has been used to study the drying characteristics and related dissolution states of silver nanomaterial suspensions in order to infer the nature of the initial solution.[116,117] Finally, the primary ion beam can be used to remove material from the specimen surface so that SIMS data can be collected in a depth profiling mode. Depending on material, this technique can reliably be utilized to provide depth information from a single nanometer to several micrometers, depending on mode of operation and targeted analyte. This technique has been applied to thin films of cobalt oxide nanomaterials and silver nanocomposites in order to determine elemental distribution throughout the film.[118,119]

3.2.4 X-ray Photoelectron Spectroscopy

X-ray photoelectron spectroscopy (XPS) is a nondestructive surface-sensitive technique which employs X-rays to eject core level electrons and outlined in Figure 3 (left).[98,102] A source of X-rays with a known, fixed energy interacts with and ejects a core level electron with a fixed binding energy (BE). The kinetic energy of the emitted photoelectron is then measured and used to determine the BE. XPS sampling depth is typically the top 10 nm of a sample and can detect surface concentrations as low as 0.1 atomic %.

Photoelectron intensities can be used to provide both elemental and chemical information for all elements, except H and He, at the sample's surface. Elemental information is given by the measured BE in the spectrum, which gives the core energy level of the emitted photoelectron (e.g., C 1s photoelectrons are typically found between 280 and 294 eV).[120] Quantitative elemental information can be obtained by measuring the peak heights or the integrated area under a given elemental peak, and then correcting these values using an experimentally derived sensitivity factor[104] or values provided by the manufacturer. Chemical information regarding the locally bonded atoms is determined by observing peak shifts associated with the element's oxidation state and the electronegativity of the surrounding elements. For example, fluorocarbons such as perfluorinated alkyl silanes have larger binding energies (C 1s $\approx$ 291.7 eV)[121] than carboxylates (C 1s $\approx$ 289.0 eV) and much higher than graphite-like CNTs (C 1s $\approx$ 284.5 eV).[122–124] Since such shifts can be quite small, this qualitative interpretation requires an energy scale that is periodically calibrated,[125] spectra that are energy and charge referenced to a known spectral feature,[98] and/or control samples. A common approach to obtain quantitative or semiquantitative information on the chemical distribution of different oxidation states involves deconvolution of the spectral profile by peak-fitting analysis. However, this process is complicated by several factors that have been previously discussed.[124]

In addition to the application of XPS for CNT characterization, it has been applied for routine characterization of quantum dots,[126] graphene,[127] and several noble metal nanomaterials such as gold[128] and silver.[129] It has also been employed to monitor photochemical surface transformations in composite materials such as silica nanomaterials within an epoxy matrix.[130] When considering a nanomaterial-containing surface, it is important to note that XPS provides an average of the surface chemical composition (typically, 10 μm × 10 μm areas or larger) and is incapable of assessing the composition of individual nanomaterials.

4. BEHAVIORAL CHARACTERIZATION

Behavioral characterization is defined as a measured response, in a specific environment, to specific stimuli. Stimuli could include electrical, changes in ionic strength, or loss of thermodynamic equilibrium. In this case, characterization of the ambient conditions of the measurement is just as important as characterization of the nanomaterial itself.

4.1 Particle Stability

The stability of nanomaterials refers broadly to either the colloidal stability, which is the persistence of individual species in the medium not interacting to from secondary structures, or the chemical stability, which is the resistance to chemical changes (e.g., oxidation, dissolution, phase changes, etc.). Importantly, stability is an ambiguous term unless a timescale is used as a reference point. For example, if the length of time in which a dispersion of nanomaterials remains intact (e.g., no dissolution or agglomeration/aggregation) after introduction into the experimental media exceeds the experimental residence time, the nanomaterial suspension would be considered as stable. However, if the nanomaterial undergoes a chemical transformation, such as sulfidation or oxidation, the nanomaterials would not be chemically stable. Distinguishing these two different stabilities is important for predicting the nanomaterial's efficacy as a delivery vector, its associated risks, and/or overall fate in environmental and biological systems.

Qualitative and semiquantitative measurements of nanomaterial stability can be made using spectroscopic and light scattering techniques. Sedimentation experiments examining the relative intensity of transmitted or backscattered light in a nanomaterial suspension can provide information about the relative stability, where the wavelength used must result in a linear response over the concentration range i.e., it can provide a general answer to whether the suspension is stable on a given timescale. Furthermore, the relative size and shape of associating nanomaterials can be estimated using the same techniques.[131] More structural information of the aggregates can be obtained through static light scattering. Sample preparation for sedimentation experiments to examine colloidal stability may require the introduction of energy (e.g., sonication) to break up weakly bound agglomerates and identify the primary nanomaterial unit (i.e., distribution of the smallest separable species present). However, the user should develop or follow protocols that limit possible artifacts from the dispersion process and allow repeatability.[132,133] Light-scattering measurements (dynamic or static) can be used as qualitative probes for assessing the change of nanomaterial distributions, but determination of the accurate size distribution requires more rigor. As mentioned previously for DLS measurements, an increase in the *z*-average size allows a confident assignment of growth to some nanomaterial population in the suspension. Transmission measurements, light-scattering measurements or monitoring other optical properties of well-characterized nanomaterial systems can be quickly assessed using widely available measurement techniques (e.g., UV-Vis).

Monitoring the optical signatures of nanomaterials can also provide information about their stability. For metallic nanomaterials that possess a localized surface plasmon resonance (LSPR) in the presence of incident radiation, the signature of the LSPR can be used to sensitively detect changes to the local dielectric environment, and more specifically the near-field electromagnetic coupling in nanomaterial aggregates.[134] Similarly, other optical properties of nanomaterials or target analytes can provide insight into changes in the local environments through aggregation-induced emission or quenching phenomena.[135,136] Interpretation of the changes in optical properties of nanomaterials in biological and environmental systems can be difficult due to the complexity of the media, and therefore observing a change in the stability should only be an indicator for a physical or chemical change if intimate knowledge of the energy transfer mechanism is unknown. Further structural information on the nanoscale can also be gained using previously outlined X-ray-based scattering and electron microscopy methods.

Nanomaterial stability in biological and environmental systems can result in distinct processing pathways for different material states (aggregates, primary materials, ions). Therefore, detecting, characterizing, and quantifying the mass distribution of the nanomaterial in different media are of principal interest for understanding their transformation pathways that could lead to more predictive models. Careful consideration and characterization of the simulated or natural media should be incorporated into experimental design when examining nanomaterial stability and fate.[137] For both biological and environmental systems, a principal question for metal and metal oxide nanomaterials (e.g., Ag, Cu, Zn, Fe) is the size-dependent role of reduced metal species (M^0), free ions (M^{z+}), and metal–ligand complexes present in these systems resulting from physical (solubility) and/or chemical processing. To separate the role of these individual contributions, quantifying the mass distribution between these disparate states is imperative.

A common method for quantifying dissolution of nanomaterials is through ultrafiltration, where the filtrates are subsequently analyzed with elemental detectors (e.g., ICP-MS or atomic absorption) to quantify the operationally defined dissolved fraction. The use of appropriate filter types (material, pore size, volume) is an important factor that should be considered to identify possible loss channels, filtering efficiencies and identifying uncertainty in the measurements, especially in environmentally relevant concentration regimes. A continuum of reduced metallic states can exist between nanomaterials with bulk crystalline structure and dissolved ions, which are commonly referred to as nanoclusters and possess distinct properties and toxicity from nanomaterial and ion matter states.[138,139] Experimental design should incorporate the characterization of the filtrate to confirm desired separation.[53]

Other methods for monitoring dissolution processes, and more specifically the free ion concentration, include **ion selective electrodes (ISE)**,[140] **Donnan membrane technique (DMT)**,[129,141] and **absence of gradients and Nernstian equilibrium stripping (AGNES)**.[142] ISE and AGNES are electrochemical techniques that measure electrical potential, which is related to ionic activity according to the Nernst equation. For ISE, interferences in the measurement of nonspecific ions require determination of the selectivity coefficients associated with the ion of interest for accurate determination of concentrations, which should be examined as quality control measures for each system. AGNES consists of two steps that include (1) a preconcentration step to equilibrium between the free metal ions at the electrode surface and the bulk solution, and (2) a stripping step. The intensity current or the total faradaic charge is measured during the stripping step and is proportional to the free ion concentration.

Distinct from ISE and AGNES, DMT contains an exchange membrane, which allows almost exclusively cationic exchange, separating a donor and acceptor solution. Based on Donnan membrane equilibrium,[143] the free ion concentration can be determined in situ by placing an unknown concentration of the analyte in the donor solution and similar ionic strength solution of a supporting electrolyte in the acceptor solution and allowing the solutions to continually recirculate until equilibrium is reached (time frame for experiment much longer than ISE or AGNES). The equilibrium cationic activity relationship of the electrolyte and the analyte of interest in donor and acceptor solutions (measured with elemental detectors) are proportional to the starting unknown free ion concentration. Comparison of the relative advantages of AGNES and ISE for specific metals has been previously summarized.[144] Potential artifacts can occur if significant electrolyte complexation occurs, changing the free ion concentration equilibrium.

4.2 Surface Charge

The surface charge of nanomaterials is one of the key factors that determine their stability in suspensions, which will affect their overall fate in natural systems. The properties of nanomaterial systems can be determined directly or indirectly by their electrical charge (or potential). According to DLVO theory, the stability of particles in a suspension is related to the sum of the potential energies from attractive, repulsive, and solvent contributions, where the latter is significant near the surface (a few nanometers) and the two former contributions remain significant over larger distances as particles interact due to Brownian motion. Because nanomaterial properties can change with decreasing size from their bulk analogs, more complete descriptions of the contributing forces are needed to accurately predict nanomaterial interactions.[145] Therefore, predicting the stability of nanomaterials in suspensions requires information about the surface that may contain coatings or adsorbed species, defect surface sites, an oxide layer, acid–base groups, etc., which represent different origins of surface charge density that can contribute to the net total overall charge on a nanomaterial.

Zeta potential measurements are a commonly used method to glean information about the surface charge. More specifically, zeta potential, ζ, is the potential difference at the shear plane, which is between the Stern layer where charge is treated as fixed near the nanomaterial surface, and a diffuse layer containing a cloud of both positive and negative charge, but this potential cannot be directly measured.[146] Experimentally, ζ is calculated by measuring the electrophoretic mobility, μ_ε, motion of particles in an applied electric field, where the velocity of the particles is proportional to their charge by the Henry equation:

$$\mu_\varepsilon = \frac{2\varepsilon\zeta}{3\eta} f(\kappa a)$$

and ε is the dielectric constant, η is the viscosity and $f(\kappa a)$ is function based on the relative size of nanomaterial, a, and the Debye length, κ, that ranges from 1.0 to 1.5. More detailed description can be found elsewhere by Hunter.[147] When reporting ζ values in aqueous solutions the pH and ionic strength, I, of the solutions should always be included. ζ values are also generally assumed to be size independent, where the surface charge density scales with surface area and results in similar μ_ε. Determination of the μ_ε can also be achieved using electroacoustic measurements that relate nanomaterial sound wave generation to their frequency-dependent electrophoretic mobility. Potential artifacts for zeta potential measurements can be introduced when a dynamic equilibrium is present and the solution conditions of the medium are dissimilar to the measured solution parameters (e.g., concentration, ionic strength etc.). The cell chosen (path length and electrode) and applied voltage also can affect the stability of the nanomaterial. A resource for more details regarding sample preparation considerations, including necessary nanomaterial concentrations for the above configuration, and data evaluation for nanomaterials commonly used in biological applications can be found elsewhere.[148]

Different measures of surface charge are also experimentally determined using potentiometric titration techniques with H^+ and OH^- activities and nonspecific ion adsorption with changes in H^+ and OH^- activities. The point where the nanomaterials do not move in the applied electric field is defined as the isoelectric point, which can be further specified by the contributing factors to net total charge in the specific set of examined system conditions. However, it is necessary to have an understanding of charge generation, which requires thorough physicochemical characterization of the nanomaterials. More detailed discussion can be found elsewhere.[149,150]

5. COMBINED PHYSICAL–CHEMICAL CHARACTERIZATION

The coupling of multiple analytical techniques that were described in the previous sections provides a route to perform orthogonal or complementary measurements on the same population of nanomaterials, which are referred to as hyphenated techniques.[151] More specifically, an instrument for separation is employed on the front end and coupled to spectroscopic or spectrometric detection that allow further physiochemical characterization and/or (semi)quantification. The ability to serially couple or hyphenate more than two instruments is possible and increases the amount of information that can be confidently assigned to specific populations of nanomaterials and can also result in higher throughput than conventional methods.[152] A number of chromatography techniques (e.g., SEC) for separation can be coupled to elemental detectors (e.g., ICP-MS), chemical detectors (e.g., Raman), light scattering (e.g., DLS), and mass detectors (e.g., MS) to name a few. Furthermore, well-studied separation methods based on mobilities (e.g., ion, aerodynamic, electrical) of aerosols[153] can be applied to nanomaterial separation and can similarly be hyphenated with detection modalities. These applied aerosol techniques can provide much higher resolution separation and size distribution information as well as insight into the contribution of hydrated surfaces.[154] Although the number of hyphenation iterations with separation and detection is large, some hyphenated techniques have been particularly adept at separation and characterization over a significant portion of the nanorange and these techniques are further discussed in subsequent chapters.

6. CONCLUSIONS

There are many different techniques available for the physical, chemical, and behavioral characterization of nanomaterials. Each characterization technique has inherent strengths and limitations, and the analyst should be familiar with both when choosing which technique to use, how to conduct the actual measurement, and how best to report the results. Depending on application, multiple characterization techniques are necessary to provide a comprehensive analysis of the specific nanomaterial.

REFERENCES

1. Boukai AI, Bunimovich Y, Tahir-Kheli J, Yu JK, Goddard WA, Heath JR. Silicon nanowires as efficient thermoelectric materials. *Nature* 2008;**451**(7175):168–71.
2. Nel A, Xia T, Madler L, Li N. Toxic potential of materials at the nanolevel. *Science* 2006;**311**(5761):622–7.
3. Yan LA, Zhao F, Li SJ, Hu ZB, Zhao YL. Low-toxic and safe nanomaterials by surface-chemical design, carbon nanotubes, fullerenes, metallofullerenes, and graphenes. *Nanoscale* 2011;**3**(2):362–82.
4. International Organization for Standardization. *TS 80004-1: nanotechnologies - vocabulary - Part 1: Core terms*. Geneva: Switzerland; 2010.
5. Stefaniak AB, Hackley VA, Roebben G, Ehara K, Hankin S, Postek MT, et al. Nanoscale reference materials for environmental, health and safety measurements: needs, gaps and opportunities. *Nanotoxicology* 2013;**7**(8):1325–37.
6. Hansen SF, Larsen BH, Olsen SI, Baun A. Categorization framework to aid hazard identification of nanomaterials. *Nanotoxicology* 2007;**1**(3). 243–U369.
7. Magdolenova Z, Collins A, Kumar A, Dhawan A, Stone V, Dusinska M. Mechanisms of genotoxicity. A review of in vitro and in vivo studies with engineered nanoparticles. *Nanotoxicology* 2014;**8**(3):233–78.
8. Warheit DB, Sayes CM, Reed KL, Swain KA. Health effects related to nanoparticle exposures: environmental, health and safety considerations for assessing hazards and risks. *Pharmacol Ther* 2008;**120**(1):35–42.
9. Tiede K, Boxall ABA, Tear SP, Lewis J, David H, Hassellov M. Detection and characterization of engineered nanoparticles in food and the environment. *Food Addit Contam Part A Chem Anal Control Expo Risk Assess* 2008;**25**(7):795–821.
10. Philip JW. Light scattering and the absolute characterization of macromolecules. *Anal Chim Acta* 1993;**272**(1):1–40.
11. Fagan JA. Carbon nanomaterials standards efforts at NIST. In: Obeng Y, et al., editors. *Graphene and emerging materials for post-CMOS applications*; 2009. p. 153–9.
12. Hackley VA, Stefaniak AB. "Real-world" precision, bias, and between-laboratory variation for surface area measurement of a titanium dioxide nanomaterial in powder form. *J Nanopart Res* 2013;**15**(6).
13. ISO TC 229, N., Working Group 3, Health, Safety, and Environment. *Nanotechnologies – guidance on physico-chemical characterization of engineered nanoscale materials for toxicologic assessment, in ISO/TR 13014:2012*. International Organization for Standardization; 2012.
14. Taurozzi JS, Hackley VA, Wiesner MR. Ultrasonic dispersion of nanoparticles for environmental, health and safety assessment – issues and recommendations. *Nanotoxicology* 2011;**5**(4):711–29.
15. Taurozzi JS, Hackley VA, Wiesner MR. A standardised approach for the dispersion of titanium dioxide nanoparticles in biological media. *Nanotoxicology* 2013;**7**(4):389–401.
16. Gorham JM, Rohlfing AB, Lippa KA, MacCuspie RI, Hemmati A, Holbrook RD. Storage Wars: how citrate-capped silver nanoparticle suspensions are affected by not-so-trivial decisions. *Journal of Nanoparticle Research* 2014;**16**(4).
17. Methner M, Hodson L, Dames A, Geraci C. Nanoparticle emission assessment technique (NEAT) for the identification and measurement of potential inhalation exposure to engineered nanomaterials-part B: results from 12 field studies. *J Occup Environ Hyg* 2010;**7**(3):163–76.
18. Baalousha M, Ju-Nam Y, Cole PA, Gaiser B, Fernandes TF, Hriljac JA, et al. Characterization of cerium oxide nanoparticles-Part 1: size measurements. *Environ Toxicol Chem* 2012;**31**(5):983–93.

19. Baalousha M, Lead JR. Characterization of natural and manufactured nanoparticles by atomic force microscopy: effect of analysis mode, environment and sample preparation. *Coll Surf A-Physicochem Eng Asp* 2013;**419**:238–47.
20. Baalousha M, Lead JR. Rationalizing nanomaterial sizes measured by atomic force microscopy, flow field-flow fractionation, and dynamic light scattering: sample preparation, polydispersity, and particle structure. *Environ Sci Technol* 2012;**46**(11):6134–42.
21. Baalousha M, Prasad A, Lead JR. Quantitative measurement of the nanoparticle size and number concentration from liquid suspensions by atomic force microscopy. *Environ Sci Proc Impacts* 2014;**16**(6):1338–47.
22. Jillavenkatesa AD, Stanley J, Lum Lin-Sien H. Particle size characterization. In: *NIST recommended practice guide*. Washington (DC): National Institute of Standards and Technology; 2001. Special Publication 960-1.
23. International Organization for Standardization. *Particle Size Analysis - Image Analysis Methods - Part 1: Static Image Analysis Methods*. 2004. IS 13322–1.
24. Baer DR, Engelhard MH, Johnson GE, Laskin J, Lai JF, Mueller K, et al. Surface characterization of nanomaterials and nanoparticles: important needs and challenging opportunities. *J Vac Sci Technol A* 2013;**31**(5).
25. Pettitt ME, Lead JR. Minimum physicochemical characterisation requirements for nanomaterial regulation. *Environ Int* 2013;**52**:41–50.
26. Ayache J, Beaunier L, Boumendil J, Ehret G, Laub D. *Sample preparation handbook for transmission electron microscopy: techniques*. Berlin: Springer; 2015.
27. Echlin P. *Handbook of sample preparation for scanning electron microscopy and X-ray microanalysis*. New York: Springer; 2015.
28. Egerton RF, Li P, Malac M. Radiation damage in the TEM and SEM. *Micron* 2004;**35**(6):399–409.
29. Goldstein J, Newbury DE, Joy DC, Lyman CE, Echlin P, Lifshin E, et al. *Scanning electron microscopy and X-ray microanalysis*. Springer; 2003.
30. Reimer L. *Scanning electron microscopy: physics of image formation and microanalysis*. Springer; 2013.
31. Giannuzzi LA, Stevie FA, editors. *Introduction to focused ion beams: instrumentation, theory, techniques and practice*. Springer; 2005.
32. Fultz B, Howe J. *Transmission electron microscopy and diffractometry of materials*. Springer; 2012.
33. Reimer L, Kohl H. *Transmission electron microscopy: physics of image formation*, vol. 36. Springer; 2008.
34. Williams DB, Carter CB. *Transmission electron microscopy: a textbook for materials science*. New York: Springer; 2009.
35. Spence JC. *Experimental high-resolution electron microscopy*. Oxford University Press; 1988.
36. Kirkland EJ. *Advanced computing in electron microscopy*. Springer; 2010.
37. Keyse RJ. *Introduction to scanning transmission electron microscopy*. BIOS Scientific Publishers in Association with the Royal Microscopical Society; 1998.
38. Pennycook SJ, Nellist PD. *Scanning transmission electron microscopy: imaging and analysis*. Springer; 2011.
39. Danilatos G. Foundations of environmental scanning electron microscopy. *Adv Electron Electron Phys* 1988;**71**(109–250).
40. Hofmann S, Sharma R, Ducati C, Du G, Mattevi C, Cepek C, et al. In situ observations of catalyst dynamics during surface-bound carbon nanotube nucleation. *Nano Lett* 2007;**7**(3):602–8.

41. Crozier PA, Wang R, Sharma R. In situ environmental TEM studies of dynamic changes in cerium-based oxides nanoparticles during redox processes. *Ultramicroscopy* 2008;**108**(11): 1432–40.
42. Klein K, Anderson I, De Jonge N. Transmission electron microscopy with a liquid flow cell. *J Microsc* 2011;**242**(2):117–23.
43. Grogan JM, Rotkina L, Bau HH. In situ liquid-cell electron microscopy of colloid aggregation and growth dynamics. *Phys Rev E* 2011;**83**(6):061405.
44. Williamson M, Tromp R, Vereecken P, Hull R, Ross F. Dynamic microscopy of nanoscale cluster growth at the solid-liquid interface. *Nat Mater* 2003;**2**(8):532–6.
45. Zheng H, Claridge SA, Minor AM, Alivisatos AP, Dahmen U. Nanocrystal diffusion in a liquid thin film observed by in situ transmission electron microscopy. *Nano Lett* 2009;**9**(6):2460–5.
46. Zheng H, Smith RK, Jun Y-W, Kisielowski C, Dahmen U, Alivisatos AP. Observation of single colloidal platinum nanocrystal growth trajectories. *Science* 2009;**324**(5932):1309–12.
47. Binnig G, Quate CF, Gerber C. Atomic force microscope. *Phys Rev Lett* 1986;**56**(9):930–3.
48. Giessibl FJ. Advances in atomic force microscopy. *Rev Modern Phys* 2003;**75**(3):949–83.
49. Starostina N, Brodsky M, Prikhodko S, Hoo CM, Mecartney ML, West P. AFM capabilities in characterization of particles and surfaces: from angstroms to microns. *J Cosmet Sci* 2008; **59**(3):225–32.
50. Starostina N, West P. *Part II: sample preparation for AFM particle characterization*. Santa Clara (CA): Pacific Nanotechnology Inc; 2006. p. 1–10.
51. Giddings JC. Field-flow fractionation – analysis of macromolecular, colloidal, and particulate materials. *Science* 1993;**260**(5113):1456–65.
52. Giddings JC, Yang FJ, Myers MN. Flow field-flow fractionation: a versatile new separation method. *Science* 1976;**193**:1244.
53. Gigault J, Pettibone JM, Schmitt C, Hackley VA. Rational strategy for characterization of nanoscale particles by asymmetric- flow field flow fractionation: a tutorial. *Anal Chim Acta* 2014;**809**:9–24.
54. Gigault J, Hackley VA. Observation of size-independent effects in nanoparticle retention behavior during asymmetric-flow field-flow fractionation. *Anal Bioanal Chem* 2013; **405**(19):6251–8.
55. Schimpf ME, Caldwell K, Giddings JC. *Field-flow fractionation handbook*. Wiley; 2000.
56. Barth HG, Jackson C, Boyes BE. Size exclusion chromatography. *Anal Chem* 1994;**66**(12): 595R–620R.
57. Sun T, Chance RR, Graessley WW, Lohse DJ. A study of the separation principle in size exclusion chromatography. *Macromolecules* 2004;**37**(11):4304–12.
58. Ju-Nam Y, Lead JR. Manufactured nanoparticles: an overview of their chemistry, interactions and potential environmental implications. *Sci Total Environ* 2008;**400**(1–3):396–414.
59. Al-Somali A, Krueger KM, Falkner JC, Colvin VL. Recycling size exclusion chromatography for the analysis and separation of nanocrystalline gold. *Anal Chem* 2004;**76**(19): 5903–10.
60. Cole JL, Lary JW, Moody TP, Laue TM. Analytical ultracentrifugation: sedimentation velocity and sedimentation equilibrium. IN: *Methods in cell biology*. Academic Press; 2008. p. 143–179.
61. Planken KL, Colfen H. Analytical ultracentrifugation of colloids. *Nanoscale* 2010;**2**(10): 1849–69.
62. Philip J,W. Submicrometer particle sizing by multiangle light scattering following fractionation. *J Colloid Interface Sci* 1998;**197**(1):9–20.

63. Filella M, Zhang J, Newman ME, Buffle J. Analytical applications of photon correlation spectroscopy for size distribution measurements of natural colloidal suspensions: capabilities and limitations. *Aquatic Colloid Surf Chem* 1997;**120**(1–3):27–46.
64. Zhao Y, Ma L. Applicable range of the Rayleigh-Debye-Gans theory for calculating the scattering matrix of soot aggregates. *Appl Optics* 2009;**48**(3):591–7.
65. Link O, Snelling DR, Thomson KA, Smallwood GJ. Development of absolute intensity multi-angle light scattering for the determination of polydisperse soot aggregate properties. *Proc Combust Inst* 2011;**33**(1):847–54.
66. Jacques DA, Trewhella J. Small-angle scattering for structural biology—expanding the frontier while avoiding the pitfalls. *Protein Sci* 2010;**19**(4):642–57.
67. Porod G. In: Glatter O, Kratky O, editors. *Small angle X-ray scattering*. London: Academic Press; 1982. p. 17–52.
68. Sun YG. Watching nanoparticle kinetics in liquid. *Mater Today* 2012;**15**(4):140–7.
69. Brunauer S, Emmett PH, Teller E. Adsorption of gases in multimolecular layers. *J Am Chem Soc* 1938;**60**(2):309–19.
70. Klobes PMK, Munro RG. *Porosity and specific surface area measurements for solid materials*. National Institute of Standards and Technology Special Publication 960–17; 2006.
71. Thomas R. Spectroscopy tutorial – a beginner's guide to ICP-MS. *Spectroscopy* 2001; **16**(4):38–+.
72. Cottingham K. ICPMS: It's elemental. *Anal Chem* 2004;**76**(1):35a–8a.
73. Bar-Ilan O, Albrecht RM, Fako VE, Furgeson DY. Toxicity assessments of multisized gold and silver nanoparticles in zebrafish embryos. *Small* 2009;**5**(16):1897–910.
74. Fabrega J, Fawcett SR, Renshaw JC, Lead JR. Silver nanoparticle impact on bacterial growth: effect of pH, concentration, and organic matter. *Environ Sci Technol* 2009;**43**(19):7285–90.
75. Cleveland D, Long SE, Pennington PL, Cooper E, Fulton MH, Scott GI, et al. Pilot estuarine mesocosm study on the environmental fate of silver nanomaterials leached from consumer products. *Sci Total Environ* 2012;**421–422**:267–72.
76. Strong KL, Anderson DP, Lafdi K, Kuhn JN. Purification process for single-wall carbon nanotubes. *Carbon* 2003;**41**(8):1477–88.
77. Pace HE, Rogers NJ, Jarolimek C, Coleman VA, Higgins CP, Ranville JF. Determining transport efficiency for the purpose of counting and sizing nanoparticles via single particle inductively coupled plasma mass spectrometry. *Anal Chem* 2011;**83**:9361–9.
78. Tuoriniemi J, Cornelis G, Hassellöv M. Size discrimination and detection capabilities of single-particle ICPMS for environmental analysis of silver nanoparticles. *Anal Chem* 2012;**84**:3965–72.
79. Mitrano DM, Lesher EK, Bednar A, Monserud J, Higgins CP, Ranville JF. Detecting nanoparticulate silver using single-particle inductively coupled plasma–mass spectrometry. *Environ Toxicol Chem* 2012;**31**:115–21.
80. Lespes G, Gigault J. Hyphenated analytical techniques for multidimensional characterisation of submicron particles: a review. *Anal Chim Acta* 2011;**692**(1–2):26–41.
81. Gigault J, Nguyen TM, Pettibone JM, Hackley VA. Accurate determination of the size distribution for polydisperse, cationic metallic nanomaterials by asymmetric-flow field flow fractionation. *J Nanopart Res* 2014;**16**(11).
82. Joy DC, Romig Jr AD, Goldstein J. *Principles of analytical electron microscopy*. Springer; 1986.
83. Vickerman JC, Gilmore IS. *Surface analysis the principal techniques*. 2nd ed. Chichester (UK): Wiley; 2009. xix, 666 p.

84. Skoog DA, Holler FJ, Crouch SR. *Principles of instrumental analysis* [Chapter 18]. 6th ed. Belmont (CA): Thomson Brooks/Cole; 2007. xv, 1039 p.
85. Gouadec G, Colomban P. Raman spectroscopy of nanomaterials: how spectra relate to disorder, particle size and mechanical properties. *Prog Cryst Growth Charact Mater* 2007;**53**(1):1–56.
86. Dresselhaus MS, Jorio A, Hofmann M, Dresselhaus G, Saito R. Perspectives on carbon nanotubes and graphene Raman spectroscopy. *Nano Lett* 2010;**10**(3):751–8.
87. Popović ZV, Dohčević-Mitrović Z, Konstantinović MJ, Šćepanović M. Raman scattering characterization of nanopowders and nanowires (rods). *J Raman Spectrosc* 2007;**38**(6): 750–5.
88. Iwasawa Y. *X-ray absorption fine structure for catalysts and surfaces. World Scientific series on synchrotron radiation techniques and applications.* Singapore; River Edge (NJ): World Scientific; 1996. xvi, 410 p.
89. Teo BK. *EXAFS: basic principles and data analysis* [Chapter 2]. Berlin: Springer; 1986.
90. Hasnain SS, editor. *X-ray absorption fine structure.* West Sussex: Ellis Horwood Limited; 1991.
91. Levard Cm, Reinsch BC, Michel FM, Oumahi C, Lowry GV, Brown GE. Sulfidation processes of PVP-coated silver nanoparticles in aqueous solution: impact on dissolution rate. *Environ Sci Technol* 2011;**45**(12):5260–6.
92. Reinsch BC, Levard C, Li Z, Ma R, Wise A, Gregory KB, et al. Sulfidation of silver nanoparticles decreases *Escherichia coli* growth inhibition. *Environ Sci Technol* 2012;**46**(13):6992–7000.
93. Sharma NC, Sahi SV, Nath S, Parsons JG, Gardea- Torresde JL, Pal T. Synthesis of plant-mediated gold nanoparticles and catalytic role of biomatrix-embedded nanomaterials. *Environ Sci Technol* 2007;**41**(14):5137–42.
94. Cullity B, Stock S. *Elements of X-ray diffraction.* 2001.
95. Hammond C. *The basics of crystallography and diffraction.* 2009.
96. Schwartz LH. *Diffraction from materials.* Elsevier; 2012.
97. Bragg WH, Bragg WL. The reflection of X-rays by crystals. *Proc R Soc London A* 1913; **88**(605):428–38.
98. Hofmann S. *Auger- and X-ray photoelectron spectroscopy in materials science: a user-oriented guide.* New York: Springer; 2012.
99. Hofmann S. *Auger and X-ray Photoelectron Spectroscopy in Material Science*, Vol. 49. London: Springer; 2013.
100. Baer DR, Engelhard MH, Johnson GE, Laskin J, Lai J, Mueller K, et al. Surface characterization of nanomaterials and nanoparticles: important needs and challenging opportunities. *J Vac Sci Technol A Vac Surf Films* 2013;**31**(5):050820.
101. Vickerman JC. Prologue: ToF-SIMS – an evolving mass spectrometry of materials. In: Vickerman JC, Briggs D, editors. *Tof-sims: materials analysis by mass spectrometry.* Manchester (UK): IM Publications LLP and Surface Spectra; 2013. p. 38.
102. Briggs D, Grant JT. *Surface analysis by Auger and X-ray photoelectron spectroscopy.* Chichester (West Sussex, UK): IM Publications; 2003. xi, 899 p.
103. Attard G, Barnes C. *Surfaces.* Oxford Science Publications. Oxford (New York): Oxford University Press; 1998. 92 p.
104. Tanuma S. Summary of ISO/TC 201 Standard: XX ISO 18118: 2004-Surface chemical analysis – Auger electron spectroscopy and X-ray photoelectron spectroscopy – guide to the use of experimentally determined relative sensitivity factors for the quantitative analysis of homogeneous materials. *Surf Interface Anal* 2006;**38**(3):178–80.

105. Rades S, Wirth T, Unger W. Investigation of silica nanoparticles by Auger electron spectroscopy (AES). *Surf Interface Anal* 2014. p. 952–56.
106. Brydson R. *Electron energy loss spectroscopy.* Oxford (UK): BIOS Scientific Publishers Ltd; 2001.
107. Egerton R. *Electron energy-loss spectroscopy in the electron microscope*. Springer; 2011.
108. Rossouw D, Botton GA. Plasmonic response of bent silver nanowires for nanophotonic subwavelength waveguiding. *Phys Rev Lett* 2013;**110**(6):066801.
109. Rossouw D, Couillard M, Vickery J, Kumacheva E, Botton G. Multipolar plasmonic resonances in silver nanowire antennas imaged with a subnanometer electron probe. *Nano Lett* 2011;**11**(4):1499–504.
110. Scholl JA, Koh AL, Dionne JA. Quantum plasmon resonances of individual metallic nanoparticles. *Nature* 2012;**483**(7390):421–7.
111. Krivanek OL, Mory C, Tence M, Colliex C. EEELS quantification near the single-atom detection level. *Microsc Microanal Microstruct* 1991;**2**(2–3):257–67.
112. Suenaga K, Tence T, Mory C, Colliex C, Kato H, Okazaki T, et al. Element-selective single atom imaging. *Science* 2000;**290**(5500):2280–2.
113. Varela M, Findlay SD, Lupini AR, Christen HM, Borisevich AY, Dellby N, et al. Spectroscopic imaging of single atoms within a bulk solid. *Phys Rev Lett* 2004;**92**(9).
114. Yu Z, Guo L, Du H, Krauss T, Silcox J. Shell distribution on colloidal CdSe/ZnS quantum dots. *Nano Lett* 2005;**5**(4):565–70.
115. Xin HL, Mundy JA, Liu Z, Cabezas R, Hovden R, Kourkoutis LF, et al. Atomic-resolution spectroscopic imaging of ensembles of nanocatalyst particles across the life of a fuel cell. *Nano Lett* 2011;**12**(1):490–7.
116. Szakal C, McCarthy JA, Ugelow MS, Konicek AR, Louis K, Yezer B, et al. Preparation and measurement methods for studying nanoparticle aggregate surface chemistry. *J Environ Monit* 2012;**14**(7):1914–25.
117. Szakal C, Ugelow MS, Gorham JM, Konicek AR, Holbrook RD. Visualizing nanoparticle dissolution by imaging mass spectrometry. *Anal Chem* 2014;**86**(7):3517–24.
118. Barreca D, Cruz-Yusta M, Gasparotto A, Maccato C, Morales J, Pozza A, et al. Cobalt oxide nanomaterials by vapor-phase synthesis for fast and reversible lithium storage. *J Phys Chem C* 2010;**114**(21):10054–60.
119. Zanna S, Saulou C, Mercier-Bonin M, Despax B, Raynaud P, Seyeux A, et al. Ageing of plasma-mediated coatings with embedded silver nanoparticles on stainless steel: an XPS and ToF-SIMS investigation. *Appl Surf Sci* 2010;**256**(22):6499–505.
120. Molder JF, Stickle WF, Sobol PE, Bomben KD. In: Chastain J, editor. *Handbook of X-ray photoelectron spectroscopy.* Eden Prairie: Perkin-Elmer Corporation; 1992.
121. Gorham JM, Stover AK, Fairbrother DH. Modification of 1H,1H,2H,2H-Perfluorooctyltrichlorosilane self-assembled monolayers by atomic hydrogen. *J Phys Chem C* 2007;**111**(50): 18663–71.
122. Ging J, Tejerina-Anton R, Ramakrishnan G, Nielsen M, Murphy K, Gorham JM, et al. Development of a conceptual framework for evaluation of nanomaterials release from nanocomposites: environmental and toxicological implications. *Sci Total Environ* 2014;**473–474**:9–19.
123. Petersen EJ, Lam T, Gorham JM, Scott KC, Long CJ, Stanley D, et al. Methods to assess the impact of UV irradiation on the surface chemistry and structure of multiwall carbon nanotube epoxy nanocomposites. *Carbon* 2014;**69**:194–205.
124. Wepasnick KA, Smith BA, Bitter JL, Howard Fairbrother D. Chemical and structural characterization of carbon nanotube surfaces. *Anal Bioanal Chem* 2010;**396**(3):1003–14.

125. Seah MP. Summary of ISO/TC 201 Standard: VII ISO 15472: 2001-surface chemical analysis – X-ray photoelectron spectrometers – calibration of energy scales. *Surf Interface Anal* 2001;**31**(8):721–3.
126. Metz KM, Mangham AN, Bierman MJ, Jin S, Hamers RJ, Pedersen JA. Engineered nanomaterial transformation under oxidative environmental conditions: development of an in vitro biomimetic assay. *Environ Sci Technol* 2009;**43**(5):1598–604.
127. Zhu S, Zhang J, Qiao C, Tang S, Li Y, Yuan W, et al. Strongly green-photoluminescent graphene quantum dots for bioimaging applications. *Chem Commun* 2011;**47**(24):6858.
128. Joseph Y, Besnard I, Rosenberger M, Guse B, Nothofer H-G, Wessels JM, et al. Self-assembled gold nanoparticle/alkanedithiol films: preparation, electron microscopy, XPS-analysis, charge transport, and vapor-sensing properties. *J Phys Chem B* 2003;**107**(30):7406–13.
129. Gorham JM, MacCuspie RI, Klein KL, Fairbrother DH, Holbrook RD. UV-induced photochemical transformations of citrate-capped silver nanoparticle suspensions. *J Nanopart Res* 2012;**14**(10).
130. Gorham JM, Nguyen T, Bernard C, Stanley D, Holbrook RD. Photo-induced surface transformations of silica nanocomposites. *Surf Interface Anal* 2012;**44**(13):1572–81.
131. Phenrat T, Saleh N, Sirk K, Tilton RD, Lowry GV. Aggregation and sedimentation of aqueous nanoscale zerovalent iron dispersions. *Environ Sci Technol* 2006;**41**:284–90.
132. Taurozzi JS, Hackley VA, Wiesner MR. Ultrasonic dispersion of nanoparticles for environmental, health and safety assessment – issues and recommendations. *Nanotoxicology* 2011;**5**:711–29.
133. Taurozzi JS, Hackley VA, Wiesner MR. A standardised approach for the dispersion of titanium dioxide nanoparticles in biological media. *Nanotoxicology* 2013;**7**:389–401.
134. Jain PK, Huang WY, El-Sayed MA. On the universal scaling behavior of the distance decay of plasmon coupling in metal nanoparticle pairs: a plasmon ruler equation. *Nano Lett* 2007;**7**:2080–8.
135. Maxwell DJ, Taylor JR, Nie S. Self-assembled nanoparticle probes for recognition and detection of biomolecules. *J Am Chem Soc* 2002;**124**:9606–12.
136. Hong Y, Lam JWY, Tang BZ. Aggregation-induced emission: phenomenon, mechanism and applications. *Chem Commun* 2009:4332–53.
137. Baalousha M, Stolpe B, Lead JR. Flow field-flow fractionation for the analysis and characterization of natural colloids and manufactured nanoparticles in environmental systems: a critical review. *Flow-Field-Flow Fractionation* 2011;**1218**(27):4078–103.
138. Pan Y, Neuss S, Leifert A, Fischler M, Wen F, Simon U, et al. Size-dependent cytotoxicity of gold nanoparticles. *Small* 2007;**3**:1941–9.
139. Mingos DMP, Slee T, Lin ZY. Bonding models for ligated and bare clusters. *Chem Rev* 1990;**90**(2):383–402.
140. Bakker E, Pretsch E. Modern potentiometry. *Angew Chem Int Ed* 2007;**46**:5660–8.
141. Temminghoff EJM, Plette ACC, Van Eck R, Van Riemsdijk WH. Determination of the chemical speciation of trace metals in aqueous systems by the Wageningen Donnan membrane technique. *Anal Chim Acta* 2000;**417**(2):149–57.
142. Galceran J, Companys E, Puy J, Cecilia J, Garces JL. AGNES: a new electroanalytical technique for measuring free metal ion concentration. *J Electroanal Chem* 2004;**566**:95–109.
143. Donnan FG. The theory of membrane equilibria. *Chem Rev* 1924;**1**:73–90.
144. Chito D, Weng L, Galceran J, Companys E, Puy J, van Riemsdijk WH, et al. Determination of free Zn^{2+} concentration in synthetic and natural samples with AGNES (Absence of Gradients and Nernstian Equilibrium Stripping) and DMT (Donnan membrane technique). *Sci Total Environ* 2012;**421–422**:238–44.

145. Boinovich LB. Long-range surface forces and their role in the progress. *Russ Chem Rev* 2007;**76**:471.
146. Bouhaik IS, Leroy P, Ollivier P, Azaroual M, Mercury L. Influence of surface conductivity on the apparent zeta potential of TiO_2 nanoparticles: application to the modeling of their aggregation kinetics. *J Colloid Interface Sci* 2013;**406**:75–85.
147. Hunter RJ. *Zeta potential in colloid science: principles and applications*. 3rd ed. San Diego: Academic Press; 1988.
148. Clogston JD, Patri AK. In: McNeil SE, editor. *Zeta potential measurement, in characterization of nanoparticles intended for drug delivery*. London: Humana Press; 2011. p. 63–70.
149. Stumm W, Morgan JJ. *Aquatic chemistry*. 3rd ed. New York (NY): John Wiley and Sons, Inc; 1996.
150. Hiemenz PC. Principles of colloid and surface chemistry. In: Lagowski JJ, editor. *Undergraduate chemistry*. New York (NY): Marcel Dekker, Inc; 1986.
151. Hirschfeld T. The hy-phen-ated methods. *Anal Chem* 1980;**52**:297A–312A.
152. Wilson ID, Brinkman UAT. Hyphenation and hypernation: the practice and prospects of multiple hyphenation. *J Chromatogr A* 2003;**1000**:325–56.
153. Hinds WC. *Aerosol technology*. 2nd ed. New York: John Wiley and Sons, Inc; 1999.
154. Elzey S, Tsai DH, Yu LL, Winchester MR, Kelley ME, Hackley VA. Real-time size discrimination and elemental analysis of gold nanoparticles using ES-DMA coupled to ICP-MS. *Anal Bioanal Chem* 2013;**405**:2279–88.

Part II

Physicochemical Characterization

Chapter 3

Size Distributions

James Ranville[1] and Manuel D. Montano
Department of Chemistry and Geochemistry, Colorado School of Mines, Golden, CO, USA
[1]*Corresponding author: E-mail: jranvill@mines.edu*

Chapter Outline

1. INTRODUCTION

The physical descriptor of nanomaterials (NMs) that is most commonly reported, and arguably most critical to understanding NM environmental behaviour, is particle size.[1,2] Particle size influences a myriad of processes relevant to the distribution and reactivity of NMs in environmental and biological systems.[3] Size influences the solubility of NMs because of the increasing prevalence of surface free energy (i.e. surface tension) brought on by the increase in the radius of curvature that accompanies decreasing size.[4,5] Additionally, NM diffusion in solution is inversely proportional to particle size and thus particle collisions increase with decreasing size.[6] This affects aggregation rates which play a large role in environmental NM transport and fate (e.g. sedimentation in surface water and air, particle capture in porous

Frontiers of Nanoscience, Vol. 8. http://dx.doi.org/10.1016/B978-0-08-099948-7.00003-8

media).[6–8] Size also plays a significant role in surface area-dependant reactions, such as those that drive many NM applications, as well as the several processes that occur in environmental and biological media.[9–12] Biological interactions such as ingestion by filter feeding organisms and transport across membranes are influenced by NM size.[13–15]

Despite being a crucial metric, NM size is often misrepresented in the literature.[16] Generally, NMs inherently possess a certain degree of size dispersity, and thus the reported size must reflect this degree of dispersity by reporting both the NM size as well as the size distribution of the material (i.e. average ± standard deviation).[17] Additionally, different analytical techniques measure different size parameters, making any comparison between studies a difficult issue.[18,19] Another important aspect that should be reported is whether the measured size is that of the primary NMs or of their aggregates. While primary particle size determines the NM's inherent properties, the aggregate particle size, which is highly dynamic, influences NM fate, transport, bioavailability and uptake.[20–23]

Despite being simple in principle, particle size can often be an insufficient descriptor of NM physical characteristics, analytically difficult to determine, and a dynamic property subject to temporal changes dependant on the dispersing media. To reflect these intricacies, this discussion includes elements of:

- general consideration for size measurement
- particle size descriptors such as 'average' size and size distribution
- measured size parameters: effects of measurement technique
- influence of sample preparation and history
- challenges to NM size characterization in environmental media.

As a result of its importance on reactivity and behaviour, reporting particle size has become an essential element of particle characterization for publishing the results of studies on the environmental implications of nanotechnology.[24] However, it is clear that great caution must be taken in the measurement, interpretation and reporting of NM size. For example, the lack of an observed effect of size of a given NM on its toxicity may not support a lack of this relationship. Rather it might reflect that size dispersity, aggregation, or dissolution during organism exposure confound the impact of NM size.[25,26]

2. GENERAL CONSIDERATIONS FOR SIZE ANALYSIS

2.1 Particle Shape

Only a spherical particle can be described by a single number; the diameter or radius. A nonspherical particle may possess multiple length measures and thus present greater complexity for characterization. Often, the size of nonspherical

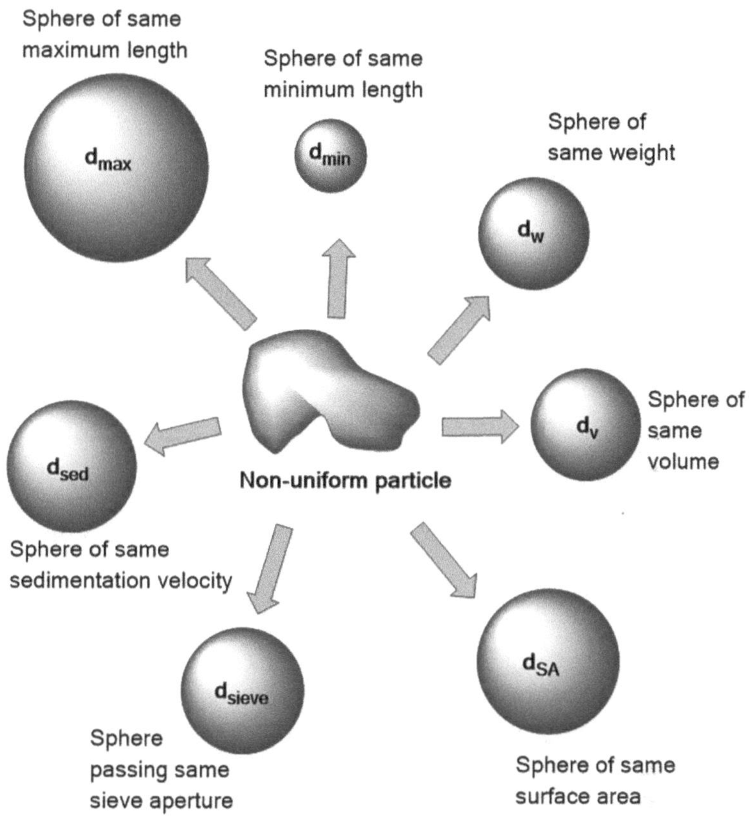

FIGURE 1 Equivalent diameters from irregular shapes.

particles is described as the 'equivalent sphere' sharing some property of the actual particle (Figure 1). This shared property might be a physical characteristic such as equivalent particle volume or may be an equivalent response of the sizing method such as cross-sectional area in an electron microscopy (EM) measurement. Simple conversions between spheres and nonspherical geometries, including cylinders and plates, follow from:

$$\frac{4}{3}\pi r^3\ (spheres) = \pi r^2 l\ (cylinders,\ discs) = l_1 l_2 l_3\ (plates) \qquad (1)$$

where

for rods, $l =$ major (length) dimension
for disks, $l =$ thickness
for plates having three nonequal dimensions, $l =$ thickness, minor length, major length

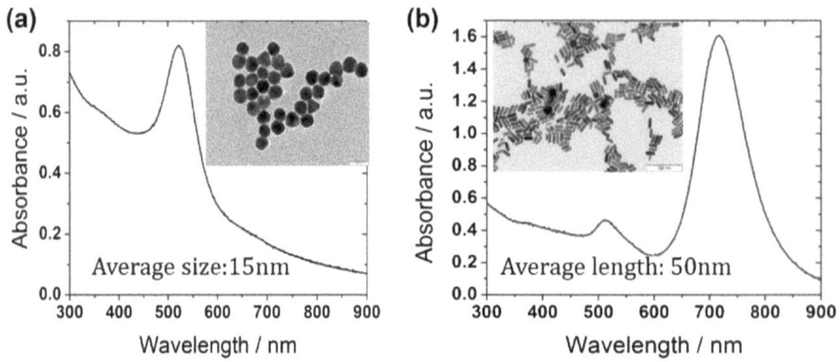

FIGURE 2 Surface plasmon resonance from Au (a) spheres and (b) rods.

Although, describing NM size as the equivalent sphere is a good first approximation, it should be noted that two particles of different shapes and sizes could possess the same equivalent diameter, but other metrics should be considered in detail if subtle differences in NM properties or effect are observed. Defining an equivalent spherical diameter may be an effective approximation for irregular particles with low aspect ratios; however, for particles with high aspect ratios it may provide a poor descriptor for assessing the size dependence of NM behaviour. Obvious examples include cylindrical carbon nanotubes (radius $<<$ length)[27] and nanoclays (thickness $<$ minor dimension $\leq$ major dimension).[28,29] These shapes will have properties such as aggregation rate or adsorption capacity that may be significantly different from that of the equivalent spherical particle. Microscopy and image analysis provide the best measurement method for shape analysis; however, other techniques have some sensitivity to shape. An example is in the UV-Vis characterization of metal NMs.[30] Surface plasmon resonance arises from the absorption of light by electrons in the near-surface region of the metal NMs and is heavily influenced by particle size and shape.[31,32] In the case of spheres, only one length is present and thus the absorption feature is monomodal (Figure 2(a)). For rods or simple dimers formed by aggregation, a notable second adsorption feature is present representing the longitudinal surface plasmon resonance (Figure 2(b)). Other size measurement techniques are influenced by particle shape, and as such shape information might be obtained from instrument response.[33]

2.2 Particle Alteration

The fewest ambiguities in particle size analysis occur when pristine (i.e. as manufactured) NMs are characterized in simple media. Yet, characterization of particle size for these materials may have little value for the purpose of most studies. Various alterations to NMs that affect their size, or the size measurement process itself, can occur and are both time variant and media dependant. NMs can be modified during incorporation into products, product use, end-of-life (disposal), biological uptake and environmental transport.[34] Sample preparation (e.g. dispersion) prior to analysis may or may not return the NM to its original size.[26] Subsequently, despite more robust size measurements being obtained from the original NM, the relevant environmental behaviour of the NM is more likely a result of its 'as sampled' size.

One of the most common alterations of NMs is aggregation, both homoaggregation (like particles)[30,35–38] and heteroaggregation (dissimilar particles),[39–42] which results in growth in particle size. Depending on the strength of interaction this process may be reversible. Ultrasonication is frequently applied as a first step in NM size characterization in order to disrupt aggregation and return the NM to its primary size. Adjustable operational parameters include applied power, sonication duration and instrument type (bath or probe sonicators). Although specific details of the sonication processes vary widely from study to study, standardized methodology is available for some NMs.[43]

Dissolution is likely process for some NMs (ZnO, Ag, etc.) resulting in a decrease in NM primary particle size.[44–47] This process is irreversible in that even if reprecipitation occurs, it is unlikely to return the NM to its original size. Quantification of the resulting dissolved ions could be used in a back calculation of the original size, but any attempt to make this correction is likely to be highly inaccurate. An exception would be very simple systems (dissolution in simple solutions) where a mass balance can be obtained for the element of interest. In these systems, the NM of interest should either be the primary (only) source of the dissolved ion or the initial concentration of the dissolved ion is known. Regardless, unlike aggregation, dissolution leads to a new particle size that may actually represent the behaviour of the NM in the media of interest, and thus the size of the altered primary particle is the relevant measurand. Dissolution complicates matters more in that, now any potential ecotoxic effect may be the result of the altered NM, the dissolved ion, or both.

Finally, the formation, loss, or alteration of surface coating on NMs alters the overall particle size.[48] Obvious examples include the formation of protein coronas in biological systems and adsorption of natural organic matter in natural waters.[9,11,49] In this case, information on the overall particle size (core + coating) should be captured. Care must be taken in choosing the sizing methodology, and in most cases application of multiple methods is advised.[19] The issue arises in whether the size analysis technique accurately responds to both the NM and the coating. For example, methods that rely on NM mass (e.g. centrifugation) may be relatively insensitive to a low-density organic coating on a dense core particle (e.g. Au) whereas a method that measures hydrodynamic size (e.g. light scattering, field flow fractionation (FFF)) may readily capture the effect of the coating on size.

2.3 Particle Dispersity

Quite often NM size is reported as an average value without any description of their dispersity. Size dispersity reflects the inherent variability in NM size, as it is practically impossible that all particles in an NM suspension have exactly the same size (see Baalousha and Lead, *Nature Nano,* **2013** for NIST definition of size dispersity that can be reported here). NM size dispersity is just as important as average size and can be a key confounding factor in understanding NM size-dependant properties and behaviours. Size dispersity can result in a several fold difference in the measured size when using different analytical techniques (e.g. atomic force microscopy (AFM), flow-FFF (Fl-FFF) and dynamic light scattering (DLS)) on the same sample.[50] This is due to differences in the dependence of instrumental response to NM size dispersity. Such variability in the measured size hampers the comparison of studies performed by different research groups, in particular when only one analytical technique is used to measure particle size.

2.4 Size Distribution Weighting

Differently weighted particle size distributions (PSDs) are generally obtained by different analytical techniques as discussed below. Specific examples include number size distribution by microscopy techniques, mass-weighted size distributions by analytical centrifugation, volume-weighted size distributions by Fl-FFF-UV-Vis and intensity-weighted size distribution by laser light scattering techniques. The weighting of the PSD must be kept in mind when selecting the analytical technique as well as when comparing size distribution measured by different analytical techniques. It is very common that sizes measured by different techniques are different (comparison and conversion of size distribution is discussed in detail below).

3. DESCRIPTORS OF SIZE DISTRIBUTIONS

3.1 Representations of PSDs

Dispersed NMs nearly always contain multiple particle sizes. If the span of diameters is narrow, the particles are considered monodispersed. NIST defines monodisperse as a PSD where 90% of the distribution lies with 5% of the median size. However the span of the PSD is large, the sample is considered polydispersed. The breadth of the distribution has obvious significance on the physical properties of the particles and on NM size measurement. A 10-fold change in diameter results in a 100-fold and 1000-fold change in surface area and volume (mass), respectively. Increasing polydispersity imposes more stringent requirements on the analytical technique, requiring a wider range of method sensitivity. The degree of polydispersity will also have a significant effect on particle reactivity.

PSDs are usually reported as tabular data and graphically presented as a histogram, or in many cases as a smoothed histogram. The two ways of representing size distributions are as a differential or a cumulative plot. In a differential size distribution the y-axis represents some measure of the amount of material (frequency) in a given size class, the most common being particle number or total mass. The y-axis is often normalized to the total amount of material measured (number, mass, etc.) in order to create a fractional frequency distribution of NM across the total size range. The size data is classified into bins and is plotted as the histogram x-variable. The bin width depends on the size resolution of the method used. Comparison of data from various techniques requires the use of bins of equal size range. In the cases where this is not possible, the problem can be overcome by dividing the y-axis frequency values by the size range (ISO 9276-1). The resulting size-normalized frequency (fraction/micron) can be plotted against nonuniform size bins. The same normalization to total amount of material is applied to a cumulative size distribution plot. In this case the y-axis is plotted as the fractional amount of material (number, mass, etc.) less than the given size bin.

3.2 Descriptors of Central Values

Three important descriptors of a PSD are the mean, median and mode of the distribution. In the case of a perfectly symmetrical PSD these descriptors have the same value. The numerical values of these size descriptors depend on the metric used to quantify the amount of NM in each size class, the possibilities usually being particle number, surface area, or mass (volume).

Mode: The mode is perhaps the most obvious feature of a differential PSD and represents the most frequently occurring size bin. It appears as the obvious 'peak' in the distribution and is thus most often the feature in the PSD that draws the most attention. It is very useful in a descriptive sense, especially if a PSD is multimodal. In this case multiple modes can be defined as the midpoints of the various peaks. It is not often used in a quantitative manner, the mean and medians being favoured. It is also not visible in the cumulative distribution.

Median: The median value is the size that divides the NM population in half. This is often represented as D50 or x50. The value is clearly dependant on what parameter is used to report the population of particles. The median is the smallest for number-based PSDs, larger for surface area-based PSDs and the greatest for mass (volume)-based PSDs. It is the central value shown on a cumulative PSD plot. Other common descriptors of the cumulative PSD are D10 (10% smaller) and D90 (90% smaller).

Mean: The mean is arguably the most valuable descriptor of the 'center' of the PSD. It represents a numerical average size value of the PSD. The mean is 'weighted' by the parameter used for reporting the amount of particles in each size bin (number, surface area, mass/volume). ASTM E799-03(2009) presents the general formula for computing the PSD mean as:

$$\overline{D}_{pq}^{(p-q)} = \frac{\sum_i D_i^p}{\sum_i D_i^q} \tag{2}$$

where

$\overline{D}$ = the average diameter
$(p - q)$ $p > q$ = the algebraic power of $\overline{D}_{pq}$
p and q = the integers 1,2,3 or 4
D_i = the diameter of the ith particle
$\sum_i$ = the summation of D_i^p or D_i^q, representing all particles in the sample
$0 = p$ and q = values of 0,1,2,3 or 4
$\sum_i D_i^0$ = the total number of particles in the sample

From this general formula, a wide variety of mean values can be obtained. In general the most important values are:

$$\text{Mean number diameter} \quad < D_{1,0} > = \left(\frac{\sum n_i D_i}{\sum n_i}\right) \tag{3}$$

$$\text{Mean surface area diameter} \quad < D_{2,0} > = \left(\frac{\sum n_i D_i^2}{\sum n_i}\right)^{1/2} \tag{4}$$

$$\text{Mean volume (mass) diameter} \quad < D_{3,0} > = \left(\frac{\sum n_i D_i^3}{\sum n_i}\right)^{1/3} \tag{5}$$

These formulae provide a representation of the PSD average value that accounts for which parameter is considered the most important: particle number, surface area, or volume (mass). The mean value will be the smallest for a number-based PSD and be the largest for the volume (mass)-based PSD. The choice of appropriate mean is also dependant on the basis of the measurement technique. For example, one can compare EM to laser diffraction: the former being a direct measure of particle number and the latter responding to particle volume. In this case $D_{1,0}$ is most appropriate for the EM and $D_{3,0}$ for the laser diffraction.

The general formula shown above illustrates one difficulty in obtaining these simple averages. Since the summed size parameters are divided by the total particle number, a large number (all) of particles need to be accurately counted. This has led to the introduction of moment means, which are considered analogous to moments of inertia. The means can be thought of as the central size point around which the PSD would rotate. Surface area and volume are usually used and generation of the moments is accomplished by adding an additional length term to these parameters. Thus two new means are commonly generated:

$$\text{Surface area} - \text{weighted mean diameter} \quad < D_{3,2} > = \frac{\sum n_i D_i^3}{\sum n_i D_i^2} \tag{6}$$

$$\text{Volume (mass)} - \text{weighted mean diameter} < D_{4,3} > = \frac{\sum n_i D_i^4}{\sum n_i D_i^3} \tag{7}$$

These means no longer have a dependence on determining particle number.

3.3 Descriptors of PSD Width

For polydisperse systems a key descriptor of the PSD is its width, and a number of approaches can be taken to define this. For macromolecules it is common to report a polydispersity index (PDI). This is obtained from the ratio of the weight average molecular weight to the number average molecular weight with values increasing from 1 as the polydispersity increases. The span of the PSD is often used and is given by:

$$Span = \frac{D_{0.9} - D_{0.1}}{D_{0.5}} \tag{8}$$

where

$D_{0.1}$ = diameter for which 10% of the NM is less than that size
$D_{0.9}$ = diameter for which 90% of the NM is less than that size
$D_{0.5}$ = median diameter

The relevant terms are easily obtained from a cumulative PSD analysis.

The graphical differential size distribution can be described numerically by the mode (i.e. peak maximum) and the distribution width defined as 'full width at half maximum'. Thus at the two points on the y-axis that represent ½ the height of the mode, the size range can be reported as a descriptor of the PSD width.

If the PSD is normally distributed around a mean value a statistical analysis can be applied to describe the PSD width. In this case the familiar statistical terms of standard deviation (σ) and variance (σ^2) are used. Thus 68.27% and 95.45% of the NM population lies within ± 1 or 2 σ of the mean respectively. Sometime the PSD may not be normally distributed when data are presented on a linear size scale. If the data span a wide range, the PSD may be best represented by examining the population against the log of the diameters. Often the PSD is normally distributed when log diameter is used (i.e. log-normally distributed) and thus σ has the same usefulness in describing the PSD width in log diameter space.

Variance: The magnitude of the spread of the PSD is given by the variance, which is the second statistical central moment (M_2) of the PSD:

$$M_2 = \frac{\sum\left(n_i(D_i - \overline{D})^2\right)}{\sum n_i} \tag{9}$$

The larger the variance, the greater is the polydispersity of the distribution.

3.4 Descriptors of PSD Shape

Often PSDs do not represent a normally distributed set of diameters. Non-symmetry and nongausssian peak shape thus need to be described by the parameters of skewness and kurtosis (Figure 3).

Skewness: PSDs can often be nonuniformly distributed around the central (e.g. mean) value. The presence of a few large particles can create a 'tail' in the larger size of the distribution and thus cause the PSD to be positively skewed. A similar effect could be seen with a tail in the lower size of the distribution resulting in a negative skew. The skewness, which is the third central statistical moment of the PSD is defined by:

$$M_3 = \frac{\sum\left(n_i(D_i - \overline{D})^3\right)}{\sum n_i} \tag{10}$$

The skewness index can be obtained by dividing by the cubed standard deviation (M_3/σ^3). The skewness of a normal distribution is zero. As the tails of the distribution become more important (greater polydispersity) the skewness can be greatly affected.

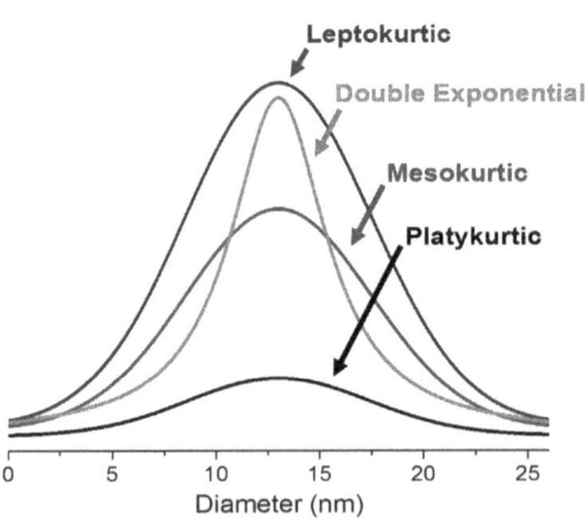

FIGURE 3 Descriptors of the shape of particle size distributions.

Kurtosis: The flatness or peakedness of a PSD is described by kurtosis (Figure 3). This is the fourth central statistical moment of the PSD and is given by:

$$M_4 = \frac{\sum\left(n_i\left(D_i - \overline{D}\right)^4\right)}{\sum n_i} \tag{11}$$

A normalized version of kurtosis is given by dividing by the standard deviation raised to the fourth power and then subtraction of three $((M_4/\sigma^4)-3)$. PSDs approximating a normal distribution, called mesokurtic, have a normalized kurtosis equal to zero. Sharper peaks, called leptokurtic, have positive values and platykurtic have negative values. One commonly used distribution obtained from analysis of light scattering data is the double exponential distribution. This PSD is characterized by a strong peak, which decays rapidly but also demonstrates significant tails. This distribution has a large positive kurtosis and may or may not be skewed.

4. MEASUREMENT METHODS

A wide range of very highly developed particle sizing methods exist in our current nanometrology toolkit. The basis and operating principles of most of these methods were described in Chapter 2. However the question remains as to how well these techniques perform on 'real world samples'. Issues include complex particle composition, particle mixtures that are chemically heterogeneous and polydisperse in size, and the complexity of the matrices that are relevant to the study of NMs in the environment. Furthermore the question about comparability between methods must be assessed as the various methods have different fundamental sensitivity to particle size and report different distributions (e.g. number-, volume-, intensity-based).[51,52] The following discussion primarily focuses on the specific issues related to how the fundamentals of each method, and the limitations to their application, affect the size information that can be obtained. Table 1 (modified from Baalousha et al.) attempts to concisely represent the relevant details of the information obtained by the methods as well as summarizing the strengths and weaknesses.[51] The following discussion briefly elaborates on information not fully contained in Table 1.

4.1 Sizing Based on Microscopy

Particle sizing by microscopy allows for the measurement of size metrics one particle at a time. Thus very polydisperse samples can be examined. Particles having nonspherical geometry can also be directly classified for additional length dimensions. However, as a single particle method, EM is very limited in its ability to perform high-throughput analysis.

Electron microscopy: High-resolution transmission electron microscopy (HR-TEM) and scanning electron microscopy (SEM) have the clear advantage

TABLE 1 Characteristics of Some Common Particle Size Characterization Techniques

Method	HR-TEM	SEM	AFM	Laser Diffraction	DLS
Measured size parameter	X-Y dimensions	X-Y-(Z) dimensions	X-Y-Z dimensions	Volume	Diffusion coefficient
Size characteristic	Morphology: Shape spherical diameter	Morphology: Shape, (height), spherical diameter	Morphology: Shape, height, spherical diameter	Spherical diameter	Spherical hydrodynamic diameter
Size distribution	Number	Number	Number	Mass	Intensity-weighted
Additional characteristics	Internal structure, atomic packing and defects	Internal structure, w/EDX: Major and minor element composition			Instrument often combined with zeta potential measurement
Operating principle	Interaction of transmitted electrons with particle while X-Y scanning across specimen surface	Interaction of scattered electrons with particle while X-Y scanning across specimen surface	Tip-particle interactions while X-Y scanning across specimen surface	Small angle light scattering of dispersed particles	Time-dependant fluctuations in light intensity due to Brownian motion of dispersed particles
Optimal size resolution	0.1–0.2 nm	5–20 nm	0.1 nm (Height)	Refractive index dependant 5–10 nm	1–5 nm
Optimized minimum size	1 nm	30 nm	1 nm	50 nm	1 nm
Sample environment	UHV	UHV	Atmospheric pressure dry conditions, liquid	Liquid	Liquid

Advantages	High resolution, particle-by-particle analysis, visual observation of the particles	Visual observation of individual particles, gives semi-3D information	High resolution, particle-by-particle analysis with morphology information	Rapid analysis is liquid media	Rapid and accurate analysis of monodisperse particles
Difficulty and limitations	TEM requires special sample preparation, requires UHV which may alter particles	Time consuming, requires UHV which may alter particles	Time consuming, requires large number of particles for representative PSD	Measures average properties of whole particle population, cannot resolve polydisperse particles	Measures average properties of whole particle population, cannot resolve polydisperse particles

Method	**DCS**	**Fl-FFF**	**NTA**	**spICP-MS**
Measured size parameter	Buoyant mass	Diffusion coefficient	Diffusion coefficient	Particle mass
Size characteristic	Spherical hydrodynamic diameter	Spherical hydrodynamic diameter	Spherical hydrodynamic diameter	Spherical diameter
Size distribution	Mass	Detector-specific mass- or intensity-weighted	Number	Number
Additional characteristics		Can recover size fractions for additional characterization	Indirect compositional information, can be combined with zeta potential measurement, solution number concentration	Major element composition, solution mass and number concentration
Operating principle	Sedimentation velocity of dispersed particles under applied gravitational field	Equilibrium between applied field and particle diffusion in a parabolic velocity gradient	Image analysis of dispersed particle velocity due to Brownian motion	Measurement of elemental mass of dispersed particles

Continued

TABLE 1 Characteristics of Some Common Particle Size Characterization Techniques—cont'd

Method	DCS	Fl-FFF	NTA	spICP-MS
Optimal size resolution	Density dependant 1–5 nm	Fl-FFF: 1 nm C-FFF: density dependant 1–5 nm	5–10 nm	Element-specific Au: 5–10 nm
Optimized minimum size	Density dependant 10 nm	1–2 nm	30–50 nm	Element- and instrument-specific 10–20 nm
Sample environment	Liquid	Liquid	Liquid	Liquid
Advantages	Very high resolution separation and rapid analysis for high-density particles	Separated fractions that can be further characterized offline or online by hyphenation	Rapid analysis on single particle basis	Rapid analysis on single particle basis with compositional discrimination possible
Difficulty and limitations	Requires knowledge of density for sizing, separated particles cannot be collected for further analysis	Optimization required to provide adequate recovery and resolution	Minimum detectable size > 20–50 nm	Minimum detectable size > 20–50 nm, composition must be known to report size

HR-TEM, high-resolution transmission electron microscopy; SEM, scanning electron microscopy; AFM, atomic force microscopy; DLS, dynamic light scattering; DCS, differential centripetal sedimentation; Fl-FFF, flow field flow fractionation; NTA, nanotracking analysis; spICP-MS, single particle inductively coupled plasma–mass spectrometry; C-FFF, centrifugal field flow fractionation.

of giving an image of each particle, from which size dimensions can be measured. SEM has the advantage over TEM in that by tilting the stage, the particle height can be imaged. Although the various irregular dimensions can be observed, size distributions are usually reported as the number-based PSDs of a single dimension i.e. 'diameter'. In studies where a limited number of size analysis techniques are applied, EM is one of the most frequently reported approaches. As with all single particle counting approaches, in order to obtain a robust PSD a large number of particles must be sized.[53] When PSDs demonstrate significant tailing (skew/kurtosis), the relatively few particles in the tails may not be statistically counted. The rather slow process of particle counting by EM limits the usefulness of the approach, although automated image analysis can free up operator time. A potentially more significant issue can arise from particle alteration by exposure to the high vacuum conditions under which EM is operated.[54,55] Although 'hard' particles may be unaffected, softer materials such as polymer coatings on particles can be significantly altered, lost or simply not imaged due to the low electron density of these materials. Although particle morphology can be directly obtained by both methods, and energy dispersive X-ray spectroscopy (EDX) can provide elemental composition information, these techniques generally cannot discriminate engineered NMs from the background of natural NMs in environmental and biological samples.

During routine application of EM methods, aggregation of particles is often observed.[54] This may or may not reflect aggregation in solution, and can easily be an artifact of sample drying.[54] It is also very difficult to translate the number of particles observed by EM to a solution particle number distribution. In general, uniform distributions of particles across the field of view do not occur. Typically, TEM samples are prepared by depositing a small drop of NM suspension on a carbon-coated TEM grid, with the anticipation that most NMs will attach to the carbon coating. This sample preparation approach results in the attachment of only few NMs on the carbon coating, depending on the interaction forces between the carbon coating and the NMs. Together the small volume typically used to prepare TEM samples, as well as the passive sampling approach, hinder the representativeness and the statistical power of the analysis. Several approaches have been recently developed to overcome these issues: (1) ultracentrifugation + surface functionalization[56] and (2) analysis of frozen NM suspension droplets using cryo-TEM.[57,58]

Atomic force microscopy: The use of AFM can in part overcome some of the issues of EM while providing very similar information including shape and number-based distributions. One of AFM's advantages over EM is a precise measurement of particle height, providing a better estimate of the equivalent spherical size of the particle. Additionally, the AFM measurement can be made in a liquid media, thus limiting possible drying artifacts. AFM shares the rather time-consuming issues of EM for counting particles as well as limitations in translating the measured particle number to solution particle concentrations. Careful sample preparation has improved on the quantitative nature of particle concentration measurements by AFM.[56] Particle discrimination in complex samples is also not possible. As such, AFM remains a rather niche technique and is not commonly utilized in studies except those in which method comparison is part of the study objectives.

4.2 Sizing Based on Diffusion Coefficient

Analysis of particle's Brownian motion allows for direct measurement of diffusion coefficient, which can then be related to particle size if certain assumptions or additional measurements are made. Given that the measurement relies on particle motion, the relevant size metric is the hydrodynamic diameter (typically represented as d_h). The simplest assumption is that of spherical particle geometry which yields the well-known Stokes law:

$$d_h = \frac{K_B T}{3\pi\eta D_T} \tag{12}$$

where

d_h = hydrodynamic diameter
K_B = Boltzmann Constant
T = temperature
η = viscosity
D_T = translational diffusion coefficient

The spherical assumption used in Stokes law is invoked to eliminate the friction coefficient, the factor that relates particle velocity to the particle's physical characteristics (e.g. mass and shape). Particle mass can be determined without assumptions about shape if sedimentation velocity and diffusion coefficient are both determined. Diffusion coefficient measurements for particle size analysis are most commonly obtained by DLS, particle tracking analysis or FFF, the latter method also providing fractionated material for further characterization.

Dynamic light scattering: Particle size measurement by DLS is perhaps the most frequently reported approach used in environmental studies (Chapter 2).[19] Unlike EM, it is not a particle-counting method but rather relies on the cumulative scattering from many thousands of particles. Its widespread popularity arises from its speed, ease of use and minimal sample preparation. Problematically, DLS results are often not carefully examined for possible artifacts such as aggregation and sedimentation of particles during these measurements. Subsequently, DLS excels in the characterization of monodisperse particles in simple aqueous media, but is limited when examining polydisperse samples or monodisperse samples in complex matrices. Furthermore, the principles behind DLS (Mie scattering) result in large particles masking the scattering intensity of smaller particles in the autocorrelation function, which can bias the measurement to larger particle sizes. As a result, spurious dust particles are particularly problematic and can make data interpretation impossible.

Light scattering depends greatly on particle size, scattering angle and the laser wavelength. With most commonly used instruments the scattering intensity is proportional to d^6 for diameters up to about 100–200 nm.[59] Above this size the dependence of scattering becomes complex, displaying size-dependant oscillations in intensity, with a general dependence roughly on d^2. Average diameters obtained from DLS are thus intensity weighted and the most commonly reported size is the z-average diameter, given by:

$$< D_Z >= \frac{\sum n_i D_i^6}{\sum n_i D_i^5} \tag{13}$$

Fluctuations in light intensity occur due to Brownian motion. The change in intensity (I) over short increments of time (τ), when integrated over some time (T) follows an exponential decay. This can be described by the autocorrelation function $G_2(\tau)$ as follows:

$$G_2(\tau) = \frac{1}{T}\int_0^T I(t)I(1+\tau)d\tau \tag{14}$$

A more useful autocorrelation function can be obtained by normalization:

$$\frac{G_2(\tau) - B}{B} = \beta e^{-2\Gamma\tau} \tag{15}$$

where B and β are constants describing the baseline and maximum normalized value of the normalized autocorrelation function respectively. The decay of the correlation function is described by Γ which is related to the translation diffusion coefficient by:

$$D = \frac{\Gamma}{q^2} \tag{16}$$

The wave vector (q) is given by:

$$q = \frac{4\pi\eta}{\lambda}\sin\left(\frac{\theta}{2}\right) \tag{17}$$

where

η = refractive index
λ = wavelength of light
θ = scattering angle

The average D_t is related to particle size through the Stokes–Einstein equation. Recognizing that since the data are intensity weighted, this corresponds to the z-average diameter. The measured correlation function must then be analyzed to obtain Γ in order to determine D_z.

For nearly monodisperse sample the method of cumulants is used. The process is described in ISO 22412:2008. In the cumulants method only the initial part of the correlation function is analyzed, usually ending the analysis when the correlation function has decayed 90% to baseline. After defining a value of the mean decay function ($\overline{\Gamma}$) a polynomial expansion of the correlation function is performed to obtain descriptors of the PSD, assuming a Gaussian distribution, as given by:

$$\text{Linear (Gamma)} \quad \ln\left[\frac{G_2(\tau) - B}{B}\right] = \ln\beta - 2\overline{\Gamma}\tau \tag{18}$$

$$\text{Quadratic (Polydispersity)} \quad \ln\left[\frac{G_2(\tau) - B}{B}\right] = \ln\beta - 2\overline{\Gamma}\tau + K_2^2\tau^2 \tag{19}$$

$$\text{Cubic (Skew)} \quad \ln\left[\frac{G_2(\tau) - B}{B}\right] = \ln\beta - 2\overline{\Gamma}\tau + K_2^2\tau^2 - \frac{K_3^3}{3}\tau^3 \tag{20}$$

$$\text{Quartic (Kurtosis)} \quad \ln\left[\frac{G_2(\tau) - B}{B}\right] = \ln\beta - 2\overline{\Gamma}\tau + K_2^2\tau^2 - \frac{K_3^3}{3}\tau^3 + \frac{K_4^4}{12}\tau^4 \tag{21}$$

Generally the skew (3rd PSD moment) and kurtosis (4th PSD moment) are not reported and polydispersity is given by the PDI which is equal to K_2/Γ^2.

For polydisperse samples several different approaches are taken. Most commonly the CONTIN approach is used. In this case a Laplace transform is preformed and a greater proportion of the autocorrelation function is analyzed. Generally the decay function is more fully analyzed and is examined to where it reaches 1% above baseline. This approach allows for resolution of multiple peaks in the distribution if size differences are sufficiently large.

Nanotracking analysis: A relatively new approach to using diffusion coefficient measurements for size analysis is nanotracking analysis (NTA). It has the advantage of DLS in that minimal sample preparation is required and analysis is relatively rapid, on the scale of a few minutes.[60,61] This approach provides several advantages over photon correlation spectroscopy (PCS), while introducing new limitations.[62,63] First, NTA measures the D_t one particle at a time by determining the track length of each particle over the period of data collection. This is accomplished by optically observing the scattering of each particle illuminated by the laser.[64–66] Given that NTA is a single particle technique, it yields number-based size distributions and has the potential to see finer features in the PSD than what could be observed by an averaging technique such as DLS. Like any counting technique, it will be biased to the smaller size of the PSD, subject to its minimum size detection limit, due to the higher likelihood of observing the greater proportion of small particles. This bias is also due to the difficulty in accurately sampling the fewer number of large particles that would dominate a volume/mass- or intensity-based PSD. Robust determination of broad PSDs thus requires acquisition of statistically significant number of particles, which increases analysis time.

The size detection limit of NTA depends on the NM composition, as the technique requires sufficient scattering by the particles to allow accurate tracking. In general its size detection limit is significantly higher than DLS.[52] Since the scattering intensity is both size and composition dependant, NTA analysis can provide some additional characterization information. If a sample contains particles of different compositions a plot of size versus intensity yields a pattern that is more complex than a plot of NMs of a single composition.[67] Although the technique does not provide element-specific information, it does allow for the detection of particle chemical heterogeneity.

Field flow fractionation: FFF has long been used to relate measured D_t to particle size.[68] Sample preparation is minimal, similar to DLS and NTA, but holds the advantage over these techniques as particles are separated by their respective D_t values. This separation opens the door to combine FFF with other techniques that can further characterize NMs. Fraction collection can be used to provide size-fractionated samples for NM characterization by EM, thus allowing for shape and morphology analysis across the PSD.[69,70] Considerable success has been achieved by online coupling of light scattering (static and dynamic) to FFF. By continuously introducing size-fractionated material into the online light scattering instruments the problems in analysing polydispersed NMs by light scattering are reduced.[71–73] Similarly the direct coupling of FFF to inductively coupled plasma–mass spectrometry (ICP-MS) provides a means of examining heterogeneity in NM composition by use of elemental analysis.[74,75]

FFF separates NMs based on the equilibrium formed between particles subjected to an applied driving force and the opposing migration due to particle diffusion. Particle diffusion dictates the position of this equilibrium within a hydrodynamic flow that displays a parabolic velocity distribution across the thin rectangular channel wherein the separation occurs. The fundamental measurement made on FFF is the retention time of a particle compared to that of the carrier fluid:

$$R = \frac{t^0}{t_R} \tag{22}$$

where t^0 is the time required for the solvent to elute and t_R is the time required for the NM. The retention ratio (R) is related to the retention parameter λ by the relationship:

$$R = 6\lambda\left[\coth\left(\frac{1}{2\lambda} - 2\lambda\right)\right] \tag{23}$$

D_t is related to λ by the relationship:

$$\lambda = \frac{l}{\lambda} = \frac{D_t}{Uw} \tag{24}$$

where

l = the average distance of the NM above the wall of the FFF channel
w = the thickness of the FFF channel
D_t = the NM diffusion coefficient
U = the migration velocity of the NM due to the applied field.

The migration velocity (U) depends on the nature of the field applied which results in different subtechniques of FFF. If a perpendicular fluid flow is used (Fl-FFF subtechnique), U is related to the volumetric flow rate and the carrier fluid viscosity. If an artificial gravitational field is used (C-FFF, centrifugal FFF subtechnique), then U is related to the applied G-force and the buoyant density of the NM. The lower limit of size resolution for FFF depends on the magnitude of the migration field that can be generated by the FFF instrumentation and on NM properties. The former depends on the either the maximum cross-flow rate generated in Fl-FFF or the maximum gravitational force generated in C-FFF. The latter depends on the NM properties of D_t or mass. For most FFF systems Fl-FFF is limited to a size of 2−5 nm whereas for C-FFF analysis of dense particles, such as gold NMs, the minimum size is about 5−10 nm.[76] The upper size limit of FFF is a result of the steric transition that results in a reversal of the influence of size on elution time. For particles greater than about 500 nm or so, the particles are not separated by their D_t but rather by their physical size.[77] While not a significant issue for NM analysis ($d < 100$ nm) micron-sized particles could elute at times similar to NMs.

The weighting of the PSDs depends on the nature of the detector used in the FFF separation. The commonly used UV-Vis detector largely responds to the total volume of particles eluting with time, giving an approximately volume/mass-based distribution. However absorption (e.g. molar extinction coefficient) is affected by size and composition so UV detection provides only an approximation of the volume-based distribution. This is particularly noticeable when examining metallic NMs, whose surface plasmon resonance is highly size dependant. Online dynamic and static light scattering provides PSDs that emphasize the larger particles due to the d^6 dependence of the scattering intensity. Direct coupling of FFF to ICP-MS gives a mass-based PSD, but the elemental composition must be known in order to convert the mass of element to mass of NM. To date, few studies have coupled particle-counting techniques directly to FFF but in principle number-based distributions could be obtained by this approach.

While the separating capability of FFF makes it a powerful NM characterization tool, several significant limitations exist. Analysis times are long, with typical analysis times for polydisperse samples exceeding 60 min in most cases, which limits the application of FFF as a high-throughput technique. Somewhat more serious is the often low recovery of NMs from the channel resulting from NM adsorption to the wall of the channel. Mass balance can be used to determine the severity of this issue. Also adsorption is size dependant, with the larger, less diffusive particles being more susceptible to this process. This results in a distortion of the PSD with the larger sizes being underrepresented. Despite significant recent improvements in FFF hardware and software, the technique remains more operationally challenging than DLS or NTA.

4.3 Sizing Based on Other Advanced Approaches

A number of additional approaches are available for determination of PSDs. Many of these have been described in Chapter 2. Thus the following discussion will be limited to new and/or relatively unused approaches that show promise for environmental nanometrology.

Single particle ICP-MS: ICP-MS has long been a powerful technique used for determining total elemental concentration due to its elemental specificity and analytical sensitivity. When coupled to a size-separation technique, ICP-MS can provide elemental specific size analysis. When combined with serial filtration or centrifugation a very low-resolution size distribution can be obtained, but serious filtration artifacts, generally the poor passage and recovery of particles smaller than the filter pore size, make interpretation problematic. Coupling ICP-MS to FFF provides high-resolution element-specific PSDs. FFF-ICP-MS provides the total elemental content of each fraction but does not directly measure the composition of each particle in the fraction.

A different approach utilizes ICP-MS to analyze samples on a particle by particle basis. Recent developments in single particle ICP-MS (spICP-MS), first introduced for aqueous samples by Degueldre[78–82] more than a decade ago, have made it possible to determine the elemental composition of individual particles in the sample. The greatest advantage of this method is its elemental specificity that may allow NM characterization in complex mixtures of particles. This technique has been used for a wide range of applications including the detection of carbon nanotubes, the analysis of silver NMs in wastewater and the release of NMs from consumer products.[83–85] Sample preparation for spICP-MS is minimal and generally is limited to dilution such that the concentration will be sufficiently low (low parts per trillion to parts per billion, depending on particle size) in order to prevent multiple particles entering the plasma at once.[63,83,85,86,87]

In spICP-MS an undigested sample is first nebulized into an argon plasma, which decomposes the sample and ionizes the constituent atoms. The ions then pass through a mass selector (e.g. quadrupole, magnetic sector) and are detected. Determining the elemental composition of single particles is achieved by reducing the dwell time (the duration for which the instrument takes a reading) to microseconds. This allows a NM to be detected as a pulse of intensity above a background of ambient ion signal. The assumption is that each pulse correlates to one NM event, which allows a particle number concentration to be calculated as the number of pulses obtained during a reading divided by the volume of sample analyzed and correcting for the sampling efficiency. Size information can also be obtained via a calibration curve of intensity for a given mass of dissolved analyte and using a standard particle to determine the efficiency of the mass flux for a given intensity. The signal intensity produced by a given particle event can then be used to calculate the equivalent mass of the particle, which can be converted into a size if the density of the material is known.[63,87] The method provides number-based PSDs and thus has the same strengths and weaknesses as other particle-counting methods.

Since spICP-MS detects the mass of a specific element, direct analysis of an equivalent spherical diameter requires knowledge of particle composition. In Figure 4 the issues regarding computation of size from spICP-MS are illustrated. For single particles of known composition (e.g. Au) the spICP-MS-computed size equals the particle size of the inorganic NM, as illustrated by the 100 and 200 nm particles used in the example. If the particle is coated with a material that was not detected by the ICP-MS (e.g. carbon) then the NM will be undersized. This issue could be addressed by performing the spICP-MS analysis on FFF fractions, the latter giving the hydrodynamic size. Alternatively multielement spICP-MS could be used if the shell material consists of an element that can be detected by ICP-MS.[88–90] A similar solution could be used to determine the size of particles that contain the element of interest as a low abundance constituent of a larger particle (example of 500 nm particle with an analyte mass equivalent to 100 nm pure NM). Finally, it has yet to be demonstrated to what degree intact aggregates can be analysed by spICP-MS. In the case where the NM of interest is heteroaggregated with dissimilar particles, multielement spICP-MS may allow

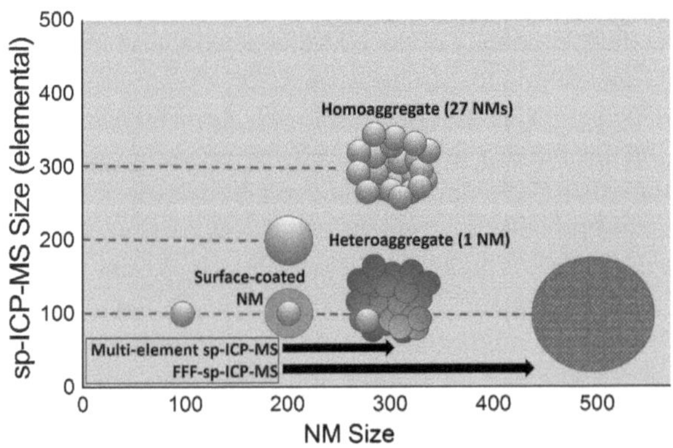

FIGURE 4 Factors importation for determining particle size distribution using single particle inductively coupled plasma–mass spectrometry (spICP-MS). FFF, field flow fractionation; NM, nanomaterial.

accurate size analysis. The homoaggregation case would require particle disaggregation in order to resolve it from a similarly massed primary particle.

Because only particles containing the element of interest are detected by the ICP-MS, other background particles do not interfere with the method. This suggests a great potential for characterizing NMs in the presence of background particles, a capability that no other technique has. Sample analysis is accomplished in less than 5 min making spICP-MS a likely candidate for high-throughput analysis.

Differential centripetal centrifugation: Recent advances in instrumentation have made differential centripetal sedimentation (DCS) an attractive means of obtaining high-resolution size information, provided the density of the particles is known. In DCS, a sample is injected into a transparent spinning disc that contains a fluid in which a density gradient has been created. The sample particles are accelerated towards the outside of the disc and pass through a beam of visible light. The resulting data (absorbance versus time) are converted into a PSD using Stokes law, assuming a spherical geometry. For high-density NMs, resolution is very high and differences in below 10% in size can be detected. Given that most instruments utilize a visible light absorption detector, the PSD is not strictly a volume/mass-based distribution but is rather like the distribution obtained by FFF with UV-Vis detection.

Depending on the particle density and degree of polydispersity, analysis times can be on the order of 1–5 min.[91] This short analysis time would make DCS amenable to high-throughput analysis. However, only a limited number of samples, on the order of a few dozen, can be injected before the analysis must be stopped, the disc drained and the fluid replaced. The major limitation of DSC as a sizing tool is the inability to analyze samples that contain particles of different densities.

5. COMPARISON AND INTERCONVERSION BETWEEN PSDs

Comparison of PSDs: As was discussed previously, PSDs and their central value (mean, median, mode) can be described in a number of ways. The most common are number-, mass- and intensity-weighted distributions. In large part this is the consequence of the inherent differences in the operating principles of the sizing methods. The differences between the average values of diameter among possible PSD-weighting factors is shown graphically in Figure 5. In this example the various averages for a mixture of 21 particles having sizes of 10 nm (3), 30 nm (5), 50 nm (9), 90 nm (3) and 150 nm (1) are shown. The number-, surface area-, volume- and z-averages are 50, 59, 104 and 145 nm, respectively. Though a monodisperse sample should give very similar averages, various analytical artifacts can distort a particular PSD. Certain analytical techniques are more sensitive to certain size distributions than others. This example illustrates the potential pitfalls of using only a single measurement to determine particle size. Two of the most commonly utilized analytical

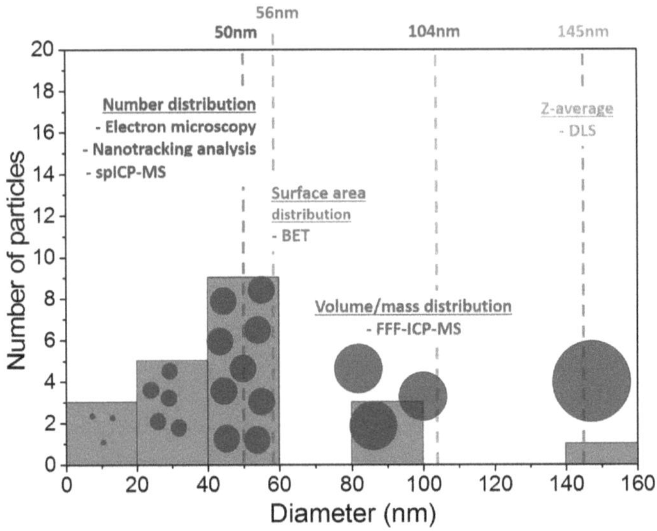

FIGURE 5 An example comparison of different weighting of the particle size distribution of polydisperse samples (note: particles are not drawn at the same scale as the x-axis). spICP-MS, single particle inductively coupled plasma—mass spectrometry; FFF, field flow fractionation; DLS, dynamic light scattering; BET, brunauer-emmett-teller.

techniques, EM and DLS, would give very different results in this scenario. As a result, it is always recommended that multiple techniques are used in the determination of size and PSD. For example, FFF-ICP-MS is capable of measuring mass-based PSDs of Au NM to a few nanometres, whereas quadrupole spICP-MS is limited to about 15—20 nm. Thus analysis of a somewhat polydisperse Au sample with a mean near 20 nm would display different PSDs by these two techniques even if the FFF-ICP-MS were converted to a number-based distribution in order to compare to the spICP-MS data.

In Figure 5 it can be seen that two averages give values below 100 nm while two give values above 100 nm. This leads to obvious issues with the established definition of NMs as being materials that have at least one dimension less than 100 nm. For simplicity, and to reflect the output of most measurement methods, the discussion is limited to reporting an equivalent spherical diameter. For this hypothetical example then, based on average size, only two distributions, number and surface area normalized distributions, would describe the sample as a NM. Currently the European Commission has recommended a definition of a NM based on number-based PSDs (Recommendation 2011/696/EU). In their recommendation "point 2" provides the following definition of a NM:

> *"Nanomaterial' means a natural, incidental or manufactured material containing particles, in an unbound state or as an aggregate or as an agglomerate and*

where, for 50 % or more of the particles in the number size distribution, one or more external dimensions is in the size range 1 nm-100 nm. In specific cases and where warranted by concerns for the environment, health, safety or competitiveness the number size distribution threshold of 50 % may be replaced by a threshold between 1 and 50 %."

The EC also specifies an alternative to this particle number-based definition as:

"A material should be considered as falling under the definition in point 2 where the specific surface area by volume of the material is greater than 60 m^2/cm^3. However, a material which, based on its number size distribution, is a nanomaterial should be considered as complying with the definition in point 2 even if the material has a specific surface area lower than 60 m^2/cm^3"

In this case the hypothetical example would meet the EC definition as a NM.

Interconversions of PSDs: Although it is mathematically relatively simple to convert between different PSDs for spherical particles, operationally there can be significant challenges. For polydisperse NMs the number-based PSDs will always be smaller than volume/mass PSDs. Thus particle-counting methods will tend to emphasis the small side of the PSD whereas mass-based analyses will emphasize the larger side of the PSD. In Figure 6 (adapted from von der Kammer, unpublished data) a hypothetical example is given to illustrate this point. For this sample the mass concentration is constant across the entire size range, an unlikely scenario yet useful for illustration purposes. Thus the particle number must exponentially decrease with increasing size. Thus a particle number-based method (TEM, NTA, spICP-MS) will have a lower probability of detecting the few large particles present in the sample,

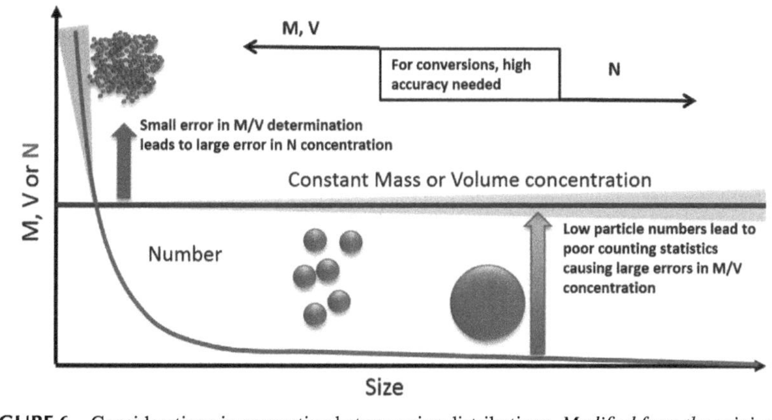

FIGURE 6 Considerations in converting between size distributions. *Modified from the original of F. von der Kammer (unpublished data).*

thus leading to considerable error in the mass-based PSD for larger sizes. Conversely, errors in a mass-based technique for small particles would lead to large uncertainties in the number concentration for that size. As an example, in FFF-ICP-MS analysis the incomplete removal of the void peak, which is often the result of dissolved constituents in the sample, would yield a nearly indetectable signal in the ICP-MS data, but due to the small size would yield large computed particle numbers. This has obvious consequences when considering the EC suggested number-based PSD for classification of NMs.

6. GENERAL CONSIDERATIONS FOR ENVIRONMENTAL SYSTEMS

Background NM population: The detection and particle size analysis of engineered NMs in the environment is complicated by ubiquitous naturally occurring colloids and NMs that are present in much larger concentrations than the engineered particles. This poses two constraints on PSD analysis in natural waters. First, the technique must be able to differentiate between natural and engineered particles.[92] In some cases this is entirely not true (DLS, DCS) or exceedingly difficult (TEM, NTA). ICP-MS-based methods (FFF-ICP-MS, spICP-MS), which respond to specific elements are likely to have the greatest potential for discriminating among particles. In some cases if the element in the engineered NM is rare (c.f. Au, Ag), ICP-MS may be able to readily detect and size the particles. If the element is present in both natural NMs and engineered NMs (c.f. Ce, Si, Ti) then detection and characterization is more ambiguous. A way forward may be to do particle-by-particle element ratio analysis that could yield particle-specific ratios that could allow differentiation.[88,92]

Even if engineered NM-specific methods are available, the feasibility of differentiating engineered NMs from natural NMs may depend in part on particle size. In natural waters, number-based PSDs (Figure 7) for natural particles have been found to follow Pareto's power law, the differential form of which is:

$$\frac{dN}{dx} = Zx^{-\beta} \tag{25}$$

where

N = the number of particles with sizes smaller than x.
Z = an empirical constant that describes the total amount of suspended particles.

β is an empirical constant generally found to be approximately 3 for natural waters,[93] although very little data exist for particles smaller than 100 nm. This relationship implies that there are 1000 times more 10-nm particles than 100-nm particles. Such a size distribution means that interference on particle-based

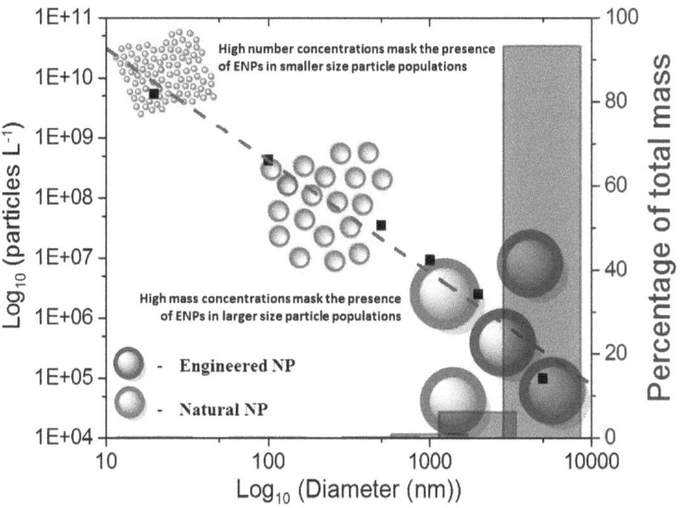

FIGURE 7 Considerations for PSDs when examining particles in natural waters. *Data from Harris (1977)*[94].

methods from an abundance of naturally occurring NMs will be more problematic as NM size decreases. However, although there are fewer particles at the upper end of the nanoscale size range, these larger particles will pose a problem for mass-based methods and in particular light scattering methods, which exhibit strong size dependence in their sensitivity.

Homo- and heteroaggregation: Given the high background of naturally-occurring particles, engineered NMs are likely to experience hetero-aggregation. With respect to environmental fate and effects, one would expect that the size dependence of these processes will depend on the aggregate size, not the original (primary) engineered NM size. However it is not clear how sampling, storage and analysis of natural samples will affect the aggregation state. Although DLS is often used in laboratory aggregation experiments, other more selective methods such as ICP-MS have not been tested to determine if aggregation state is preserved. As such rigorous and standardized methods of sample dispersion, followed by analysis of primary particle size will yield the most stable, although perhaps not most relevant, PSD characterization. If heteroaggregation is to be understood in natural systems, uniform methods of sample collection, storage and analysis must be developed and tested for repeatability and accuracy.

7. SUMMARY

Because size and size distribution play an important role in the environmental fate and impacts of NMs, accurate determination of PSDs remains a priority need. Despite a diverse range of methods to measure PSD, several challenges remain in the forefront. At the most fundamental level, the different ways of presenting the central size value, and the distribution of particles around this value, lead to potential for ambiguity. Number-, area-, mass- and intensity-based distributions will all appear different for particles other than very monodisperse samples. This may lead to the need to interconvert these distributions. Size-dependant analytical limitations of each method may lead to errors in this conversion, especially at the tails of the distributions.

The presence of interfering particles in natural samples leads towards the need to develop tools with some degree of specificity that can lead to discrimination between engineered and naturally occurring particle types. ICP-MS-based methods seem to hold significant promise in this regard but further method development is needed.

Finally it is important to recognize the inherent instability of NMs in solution that may lead to aggregation and alter the representation of particle size. This process may be at least in part reversible. Thus PSD is not a static number but is likely very dynamic changing in time and space. Since most methods for PSD analysis have not been validated for aggregates, considerable variability in results from different research groups is to be expected. Until the degree to which aggregation state can be maintained through sampling, storage and analysis is determined, rigorous disaggregation and PSD analysis of primary particles will provide the most robust and comparable size data.

REFERENCES

1. Bleeker EA, de Jong WH, Geertsma RE, Groenewold M, Heugens EH, Koers-Jacquemijns M, et al. *Regul Toxicol Pharm* 2013;**65**:119–25.
2. Klaine SJ, Alvarez PJJ, Batley GE, Fernandes TF, Handy RD, Lyon DY, et al. *Environ Toxicol Chem* 2008;**27**:1825–51.
3. Handy R, Owen R, Valsami-Jones E. *Ecotoxicology* 2008;**17**:315–25.
4. Roduner E. *Chem Soc Rev* 2006;**35**:583–92.
5. Ma R, Levard Cm, Marinakos SM, Cheng Y, Liu J, Michel FM, et al. *Environ Sci Technol* 2011;**46**:752–9.
6. Zhang Y, Chen Y, Westerhoff P, Hristovski K, Crittenden JC. *Water Res* 2008;**42**:2204–12.
7. Hotze EM, Phenrat T, Lowry GV. *J Environ Qual* 2010;**39**:1909–24.
8. Areepitak T, Ren J. *Environ Sci Technol* 2011;**45**:5614–21.
9. Cedervall T, Lynch I, Lindman S, Berggård T, Thulin E, Nilsson H, et al. *Proc Natl Acad Sci* 2007;**104**:2050–5.
10. Lundqvist M, Stigler J, Elia G, Lynch I, Cedervall T, Dawson KA. *Proc Natl Acad Sci* 2008;**105**:14265–70.
11. Monopoli MP, Walczyk D, Campbell A, Elia G, Lynch I, Baldelli Bombelli F, et al. *J Am Chem Soc* 2011;**133**:2525–34.
12. Walkey CD, Olsen JB, Guo H, Emili A, Chan WC. *J Am Chem Soc* 2012;**134**:2139–47.
13. Jiang W, Kim BY, Rutka JT, Chan WC. *Nat Nanotechnol* 2008;**3**:145–50.
14. Chithrani BD, Ghazani AA, Chan WC. *Nano Lett* 2006;**6**:662–8.
15. Ward JE, Kach DJ. *Marine Environ Res* 2009;**68**:137–42.
16. Boverhof DR, David RM. *Anal Bioanal Chem* 2010;**396**:953–61.

17. Baalousha M, Lead JR. *Nat Nanotechnol* 2013;**8**:308–9.
18. Domingos RF, Baalousha MA, Ju-Nam Y, Reid MM, Tufenkji N, Lead JR, et al. *Environ Sci Technol* 2009;**43**:7277–84.
19. Hassellov M, Readman J, Ranville J, Tiede K. *Ecotoxicology* 2008;**17**:344–61.
20. Jassby D, Farner Budarz J, Wiesner M. *Environ Sci Technol* 2012;**46**:6934–41.
21. Jassby D, Wiesner M. *Langmuir* 2011;**27**:902–8.
22. Hotze EM, Bottero JY, Wiesner MR. *Langmuir* 2010;**26**:11170–5.
23. Liu J, Aruguete DM, Murayama M, Hochella Jr MF. *Environ Sci Technol* 2009;**43**:8178–83.
24. Stefaniak AB, Hackley VA, Roebben G, Ehara K, Hankin S, Postek MT, et al. *Nanotoxicology* 2013;**7**:1325–37.
25. Montes-Burgos I, Walczyk D, Hole P, Smith J, Lynch I, Dawson K. *J Nanopart Res* 2010;**12**:47–53.
26. Petersen EJ, Henry TB, Zhao J, MacCuspie RI, Kirschling TL, Dobrovolskaia MA, et al. *Environ Sci Technol* 2014;**48**:4226–46.
27. Dai H. *Acc Chem Res* 2002;**35**:1035–44.
28. Hochella MF. *Elements* 2008;**4**:373–9.
29. Hochella MF, Lower SK, Maurice PA, Penn RL, Sahai N, Sparks DL, et al. *Science* 2008;**319**:1631–5.
30. Diegoli S, Manciulea AL, Begum S, Jones IP, Lead JR, Preece JA. *Sci Total Environ* 2008;**402**:51–61.
31. Eustis S, El-Sayed MA. *Chem Soc Rev* 2006;**35**:209–17.
32. Link S, El-Sayed MA. *J Phys Chem B* 1999;**103**:8410–26.
33. Mock JJ, Barbic M, Smith DR, Schultz DA, Schultz S. *J Chem Phys* 2002;**116**:6755–9.
34. Nowack B, Ranville JF, Diamond S, Gallego-Urrea JA, Metcalfe C, Rose J, et al. *Environ Toxicol Chem* 2012;**31**:50–9.
35. Chen KL, Mylon SE, Elimelech M. *Environ Sci Technol* 2006;**40**:1516–23.
36. Pallem VL, Stretz HA, Wells MJM. *Environ Sci Technol* 2009;**43**:7531–5.
37. Petosa AR, Jaisi DP, Quevedo IR, Elimelech M, Tufenkji N. *Environ Sci Technol* 2010;**44**:6532–49.
38. Stankus DP, Lohse SE, Hutchison JE, Nason JA. *Environ Sci Technol* 2010;**45**:3238–44.
39. Cerbelaud M, Videcoq A, Abelard P, Pagnoux C, Rossignol F, Ferrando R. *Langmuir* 2008;**24**:3001–8.
40. Lin W, Kobayashi M, Skarba M, Mu C, Galletto P, Borkovec M. *Langmuir* 2005;**22**:1038–47.
41. Yates PD, Franks GV, Biggs S, Jameson GJ. *Colloids Surf A Physicochem Eng Aspects* 2005;**255**:85–90.
42. Yates PD, Franks GV, Jameson GJ. *Colloids Surf A Physicochem Eng Aspects* 2008;**326**:83–91.
43. Taurozzi J, Hackley V, Wiesner M. *NIST Special Publication* 2012;**1200**:3.
44. Levard C, Mitra S, Yang T, Jew AD, Badireddy AR, Lowry GV, et al. *Environ Sci Technol* 2013;**47**:5738–45.
45. Li X, Lenhart JJ. *Environ Sci Technol* 2012;**46**:5378–86.
46. Liu G, Wang D, Wang J, Mendoza C. *Sci Total Environ* 2011;**409**:2852–7.
47. Mitrano D, Ranville J, Bednar A, Kazor K, Hering A, Higgins C. *Environ Sci Nano* 2014;**1**:248–59.
48. Levard C, Hotze EM, Colman BP, Dale AL, Truong L, Yang XY, et al. *Environ Sci Technol* 2013;**47**:13440–8.
49. Shannahan JH, Lai X, Ke PC, Podila R, Brown JM, Witzmann FA. *PLoS One* 2013;**8**:e74001.

50. Baalousha M, Lead J. *Environ Sci Technol* 2012;**46**:6134–42.
51. Baalousha M, Stolpe B, Lead JR. *J Chromatogr A* 2011;**1218**:4078–103.
52. Boyd RD, Pichaimuthu SK, Cuenat A. *Colloids Surf A Physicochem Eng Aspects* 2011;**387**:35–42.
53. Bootz A, Vogel V, Schubert D, Kreuter J. *Eur J Pharm Biopharm* 2004;**57**:369–75.
54. Tiede K, Tear SP, David H, Boxall ABA. *Water Res* 2009;**43**:3335–43.
55. Tiede K, Hassellöv M, Breitbarth E, Chaudhry Q, Boxall ABA. *J Chromatogr A* 2009;**1216**:503–9.
56. Baalousha M, Prasad A, Lead J. *Environ Sci Process Impacts* 2014;**16**:1338–47.
57. Tiede K, Boxall AB, Tear SP, Lewis J, David H, Hassellöv M. *Food Addit Contam* 2008;**25**:795–821.
58. Hondow N, Brydson R, Wang P, Holton MD, Brown MR, Rees P, et al. *J Nanopart Res* 2012;**14**:1–15.
59. Brar SK, Verma M. *Trends Anal Chem* 2011;**30**:4–17.
60. Dragovic RA, Gardiner C, Brooks AS, Tannetta DS, Ferguson DJ, Hole P, et al. *Nanomed Nanotechnol Biol Med* 2011;**7**:780–8.
61. La Rocca A, Di Liberto G, Shayler P, Parmenter C, Fay M. *Tribol Int* 2014;**70**:142–7.
62. Filipe V, Hawe A, Jiskoot W. *Pharm Res* 2010;**27**:796–810.
63. Pace HE, Rogers NJ, Jarolimek C, Coleman VA, Gray EP, Higgins CP, et al. *Environ Sci Technol* 2012;**46**:12272–80.
64. Hole P, Sillence K, Hannell C, Maguire CM, Roesslein M, Suarez G, et al. *J Nanopart Res* 2013;**15**:1–12.
65. Gallego-Urrea JA, Tuoriniemi J, Hassellöv M. *Trends Anal Chem* 2011;**30**:473–83.
66. Saveyn H, De Baets B, Thas O, Hole P, Smith J, Van Der Meeren P. *J Colloid Interface Sci* 2010;**352**:593–600.
67. Mahl D, Diendorf J, Meyer-Zaika W, Epple M. *Colloids Surf A Physicochem Eng Aspects* 2011;**377**:386–92.
68. Giddings JC. *Anal Chem* 1995;**67**:592A–8A.
69. Beckett R, Murphy D, Tadjiki S, Chittleborough DJ, Giddings JC. *Colloids Surf A Physicochem Eng Aspects* 1997;**120**:17–26.
70. Baalousha M, Lead J. *Environ Sci Technol* 2007;**41**:1111–7.
71. Baalousha M, Kammer FVD, Motelica-Heino M, Hilal HS, Le Coustumer P. *J Chromatogr A* 2006;**1104**:272–81.
72. Fraunhofer W, Winter G, Coester C. *Anal Chem* 2004;**76**:1909–20.
73. Hupfeld S, Holsaeter AM, Skar M, Frantzen CB, Brandl M. *J Nanosci Nanotechnol* 2006;**6**:9–10.
74. Lesher EK, Ranville JF, Honeyman BD. *Environ Sci Technol* 2009;**43**:5403–9.
75. Gray EP, Bruton TA, Higgins CP, Halden RU, Westerhoff P, Ranville JF. *J Anal At Spectrom* 2012;**27**:1532–9.
76. Kowalkowski T, Buszewski B, Cantado C, Dondi F. *Crit Rev Anal Chem* 2006;**36**:129–35.
77. Giddings JC, Myers MN. *Sep Sci Technol* 1978;**13**:637–45.
78. Degueldre C, Favarger P-Y. *Colloids Surf A Physicochem Eng Aspects* 2003;**217**:137–42.
79. Degueldre C, Favarger P-Y. *Talanta* 2004;**62**:1051–4.
80. Degueldre C, Favarger PY, Bitea C. *Anal Chim Acta* 2004;**518**:137–42.
81. Degueldre C, Favarger PY, Wold S. *Anal Chim Acta* 2006;**555**:263–8.
82. Degueldre C, Favarger P-Y, Rosse R, Wold S. *Talanta* 2006;**68**:623–8.
83. Mitrano DM, Lesher EK, Bednar A, Monserud J, Higgins CP, Ranville JF. *Environ Toxicol Chem* 2012;**31**:115–21.

84. Mitrano DM, Rimmele E, Wichser A, Erni R, Height M, Nowack B. *ACS Nano* 2014; **8**:7208–19.
85. Reed RB, Goodwin DG, Marsh KL, Capracotta SS, Higgins CP, Fairbrother DH, et al. *Environ Sci Process Impacts* 2013;**15**:204–13.
86. Benn TM, Westerhoff P. *Environ Sci Technol* 2008;**42**:4133–9.
87. Pace HE, Rogers NJ, Jarolimek C, Coleman VA, Higgins CP, Ranville JF. *Anal Chem* 2011;**83**:9361–9.
88. Montaño MD, Badiei HR, Bazargan S, Ranville J. *Environ Sci Nano* 2014;**1**:338–46.
89. Borovinskaya O, Gschwind S, Hattendorf B, Tanner M, Günther D. *Anal Chem* 2014;**86**:8142–8.
90. Borovinskaya O, Hattendorf B, Tanner M, Gschwind S, Gunther D. *J Anal At Spectrom* 2013;**28**:226–33.
91. Walczyk D, Bombelli FB, Monopoli MP, Lynch I, Dawson KA. *J Am Chem Soc* 2010;**132**:5761–8.
92. Montano M, Lowry G, von der Kammer F, Blue J, Ranville J. *Environ Chem* 2014; **11**:351–66.
93. Buffle J, Leppard GG. *Environ Sci Technol* 1995;**29**:2169–75.
94. Harris JE. *Deep-Sea Research* 1977;**24**:1055–61.

Chapter 4

Analytical Transmission Electron Microscopy and Scanning Transmission Electron Microscopy Techniques for the Characterization of Nanomaterial Composition, Phase and Crystallinity

Bojeong Kim*[,1] **and Michael F. Hochella, Jr.**[§,¶]

**Department of Earth and Environmental Science, College of Science and Technology, Temple University, Philadelphia, PA, USA;* [§]*Department of Geosciences, Virginia Tech, Blacksburg, VA, USA;* [¶]*Institute for Critical Technology and Applied Science, Environmental Nanoscience and Technology Laboratory, Virginia Tech, Blacksburg, VA, USA*

[1]*Corresponding author: E-mail: bkim@temple.edu*

Chapter Outline

1. INTRODUCTION

Nanotechnology is one of the most important new technologies of the twenty-first century. With strong demands of nanotechnology initiatives in the past few decades, much attention has been given to the analytical methods and the method developments for the characterization of chemical and structural properties of NMs. Among these methods, analytical transmission electron

Frontiers of Nanoscience, Vol. 8. http://dx.doi.org/10.1016/B978-0-08-099948-7.00004-X

microscopy (analytical TEM) is a unique tool for materials characterization at the nanoscale, yielding information of their size, morphology, aggregation states, chemical composition and crystal structure with unsurpassed spatial resolution.[55] Modern TEM with scanning coils is also capable of scanning transmission electron microscopy (STEM) mode. The combination of STEM with X-ray and electron detectors for chemical analyses has led to an atomic-scale understanding of how the crystal structure and structural variations are related to physical and chemical properties. While the TEM and STEM techniques are widely distributed in laboratories serving the fields of materials and earth sciences, as well as the biological sciences, for the characterization of NMs and nanoprocesses, their application is still somewhat limited relative to what one might expect for such powerful tools. This is mainly because TEM and STEM techniques require high degrees of expertise in sample preparation, instrument operation and data analysis and interpretation. This area of research is also a challenge due to the high cost of instrument acquisition and maintenance.

The primary objective of this chapter is to make these instruments and the techniques much more intellectually accessible and familiar to a wider variety of potential users. In this regard, we first present basic operation principles and capabilities of TEM and STEM instruments, followed by their applications and limits, data acquisitions and analyses, as well as sample preparation procedures. Our hope is to see increased usage of TEM and STEM techniques in the studies of environmental nanoscience and nanotechnology, as well as ecotoxicology, to obtain multiple, but complementary, perspectives in these emerging fields.

2. TEM AND STEM INSTRUMENTS

TEM is a microscopy technique that uses a ~100–300 kV electron beam generated from either tungsten or lanthanum hexaboride filaments or field-emission guns for (high-resolution) images and electron diffraction patterns of materials. Electrons also can be used for chemical analyses of materials in the TEM: when electrons are transmitted through a material, many of them undergo inelastic scattering processes, emitting characteristic X-ray quanta, Auger electrons and light.

There are two different modes in TEM: a 'fixed beam' mode (conventional TEM) and a rastered beam mode (STEM). Modern TEMs equipped with scanning coils are capable of both modes of operation. Both TEM and STEM modes will be discussed in detail in the following section, especially focussing on their abilities to perform phase identification, crystal structure determination, as well as compositional analyses of nanoscale materials. Hereafter, (S)TEM is used to denote both the conventional TEM and STEM modes, whereas TEM and STEM are used individually when there is a difference between those two modes.

Basic operation principle: In conventional TEM, a broader (~several micrometres), (almost) parallel electron beam is typically used to illuminate a thin sample. In Figure 1, the first two condenser lenses (C1 and C2) are used to form a parallel beam of electrons; the C1 lens is typically fixed by the manufacturer for the parallel-beam TEM mode, whereas the C2 lens is

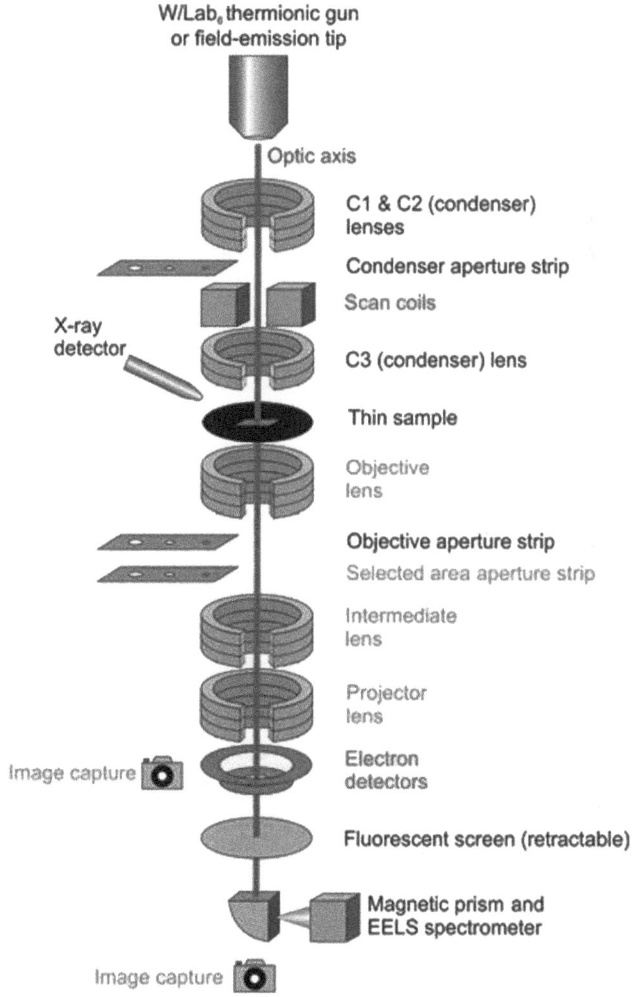

FIGURE 1 Diagram showing the relative positions of the main components of a microscope that can operate both in transmission electron microscopy (TEM) and scanning transmission electron microscopy (STEM) modes. Those components used for both TEM and STEM work are named in black, whereas components used in TEM, only, are in blue (light gray in print versions) and those in STEM, only, are in red (gray in print versions). Note that dedicated STEM instruments may differ significantly in their configuration. *Diagram and captions reproduced from Ref. 29; with permission from the Mineralogical Society.*

adjustable to make the beam parallel to the optic axis. In addition, a smaller C2 aperture decreases the convergence angle of the beam (α), producing a more parallel beam. The parallel-beam mode is often used for TEM imaging and diffraction. The objective lens is then used to form a magnified image or diffraction pattern of the illuminated area, which is then further magnified by a sequence of projection lenses beneath the sample, onto the image detector. The first lens of the projector system is sometimes called the 'intermediate' or 'diffraction' lens. Therefore, in TEM, one has to select between imaging mode and diffraction mode. Once the adjustment of the projector system for magnification is properly made, the image or diffraction pattern is then viewed on a fluorescent screen or electronically by a camera located above or below the screen.

A focussed convergent beam can be used for both (S)TEM modes, but more often in the STEM mode for imaging, microanalysis and micro-diffraction. In TEM, a focussed convergent beam, formed by adjusting C2, is most useful when the sample is thick and electron transmission is poor, as it delivers the highest intensity of illumination on a specific area of the sample. In STEM, the electron beam can be focussed to a diameter as small as 0.1 nm by the condenser lenses and rastered over a thin sample, i.e. the convergent beam is a scanning probe, and images are formed using electron detectors (Figure 1). In STEM, C1 is adjustable, and used to directly form the probe, whereas C2 is switched off or weakened. A stronger C1 gives a smaller probe size on the thin sample. Although C2 is off or weakened, the C2 aperture still controls α on the sample, which is particularly important for operations in STEM for imaging and microanalysis. In addition to C1 and C2, STEM needs a condenser–objective lens (C3), which is the upper pole piece of the objective lens converted into a third condenser lens. This is the most critical component for the STEM, as its ability to achieve the finest possible probe size, with the maximum available current, will ultimately determine the quality of the image and the resolution of the STEM. Having a well-defined, round, and non-astigmatic probe, with the smallest possible diameter and the maximum possible probe current, would be ideal, yet this had been adversely impacted from the lenses aberrations. Among the lenses aberrations, the spherical aberration (aberration coefficient, Cs) was the strongest aberration. Development and implementation of multipole-based aberration (Cs) correctors to STEM have therefore provided a much better control of lens aberrations, with an increase of the beam current. As a result, 0.5 Å probe sizes are now possible.[17,22,23,34] This advance has allowed the STEM to achieve the atomic-scale resolution for imaging and chemical analysis, which is crucial for the characterization of NMs and nanocomposites, as well as the characterization of advanced functional materials such as superconductors.[24,37,38,40,45,51,52]

In the TEM mode with aberrations corrected, a much more straightforward interpretation of high-resolution TEM (HRTEM) images can be achieved. For example, Urban et al.[50] successfully demonstrated how the application of a negative Cs value could simplify the detection of light atoms in the interfaces between two different oxides, as well as in interfacial dislocations and defects by reducing contrast artefacts in the HRTEM mode.

Imaging and electron diffraction with (S)TEM: The (S)TEM images are formed by filtering electrons that have been scattered over a certain angular range in one of three ways: (1) bright-field (BF) imaging, (2) dark-field (DF) imaging and (3) high-resolution imaging. First, a BF image is formed from the direct beam that contains unscattered and low-angle forward-scattered electrons. Therefore, regions in the sample where it is thicker and/or has higher atomic number (Z) elements will appear relatively dark on BF images, as a high proportion of incident electrons are being scattered, and thereby, lowering the intensity of the direct beam. These electrons are accepted by the objective aperture in TEM, as opposed to by an electron detector lying on the optic axis ($<0.5°$) in STEM (Figure 2).

Second, a DF image is formed by accepting forward-scattered electrons. In TEM, the incident beam is tilted so that scattered electrons are intercepted by the objective aperture (only a single-Bragg-diffracted beam for crystalline materials). On the other hand, in STEM, DF imaging uses annular (ring-shaped) detectors that can capture all the electrons scattered over certain angular ranges. For instance, low-angle annular dark-field images (LAADF) can be formed by collecting electrons in a semiangle of $>\sim 0.5-3°$, whereas

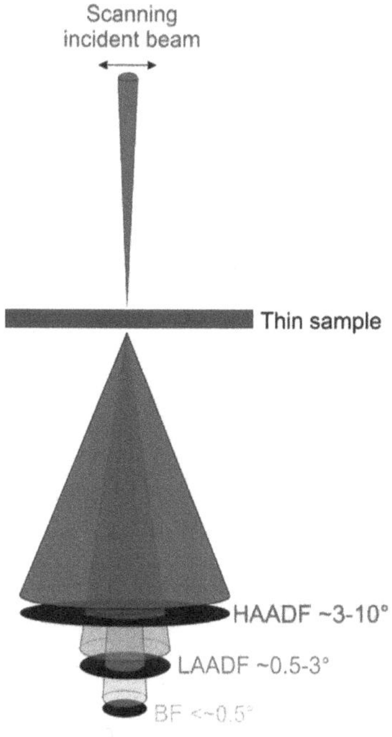

FIGURE 2 Illustration of the collection semiangles used by the different scanning transmission electron microscopy electron detectors. Note that the bright-field (BF) detector may be integrated into the electron energy-loss spectroscopy system. HAADF, high-angle annular dark-field images; LAADF, low-angle annular dark-field images. *Illustration and captions reproduced from Ref. 29; with permission from the Mineralogical Society.*

high-angle annular dark-field images (HAADF) can be formed in a semiangle of ~3–10° (Figure 2). Due to the strong Z-dependence of high-angle scattering, HAADF images can be interpreted semi-quantitatively in the same way as those obtained from a scanning electron microscopy (SEM) backscattered electron (BSE) detector. HAADF imaging and SEM BSE modes, therefore, can be exceptionally useful when studying environmental samples with a particular interest on identifying NMs with higher Z. Successful examples of such applications, including nanosilver and zinc sulphides, and nano titanium oxide particles, can be found elsewhere.[19–21]

Finally, a high-resolution image is formed from both primary and diffracted beams, simultaneously, generating an interference pattern where lattice fringes are visible for crystalline materials. Each set of fringes in the lattice images represents a set of diffracting planes whose Bragg-scattered electrons have been collected by the aperture. Therefore, lattice fringe images can be used to identify the crystal structure and crystallographic orientation of the sample. In addition, lattice fringe images can also be used to determine grain boundaries and intracrystalline features, such as dislocations, stacking faults, twin planes and exsolution lamellae. For those specific purposes, however, the area of interest should be less than 10 nm in thickness (ideally <~8 nm for imaging atoms in a medium Z materials at 200 kV[55]) and oriented with a zone axis exactly on the optic axis of the microscope. Even under ideal conditions, using high-resolution images for crystal structure analysis can be challenging due to sample damage which may occur during high magnification observation. Lack of experience or the level of skill of the user, as well as lack of knowledge of the sample, may also result in undesirable outcomes.

Often, fast Fourier transforms (FFT) of high-resolution (S)TEM images are used to determine the crystallinity, structure and orientation of the sample. By using this technique, it is possible to convert the periodicity of a high-resolution digital image into a diffractogram that can be indexed in the same manner as a selected area electron diffraction pattern (SAED; see below for the detail). Diffractograms can be very useful since they greatly reduce the degree of complexity of a high-resolution image and present (somewhat) clearly the geometric features of those images that otherwise are hard to see. For instance, a single crystal that is oriented with a zone axis at the Bragg angle relative to the incident beam will form periodic arrays of spots in an FFT-computed pattern (Figure 3(1), Ag_2S). In effect, spot patterns generated using FFT are a two-dimensional area of the reciprocal lattice of the crystal. The positions of spots are determined by the orientation of atomic planes and their d-spacing. An amorphous sample, or a selected region with amorphous nature, displays a diffuse halo or halo rings on FFT. On the other hand, a region that contains many crystals with their random orientations will form continuous ring patterns with intensity variations on FFT (Figure 3(2), ZnS). In addition to the spacings of spots, angles between diffracting planes and pattern symmetry are also important parameters to determine the crystal structure of the sample.

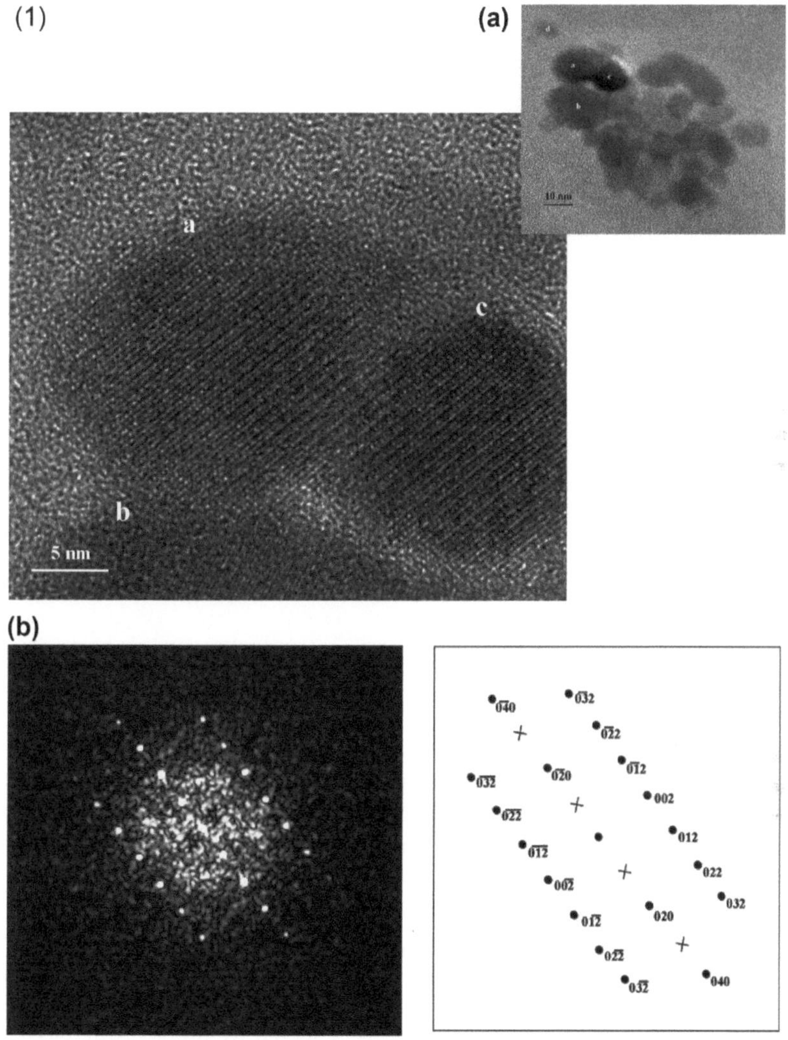

FIGURE 3 (1) Ag_2S structure analysis. (a) High-resolution transmission electron microscope (HRTEM) image of selected Ag NMs. The inset shows a bright-field transmission electron microscope (BF TEM) image of the entire NM aggregate. (b) A fast Fourier transforms (FFT) pattern, obtained from NM c, and a key diagram showing the calculated [1 0 0] pattern for α-Ag_2S. The crosses indicate positions of spots on the FFT pattern that should be systematically absent, but are present due to a defocus of the FFT image. *Images and captions reproduced from Ref. 19; with permission from American Chemical Society.* (2) ZnS structure analysis. (a) HRTEM image of ZnS sphalerite NMs. The small TEM image to the left (b) is a BF of the aggregate, with the white box marking the position of the HRTEM image, and the image to the left and bottom (c) is the resulting FFT pattern of the HRTEM image. *Images and captions reproduced from Ref. 21; with permission from American Society of Agronomy, Crop Science Society of America, and Soil Science Society of America.*

(2)

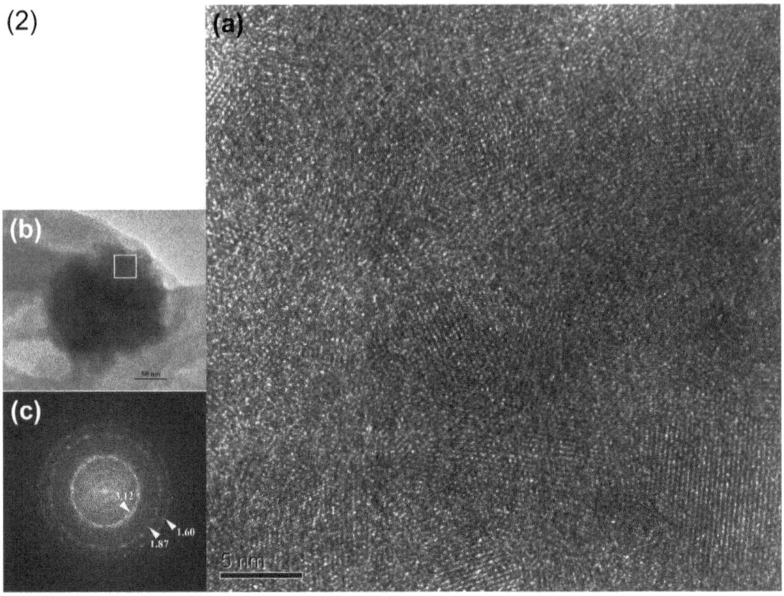

FIGURE 3 cont'd

Figure 3 showcases how to use HRTEM images for the structure analyses of α-Ag_2S and ZnS found in sewage sludge materials, i.e. by-products of wastewater treatment processes.[19,21] Alternatively, this can be done computationally, by using commercial software. Other good examples of using HRTEM images for crystal structure analyses on a sample that was collected from natural environmental settings can be found elsewhere.[33]

In addition, SAED, convergent-beam electron diffraction (CBED) and electron nanodiffraction are (S)TEM-based techniques, also often applied to identify the crystal structure and crystal defects of a sample. The SAED technique uses a broad and parallel electron beam and a selected area aperture to define the region on the sample from where the diffraction pattern is obtained. The smallest area that can be selected by the aperture is (typically) approximately 500 nm in diameter. Like FFT, SAED patterns also yield either arrays of spots or ring patterns, which can be indexed in the same way as FFT patterns for the structure analysis of the sample. Therefore, the accuracy and precision of d-spacing measurements in SAED patterns are critical for structure determinations, which require both careful calibration of the camera factors, such as camera length and camera constant (essentially for the magnification of the SAED pattern) and correction for elliptical distortion. This can be done by using a standard material, such as gold or platinum, sputtered and deposited onto the thin TEM sample, which can provide ring patterns of known d-spacings.[36] Ideally, the size and thickness of the standard material is similar to that of the sample for the analysis.

CBED technique uses a convergent, focussed electron beam, and therefore, the size of the probe determines the diffracting volume of the sample. CBED patterns can be obtained from regions approximately a few nanometres in diameter. The patterns contain discs of intensity (rather than spots as with SAED), and the CBED intensities provide phase information of crystal structure factors due to the strong dynamical diffraction effects. The structure of the discs contains other crystallographic information such as symmetry (i.e. point and space groups) and symmetry changes due to very small atomic displacement.[48] CBED also enables the refinement of crystal structure parameters.[49] Thus, CBED is a powerful technique for the crystal structure analysis of nanosized materials, especially when they are polyphasic (i.e. mosaic), crystals and/or polycrystalline, with many lattice defects or dislocations in the structure.

In STEM mode, with a very small probe available, electron nanodiffraction can be obtained from sub-nanometre-sized regions of a sample[8] by scanning the probe over the sample and acquiring electron nanodiffraction patters at each point. Electron nanodiffraction technique also produces discs of intensity as opposed to spots as with SAED, but they can be indexed in the same way as the SAED patterns. Janney et al.[18] compared diffraction characteristics of 2- and 6-line ferrihydrite, naturally occurring iron oxide nanominerals, estimated by both electron nanodiffraction and X-ray diffraction methods, and concluded that both were within less than 5% error in those measurements. Since the crystals cannot be tilted during experiments, a polycrystalline sample may be problematic, as many of the patterns acquired from the sample will not be oriented precisely on a zone axis. Electron beam damage of the sample is another severe drawback for such application. For example, kaolinite, a silicate mineral, is transformed to an amorphous material during less than 1 s of electron beam exposure while acquiring its electron nanodiffraction patterns.[8]

In any type of diffraction experiment in electron microscopy, care must be taken not to unknowingly disturb the sample. For example, dehydration by vacuum, or electron beam damage, may alter the original d-spacings of the sample, or cause the sample to become amorphous. For instance, dehydration of smectite clay minerals in the vacuum of a TEM typically cause changes in its (001) layer spacings from 1.5−1.2 to 1 nm.[36]

3-D Imaging: Interpretation of 2-D projected images of 3-D samples has been one of the disadvantages of the (S)TEM techniques. Recently, by using tilt series tomography in electron microscopy, significant progress has been made in obtaining 3-D information from NMs. A series of images, taken every few degrees of tilt, is reconstructed into a single 3-D data set, which allows a 3-D view of the sample in any desired orientation. For tomography tilt series acquisition, annular DF imaging in STEM mode is preferred over BF and high resolution in TEM. This is because, for the back-projection reconstruction procedure in tomography, images need to be good representations of the projected mass thicknesses of the sample, which requires effective suppressing of phase and diffraction contrast effects on the images. Annular DF imaging in STEM also reduces the edge-enhancing artefacts that are often found in BF and HRTEM imaging modes. One constraint with current reconstruction algorithms assumes the incident electron beams to be parallel. This assumption may not work when the desired spatial resolution increases and an electron beam becomes focussed (convergent). In such a case, the projection becomes sensitive to the focus, and therefore, different areas of the sample need to be in focus at different settings. At present, spatial resolution in electron tomography has been demonstrated at the angstrom level by using a gold NM.[44] Figure 4 presents an example of STEM tomography on a catalyst, Pt_2Au_4 NMs supported on TiO_2. This 3-D reconstruction makes possible the identification of the morphology of the TiO_2 support, as well as the location of the metal nanoclusters as elongated ellipsoids.[5] Another example of HAADF-STEM tomography on NMs of hematite (α-Fe_2O_3), one of the most common iron oxides in nature, is given in Figure 5. The tomographic image allows us to identify nanoporous structure within rhombohedral hematite nanocrystals, and therefore, helps us to understand both the external and internal structure of hematite NMs more accurately.[11]

Electron beam damage: Electron beam-induced sample deteriorations are considered to be the limiting factor for further applications of the (S)TEM techniques. Radiolysis, and heating and knock-on (or replacement) damages are three mechanisms to explain why a thin sample can be damaged during the (S)TEM work (Figure 6). Radiolysis involves the ionization of sample atoms during inelastic scattering and breaks chemical bonds, resulting in forming defects, inducing amorphization, sputtering atoms from the surface, as well as making changes in chemistry. Heating involves the transfer of energy from incident electrons to sample atoms during inelastic scattering, resulting in breaking bonds, and inducing vapourization and potential recrystallization. For the samples sensitive to beam damage by radiolysis and heating, cooling them with liquid nitrogen, or combining the increase of the accelerating voltage with a broader beam (in TEM) or rastered beam (in STEM), as well as reducing acquisition time for the analysis, may help to reduce unfavourable beam damage.

Knock-on damage involves the transfer of momentum from incident electrons to sample atoms during high-angle elastic scattering, resulting in displacing internal atoms (especially those with low Z) and sputtering weakly bonded surface atoms. Hydrocarbon contamination of a sample can also occur via hydrocarbon compounds that may be deposited during the sample preparation steps (ion milling and coating) or derived from vapour present in the electron microscope column. Sample deterioration due to hydrocarbon contamination is most problematic for the compositional (S)TEM analyses, such as energy-dispersive X-ray spectroscopy (EDS, EDX or XEDX) and electron energy-loss spectroscopy (EELS). Both EDX and EELS techniques will be discussed in the following section.

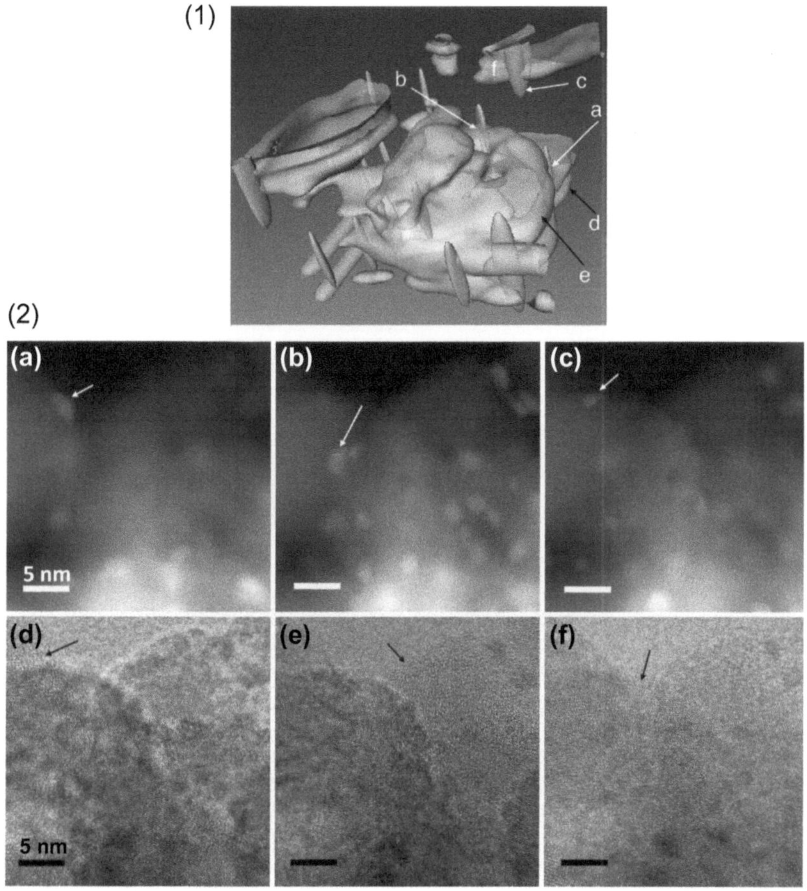

FIGURE 4 (1) 3-D rendering of the depth series of a (Pt, Au)/TiO_2 catalyst sample. Yellow (light gray in print versions) surfaces enclose the areas of high HAADF intensity and thus represent metal (Pt, Au) particles. White surfaces enclose the areas of high diffraction contrast (from bright-field, BF series) and thus represent TiO_2 substrate particles; (2) Selected frames from the depth series (raw data) collected from on a (Pt, Au)/TiO_2 catalyst sample. (a–c) HAADF images (detector inner angle, 50 mrad) taken at a distance of −20 nm (a), −88 nm (b) and −140 nm (c) from initial defocus. (d–f) the corresponding simultaneously acquired BF images (detector outer angle, 1 mrad) at defocus values of −20 nm, −88 nm and −140 nm respectively. A scale bar noted on all these images is 5 nm. Arrowed features (a–f) in Figure 4(1) match to the arrowed features in Figure 4(2) (a–f) respectively. HAADF, high-angle annular dark-field. *Images and captions reproduced from Ref. 5; with permission from National Academy of Sciences.*

Compositional analysis by (S)TEM: EDX and EELS are analytical techniques used in the electron microscopy to determine elemental composition of a sample. The combination of image and diffraction analyses, as well as spectrometry, makes (S)TEM, by far, the most powerful, single instrument for NMs characterization. Specifically, EDX uses characteristic X-rays produced

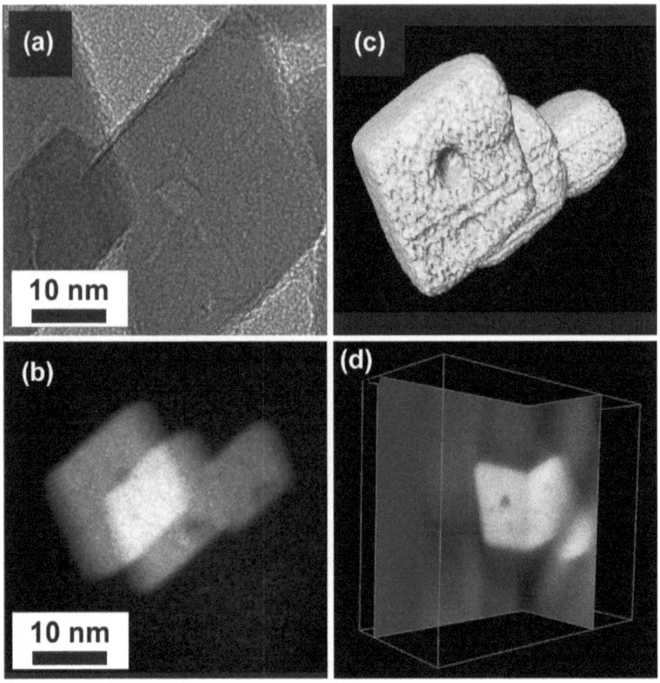

FIGURE 5 (a) High-resolution transmission electron microscope image of hematite nanocrystal showing a nanopore perpendicular to the page; (b) high-angle annular dark-field-scanning transmission electron microscopy (HAADF-STEM) image of hematite nanocrystal showing nanopores running through the crystals. (c) Isosurface image of hematite nanocrystal shown in (b). Nanoscale dimple on the surface of crystal was found by HAADF-STEM tomography. (d) Two slices perpendicular to each other taken from the tomography image shown (c). Nanopores run inside the crystal. *Images and captions reproduced from Ref. 11; with permission from the Mineralogical Society of America.*

and detected during ionization of sample atoms by high-energy, incident electrons, whereas EELS uses the changes in kinetic energy of incident electrons due to inelastic scattering, while interacting with sample atoms. Both EDX and EELS are qualitative and quantitative analytical techniques, but EELS can further examine the oxidation states and interatomic bonds of the sample, and is therefore capable of both chemical and structural analyses of the sample. For the detection of characteristic X-rays emitted by a sample, lithium-drifted silicon detectors are now commonly used for the EDX. With a super ultrathin window in front of the detector, the EDX can detect, in principle, all the elements with $Z > 3$, yet detection of lighter elements by the EDX can be challenging due to absorption effects in the sample, as well as by the detector window. EELS is the preferred method for lighter elements ($Z < 10$) due to its higher sensitivity compared with EDX. The EELS detector, fitted with a magnetic prism, is positioned beneath the sample to obtain

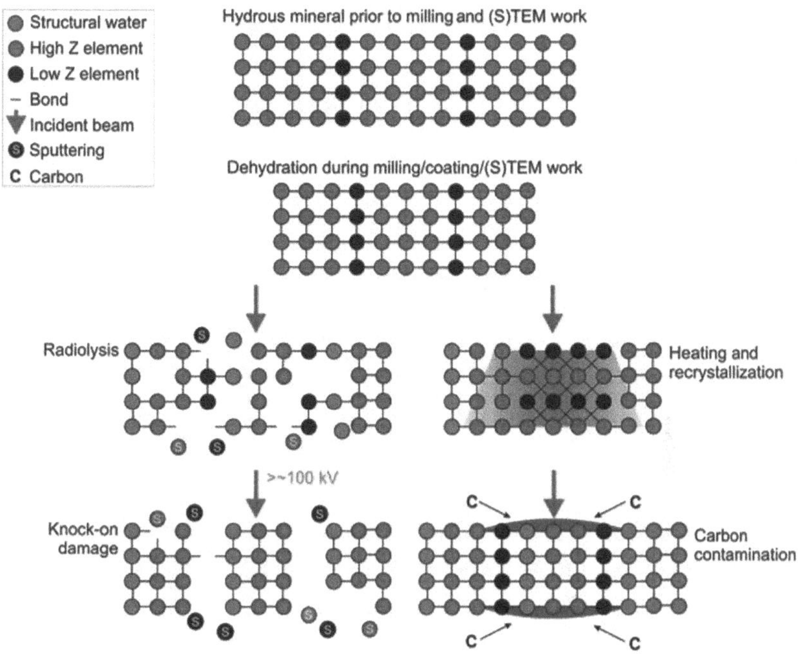

FIGURE 6 Illustrations of the various mechanisms by which a mineral can be damaged during (S)TEM work. *Illustration and captions reproduced from Ref. 29; with permission from the Mineralogical Society.*

unscattered and low-angle (<1°) scattered electrons (Figure 1). Once collected, electrons are further separated according to their kinetic energy by the magnetic prism; that is, their incident energy minus any energy loss which occurs in the sample. Before and after the EDX and EELS analyses, (S)TEM images of the area where the analyses are performed need to be taken to determine if sample drift has occurred during analysis.[6] This chemical analysis region can be a spot or area in the sample; both EDX and EELS spot analyses can be done in (S)TEM modes, whereas elemental and spectral maps can be done only in STEM mode. Furthermore, LAADF or HAADF imaging in STEM, with the availability of EELS signals, make atomic-resolution spectroscopic imaging of the sample feasible. EDX and EELS will, therefore, provide different but complementary information of the sample's chemical make-up.

EDX produces a spectrum that consists of a plot of X-ray counts detected versus X-ray energy. There are two types of X-rays generated and collected by an energy-dispersive X-ray detector during the ionization event; Bremsstrahlung and characteristic X-rays. The former is a result of electrons interacting with the nucleus, producing a continuous spectrum. The latter, on the other hand, is a result of electrons ionizing an atom, emitting the characteristic X-rays that are unique to the ionized atom, and this produces Gaussian-shaped peaks. Thus, in general, both appear in the EDX spectrum, Gaussian-shaped, characteristic X-ray peaks superimposed on a continuous background of Bremsstrahlung X-rays (Figure 7). Characteristic X-ray peaks can be further defined by (1) a capital Roman letter indicating the shell (K, L or M) that has the vacancy generated by electron bombardment, (2) a Greek letter indicating the group to which the line belongs in descending importance order (e.g. α, β, etc.), and (3) a number indicating the intensity of the line within the group in descending importance order (e.g. 1, 2, etc.). Therefore, the most intense lines of K and L are $K_{\alpha 1}$ and $L_{\alpha 1}$, and the total X-ray intensity of a given shell is divided among those several lines. For the K shell, the majority of the intensity is contained in the $K_{\alpha 1,2}$ lines. L shell contains at least 12 lines, much more complicated than the K shell, but except for $L_{\alpha 1}$, $L_{\beta 1}$ and $L_{\beta 2}$, many of them are quite weak to use for the peak identification. The observable M lines include only the M_{α} and M_{β} series. Qualitative elemental analysis based on the identification of the lines in the spectrum is, therefore, quite straightforward due to the simplicity of X-ray spectra. However, quantitative

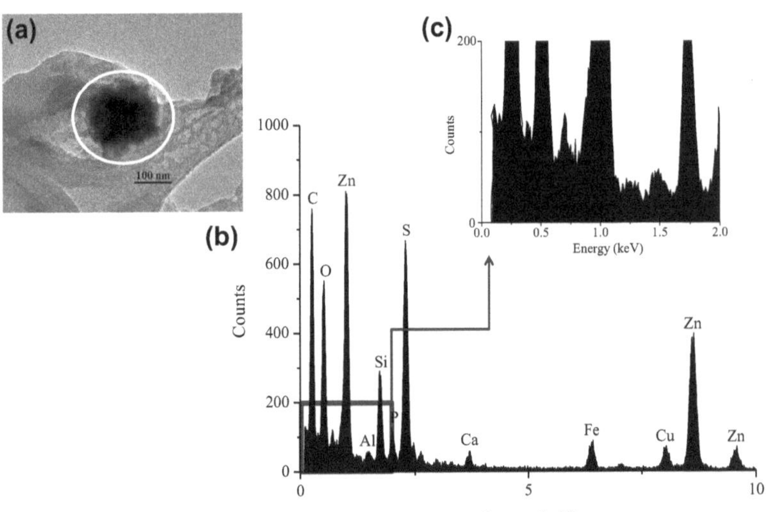

FIGURE 7 A representative energy-dispersive X-ray spectroscopy (EDX) spectrum. (a) Bright-field (BF) image of a ZnS NM aggregate with the white circle marking the position of the entire aggregate; (b) EDX spectrum of the entire aggregate in (a). Note that the peaks of C, O and Cu, appearing in the EDX spectrum, are presumably from the C support film and/or the surrounding residual sewage sludge organics and/or from the Cu grid sample holder. The remaining elements, including Fe, Al, Si, P and Ca, are most likely only from the sludge matrix. (c) The same EDX spectrum, but enlarged in the lower eV range to more clearly show continuous background, originated from Bremsstrahlung X-rays. *Images and captions reproduced from Ref. 21; with permission from American Society of Agronomy, Crop Science Society of America and Soil Science Society of America.*

elemental analysis to determine the concentrations of the elements present in the sample can be challenging for the following reasons: first, background corrections may create artefacts. As mentioned above, the continuous background originates mainly from Bremsstrahlung X-rays. If it is high as a result of the bombarding high-energy electrons, it will limit the detectability of small peaks and, as a result, the background removal or correction can be quite difficult. If it is relatively low, the background removal or correction can be done either by using the second derivative of the spectrum or by fitting an analytical function to the background. Peak intensities can be then calculated by fitting either a Gaussian function or peaks from library standards. Second, the contribution of X-rays generated from areas that are not illuminated by the electron beams may also add artefacts for the EDX spectrum. This includes the pole piece of the microscope lens and/or the grid material, which can be also excited by BSEs or by X-ray fluorescence. This may be partially overcome by acquiring a hole spectrum of the TEM sample holder when the incident electron beam passes through the hole. Finally, the increase of electron beam exposure times on the sample to compensate for low X-ray count rates in the EDX spectrum can damage the sample, which may result in the loss of elements. The electron beam will be drilling a hole in the sample, so accuracy of the quantitative analysis can be questioned when severe sample damage occurs during the EDX analysis. One of the ways to resolve some of the above issues is to have a standard material of known composition with a similar size and thickness, positioned adjacent to the spot/area of interest in the (S)TEM samples analyzed, and use its information to correct the obtained spectrum for the quantitative elemental analysis.

EELS produces a spectrum which consists of a plot of electrons collected per channel versus energy loss energy (eV). An example of an iron fluoride film (FeF_2) EELS spectrum is given in Figure 8, which possesses typical features of an energy-loss spectrum.[12] Specifically, the spectrum has three energy-loss regions: (1) the zero-loss peak, the first peak which appears around 0 eV, contains the direct beam and electrons that have undergone elastic scattering with an energy loss too small to measure. The width of the zero-loss peak is typically in the range of 0.2−5 eV, presumably reflecting the energy distribution of the electron source (Figure 8(a)); (2) the low-loss region contains electrons that have undergone inelastic scattering by conduction or valence electrons, and therefore, reflect ionization of outer shell electrons of sample atoms. This region is in the range of 5−50 eV. Figure 8(a) shows two prominent peaks in the low-loss region, including one around 22 eV and the other around 54 eV. The former results from a plasma resonance of the valence electrons, whereas the latter results from inelastic scattering from M_2 and M_3 subshells ($3p_{1/2}$ and $3p_{3/2}$ electrons) of Fe atoms. Its shape, and rapid increases in energy, followed by a gradual decrease in energy, is called an ionization edge. The shapes of the edges, therefore, provide information on the element, as well as the environment of the excited atom (e.g. bonding and valence

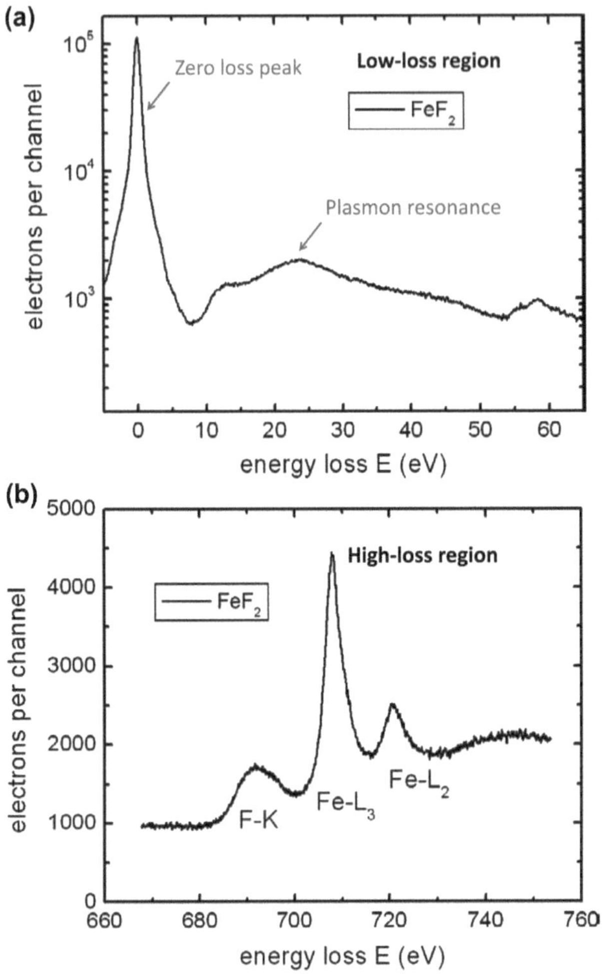

FIGURE 8 Energy-loss spectrum of iron fluoride film: (a) low-loss region with a logarithmic intensity scale and (b) part of the core-loss region, with linear vertical scale. *Spectra and captions reproduced from Ref. 12; with permission from IOP Publishing.*

states); and (3) the high-loss region reflects excitation of inner shell (core) electrons of sample atoms. Figure 8(b) has three peaks in the high-loss region, including an F-K edge (around 690 eV, as a result of excitation of 1s electrons), and Fe L_3 and L_2 edges (around 710 eV and 722 eV, as a result of excitation of Fe $2p_{3/2}$ and $2p_{1/2}$ electrons, respectively). Studies in the measurements of electrical and mechanical properties of advanced functional NMs,[9,26,57] as well as in the measurements of magnetic properties of materials,[43] have greatly benefited from EELS analysis and its ability to determine bonding and valence states of the element of interest.

Since the intensity of the core edge of a given element corresponds to the number of its atoms that are present in the illuminated volume, EELS can be also used for quantitative elemental analysis. Similarly to EDX, absorption edges in EELS are also superimposed to a background signal, originating from excitations of electrons to high vacant energy levels or in thicker samples from multiple energy-loss events. For quantification, this background has to be removed or corrected, yet the background removal or correction is much more different in EELS than in EDX because of the complexity and heterogeneity of the shapes of the various edges. Having standard materials to obtain information of approximate edge shapes and intensities, and using such calculated information to the experimental spectrum for the background removal or correction, has been proposed to partially overcome this problem, and successfully demonstrated.[7,25,53]

3. SPECIMEN PREPARATION: ION-MILLED FOILS, LOOSE GRAINS, ULTRAMICROTOME SLICES AND CRYOFIXED SAMPLES

In order to obtain scientifically valuable results by the analytical (S)TEM, making good (S)TEM specimens is the most important prerequisite step. There are two main challenges in the preparation of samples: the uniformly, 'electron transparent' thickness and the representativeness (i.e. specimens results related back to the bulk sample or property). The sample for the analytical (S)TEM analysis needs to be suitably thin enough that the incident electrons can pass through the sample without being scattered too highly, or losing too much of the initial energy, when irradiated with a 100–300 kV electron beam.

The target specimen thickness is dependent on the type of material and the analysis, but typically lies between $\leq\sim 10$ and 100 nm. For EELS analysis, the specimen should contain areas thinner than 50 nm, and for HRTEM imaging (at the atomic-scale, intended use for crystal structure analysis) areas less than 10 nm. Ultramicrotomed samples are typically in the thickness ranges of 60–90 nm. Herein, we present four different sample types, including loose grains, ion-milled foils, ultramicrotomed as well as cryofixed slices of solid-state materials for the specimen preparations in electron microscopy.

Ion-milled foils. There are two ion-milling sample preparation techniques for electron microscopy samples, argon (Ar) ion milling and focussed ion beam (FIB) techniques. For Ar ion milling, the sample is attached to a washer, loaded in the ion mill and bombarded at low angles from above and below, typically with a 5 kV Ar ion beam. If possible, the initial sample needs to be polished as thin as possible, mechanically, prior to milling processes. Ar ion-milling works reasonably well for nonmetallic compounds. Even with the optimization of ion-milling rates, however, Ar ion bombardment on solid samples is not site-specific and produces a wedge-shaped sample around holes with varying thickness. Using the wedge-shaped sample with nonuniform thickness, for elemental mapping or line scan in electron microscopy, will be extremely difficult. Furthermore, the holes in the specimen can appear randomly, especially for a polycrystalline/polyphasic sample. This is because ion-induced damage and artefacts by sputtering will preferentially occur in the regions with the hardness contrast among crystals and phases, the interfaces, disorders, as well as many point defects in the polycrystalline/polyphasic sample.

FIB technique uses high-energy heavy ions, typically 30 keV of ionized gallium (Ga^+), and cuts pairs of trenches into the surface of a sample to leave thin relict walls that then can be cut free, and mounted as a (S)TEM thin foil sample. Referring to the geometry shown in Figure 9, this technique is often called the H-bar technique. Later, this conventional technique was improved by its combination with tripod polishing,[1–3] which reduced the initial sample thickness to only 2–3 μm or less, lessening FIB milling time effectively.

(1)
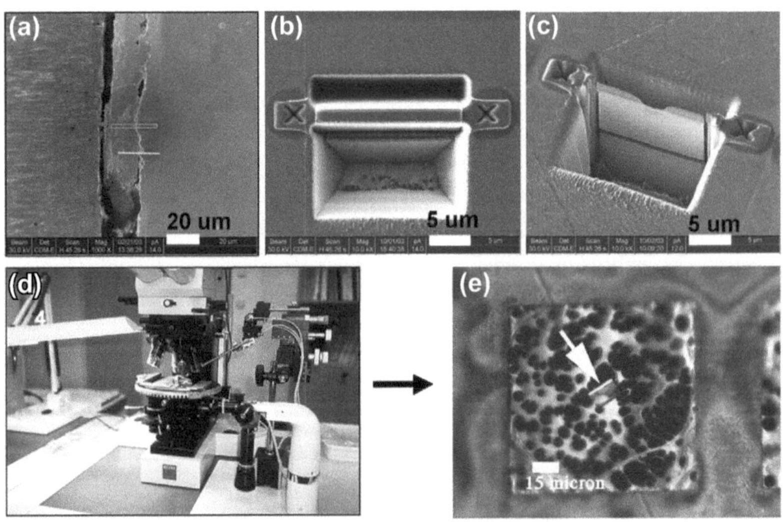

FIGURE 9 Focussed ion beam-based transmission electron microscopy (TEM) foil preparation. (1) Ex situ lift-out: (a) Secondary electron (SE) image of the sample surface with the location of the foil to be cut indicated (shown as the yellow (light gray in print versions) line on the image). The rectangular white box represents where platinum to be deposited. (b) SE image of the platinum strip deposited on the surface with the two reference crosses milled. Two trenches, one in front of the foil and a smaller one behind the foil, are already sputtered. The bright contrast is due to redeposited and sputtered material. (c) SE image of the foil ready for ex situ life-out. (d) Optical microscope with micromanipulator for lift-out of the foil. (e) TEM foil placed onto a perforated carbon film on a standard TEM copper grid; (2) In situ lift-out. (a) The tungsten tip of the manipulator is welded to the foil. (b) The foil is removed from its excavation site. (c) and (d) The tip is retracted and moved towards the TEM grid. (e) SE image of the foil welded to the TEM grid by platinum deposition between the foil and the grid. *Images and captions are reproduced from Ref. 56; with permission from Elsevier B.V.*

(2)

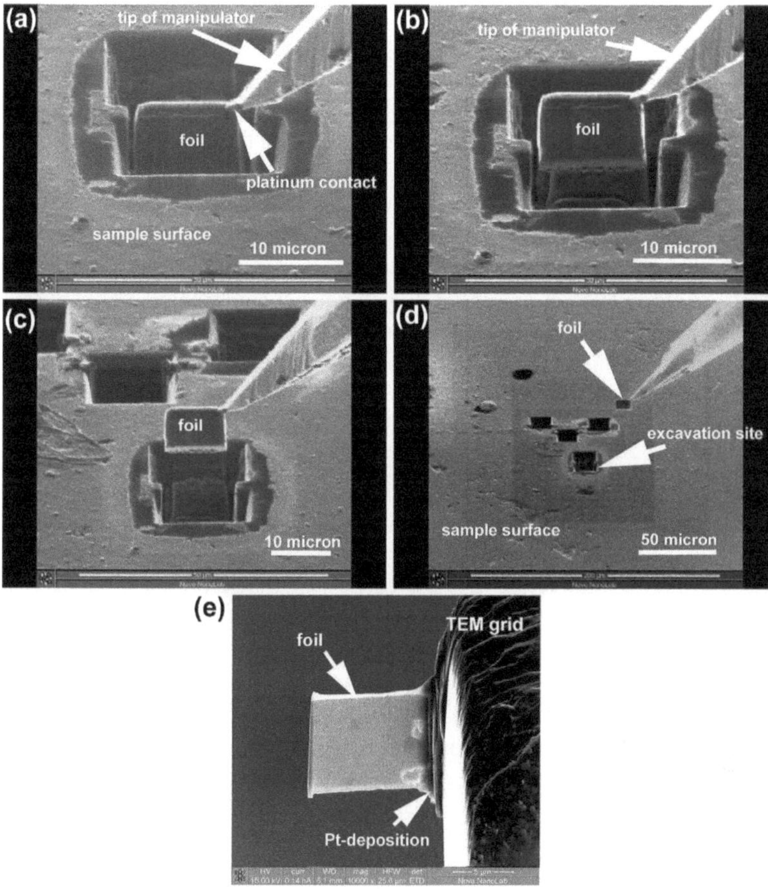

FIGURE 9 cont'd

The final sample extraction procedure, i.e. taking the foil out of the excavation site, is called the 'lift-out' technique. Lift-out techniques were first proposed by Overwijk et al.[41]; and were further improved by Giannuzzi and Stevie.[15] FIB techniques can produce TEM foil samples typically with the dimensions of 15−20 μm (width) × 10−15 μm (height) × 0.10−0.15 μm (thickness) from metals, alloys, ceramics, minerals, rocks, glasses, powdered samples, organic materials (polymers, biological materials) and even frozen liquids.[30,56]

Basic operation principle: Of the existing liquid metal ion sources, Ga is the most widely used due to its low melting temperature and low vapour pressure. The low melting temperature (29.8 °C) makes the ion source easy to design and operate, and low vapour pressure (even at high temperature) makes it stable from evaporation for long periods of time. In addition, Ga does not react with a tungsten (W) needle used to heat the Ga. Therefore, an assembly of a small Ga reservoir, in contact with the W needle, becomes a basic, essential component for an FIB instrument to produce an ion beam. During operation, the solid Ga in the reservoir is heated to its melting point and the liquid Ga flows by surface tension to the needle tip, where Ga is extracted by field emission. A large negative potential between the needle and an extraction electrode generates a strong electric field (10^{10} V/m) at the tip of the needle. The balance between the electrostatic forces and the Ga surface tension wetting the tapered W needle geometry forms a single Taylor cone at the tip of the needle. The tip radius of this cone is approximately 2−5 nm. The strong electric field applied at this small tip generates ionization and field emission of the Ga atoms. The Ga^+ ions are then accelerated in an electrical field (up to 30 keV). A continuous flow of liquid Ga to the tip replaces the extracted Ga^+ ions, producing a constant ion current, which makes a sputtering process possible. An ion beam column therefore consists of Ga-W at the source end and two electrostatic lenses (a condenser and objective lens) to define the ion beam and then focus it on the sample.

Several interactions can occur when Ga^+ ions bombard solid-state materials (Figure 10 for summary).[15,39,42] Sputtering is a process whereby surface atoms from a solid target material are ejected (as neutral atoms) through momentum transfer from the incident Ga^+ ion to the target atoms (elastic collision). This only occurs when a surface atom receives enough kinetic energy to overcome the surface binding energy (≥1 eV). Inelastic scattering of the incident Ga^+ ions with the target material generates the emission of secondary electrons, phonons, plasmons and X-rays.[56] Secondary electrons are

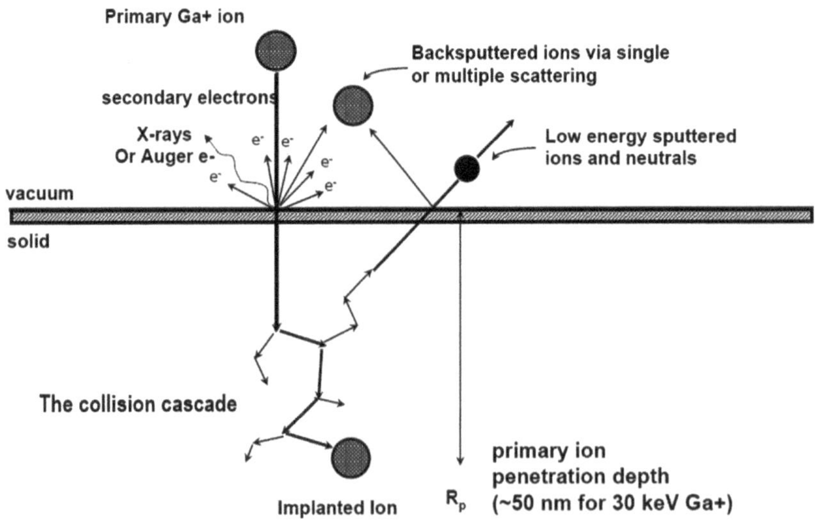

FIGURE 10 Ga^+ ion-solid-state material interactions. *Illustration reproduced from Ref. 15; with permission from Springer.*

used for imaging (with a resolution of ~7 nm) by collecting the electrons with a continuous dynode electron multiplier (CDEM) detector in a single-beam FIB mode.[27] Details of similarities and differences between single- and dual-beam FIB modes are provided below.

Incident Ga^+ ions may also lose their kinetic energy through the collision cascade that involves a series of independent binary collision processes on a solid (the linear collision cascade regime).[46,47] Schematic illustration of the collision cascade model is provided in Figure 11. In brief, once the target atom is knocked out of its original site during a collision, by receiving sufficiently higher energy to overcome its threshold displacement energy, an interstitial-vacancy pair will be created in the sample. This primary recoil atom may also generate secondary recoil atoms, and so on and so forth, resulting in having larger numbers of atoms with excess kinetic energy in the crystal. A grey area, in Figure 11, illustrates the collision cascade volume, created by 30 keV incident Ga^+ ions on a crystalline sample.[14] For comparison, the displacement energy is typically in the order of 20 eV, much higher than the surface binding energy for the atoms (≥1 eV). If a displacement collision occurs near the surface, the recoil atom may be emitted from the solid and cause sputtering. After approximately 10^{-11} s, the collision cascade will come to an end, as (1) an ion has rested in the crystal, a certain distance below the surface of the target material (i.e. ion implantation) and (2) all participating particles have energies below the displacement energy. Therefore, this process is always in company with radiation, ion beam damage such as lattice defects (i.e. vacancies, interstitials and dislocations), ion implantation and heat, and all of them may further interact and evolve.[54] For instance, the penetration depth of the ion is estimated to be between 10 and 100 nm (projected, R_p) and between 5 and 50 nm (lateral, R_l) when using the 30 keV Ga on a crystal sample (Figure 10). If the incident Ga^+ ion is backscattered out of the target surface, it will come to rest on the surface of the target material, generating Ga contamination on the sample surface. For normal incidence of 30 keV Ga, sputtering yields (sputtered target atoms per incoming ion) are between 1 and 20, but the yields depend on the material (i.e. Z, surface structure and geometry, and chemistry), crystal orientation, ion beam incidence angle, as well as the extent of redeposition[4] (Figure 12).

Currently, there are two FIB modes in use: single-beam and dual-beam modes. Dual-beam mode has the FIB column, with an additional SEM column in one single apparatus, and therefore, it can use both the ion and electron beams to interact with the solid samples. Furthermore, all of the techniques associated with SEM, including EDX, BSE imaging and electron backscatter diffraction can be used in dual-beam modes. Foil manufacture procedures are similar for both modes, but lift-out can be different (i.e. ex situ vs in situ micromanipulator).

Foil manufacture: Semiconductors and insulators require a thin layer of coating with a conducting material such as C or Au prior to the use of FIB

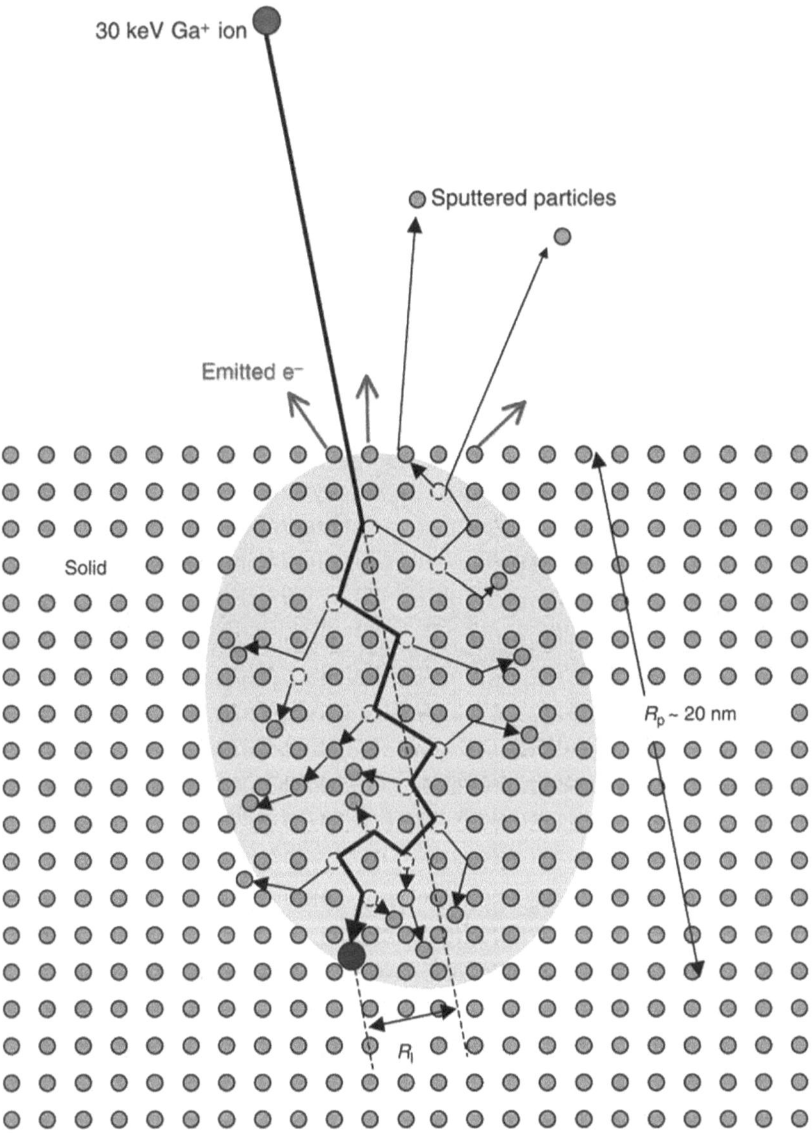

FIGURE 11 Schematic illustration of a collision cascade generated by a 30 keV Ga^+ ion incident on a crystal lattice, showing the damage created in the collision cascade volume, and the projected range R_p and lateral range R_l of the implanted ion. *Illustration and captions reproduced from Ref. 54; with the permission from Material Research Society.*

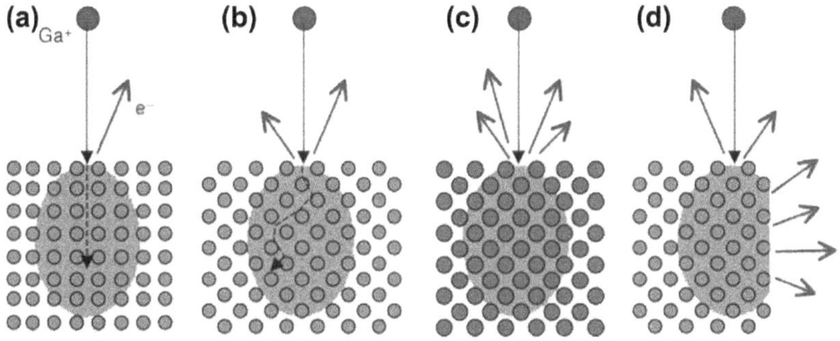

FIGURE 12 Schematics showing the influence of (a), (b) crystal orientation, (c) atomic mass and (d) surface geometry on 30 keV Ga^+ collision cascades and sputtering yields. The orange (dark gray in print versions) atoms in (c) are more massive than the yellow (light gray in print versions) atoms in (a), (b) and (d). *Illustration and captions reproduced and modified from Ref. 54; with the permission from Material Research Society.*

techniques. Figure 9 presents a series of images for the FIB-TEM foil preparation steps. Two reference markers are milled on the sample surfaces (e.g. crosses in Figure 9(1-b)), which are cut into the surface near the milling site. Then, a thin layer (1−2 μm) of a protective Pt (or W) strip is deposited onto the surface area where the foil should be cut and the two reference markers have been milled. This Pt (or W) layer protects the foil from being sputtered by the Ga^+ ion beam, especially for the final stage of the foil preparation when its thickness is much less than 500 nm. Furthermore, having this protective Pt (or W) strip is also preventative of beam shift (or sample drift), as beam shift can occur by charging of the specimen due to insufficient conductivity of the milled trenches during the FIB operation. The importance of having a sufficiently thick Pt layer, when investigating the nearby surface of a sample, is well demonstrated by Lee et al.[28]

The milling procedure is provided in great detail elsewhere.[28,58,59] In brief, a large trench is milled in from the Pt-strip and a smaller one in the back at the beginning of the milling process (Figure 9(1-b)). The progress of the sputtering process is carefully monitored in real time using a CDEM detector by acquiring a secondary electron image from the sample. After the large trenches are made, milling is continued but with a smaller beam size and ion current (350 pA). A small rectangular area is selected and material is sputtered from the sample for a given time, which is dependent on the size of the foil. The pattern is shifted towards the foil and the process is repeated until it reaches a foil of approximately 500 nm in thickness. The sample is then tilted at about 45° with respect to the ion beam and the foil is cut free on both sides and at its base with only a narrow strip of Pt-material on the top of the foil to fix it in its position to continue polishing. After tilting the sample back to its original position, milling is continued, but with a much-reduced beam and Ga^+ ion current (70 pA) to achieve the final sample thickness. At the end of the foil preparation process, the Pt-strip is completely removed and the foil is ready for lift-out. Lift-out of the foil from its excavation site is often done with an optical microscope using a special manipulator (Figure 9(1-d)). The foil is then placed onto the perforated membrane on a TEM grid. It is important to gently press the foil down with the glass fibre so that the adhesive forces keep the foil in position.

The sputtering process always results in the implantation of Ga^+ into the sample surface, and therefore, generating an amorphous layer on both sides of the foil. The thickness of the amorphous layer depends on the material, the ion current and the angle of incidence. Under optimal conditions, the thickness of the amorphous layer is in the size range of approximately 5–15 nm on both surfaces, but it varies depending on the material, the ion current and the angle of the incident beam.

A single-beam model uses ions for both imaging and milling, and therefore, more Ga^+ contamination on the foil sample will be expected for this model than the dual-beam model. Especially for the HRTEM imaging or EELS analyses, the foil needs to be very thin, ideally less than 50 nm. In a single-beam mode, it is not possible to prepare a foil having 15–20 μm wide and 10–15 μm high, but only a thickness of 50 nm (or less) due to damages occurred on the foil by sputtering, but this can be possible in a dual-beam mode (see below for the detail). Furthermore, in the single-beam mode, it may be difficult to identify the sputtering site under an optical microscope for ex situ lift-out of the foil, using a micromanipulator when the sample surfaces are unpolished.

On the other hand, dual-beam models use the electron column for imaging, and therefore, the foil can be imaged at high resolutions without concern of the target surface contamination/damage during imaging.[16] The ion-milling process can be monitored and controlled with the SEM, which provides a better precision of sputtering the target material. This is especially important when the sputtering rate by the electrons is very low, for example, during the final milling process of thin TEM foils. Dual-beam models may have an in situ micromanipulator that allows for in situ TEM foil lift-out (Figure 9(2)). Therefore, the dual-beam mode gives much better control over the milling process, as well as the sample handling procedure (even at 1000 nm of sample thickness).

Loose grains. Powder type samples (e.g. NMs) can be dispersed in a solution by ultrasonication. The solution chosen for the NM dispersion (1) should not react with the NMs of interest, (2) should not leave carbon residue on the surface after evaporation and (3) should evaporate in a reasonable time frame at room temperature. In general, hexane or methanol is often a choice for the suspending solution, but in environmental science studies, water can be also used (a longer evaporation time is expected). For environmental samples with high levels of dissolved organics, however, organic solvents such as methanol will work better than water.[19] Sewage sludge materials, the by-products of wastewater treatment plants, are good examples of this type of sample, as they contain 50–70% organics by weight. Once NM suspensions are made, the whole mount TEM sample can be prepared by dipping the grid in the solution gently and withdrawing horizontally (on edge), or by placing a few drops of the solution directly on a carbon-coated TEM grid.

Ultramicrotome slices. Embedding NMs in a medium can be very useful (1) when they possess magnetic nature and (2) when there is a need to preserve all spatial information of the particles in the sample. For the former, when the magnetic NMs are placed between pole pieces of the objective lens of the TEM, they may become magnetized. Therefore, without a medium that holds particles together and makes them stable for TEM investigations, particles can be attracted to the pole pieces and even contaminate them over time. Magnetic particles can also drift under the electron beam during analysis, thus making it difficult to perform HRTEM studies.

For the latter, field-collected environmental samples are very complex and heterogeneous, having both organic (including biological samples) and inorganic solid fractions. It is, therefore, important to make TEM samples that preserve both minerals and biological samples, as well as keep their heterogeneous spatial distribution as close as possible to the environmental conditions. For this specific purpose, TEM sample preparation needs three parts: fixative solution, resin and hardener. A field-collected solid sample can be preserved onsite by immediate immersion in a primary fixative solution (2% glutaraldehyde in 0.1 mol L^{-1} Na cacodylate buffer, pH adjusted to 7.0) in a centrifuge tube. The Na cacodylate buffer is typically used for the preparation and fixation of biological samples in TEM studies. The preserved solid samples are washed at least twice with 0.1 mol L^{-1} Na cacodylate buffer for 15 min each time. The samples are then fixed again in 1% OsO_4 in 0.1 mol L^{-1} Na cacodylate buffer for 1 h. Here, OsO_4 serves as a staining agent to provide contrast in TEM images, as well as a fixative to preserve biological samples. After discarding the OsO_4 solution, the resulting samples are washed twice for 10 min each with 0.1 mol L^{-1} Na cacodylate buffer. They are then dehydrated by applying a series of ethanol solutions starting at 15% and going to 100% ethanol. After the ethanol solutions, the samples are submersed in 100% propylene oxide for 15 min, followed by overnight infiltration with a mixture of propylene oxide and Poly/Bed 812 (50:50 v/v). In this context, propylene oxide and Poly/Bed 812 mixtures are resin and hardener, respectively. This is because they preserve biological cell structures and provide hardness for the ultramicrotome-cut sections that are required for TEM analyses. On the following day, the samples are again subjected to overnight infiltration with 100% Poly/Bed 812. The samples are then embedded in a flat embedding mould using freshly prepared 100% Poly/Bed 812. Moulds are placed in an oven that is set at 60 °C for at least 2 days to cure. Ultrathin sections are obtained by ultramicrotomy using a diamond knife, and sections approximately 60–90 nm thick are collected on a TEM grid for further analysis. Its successful application to various environmental samples, including the ferric oxyhydroxysulphate nanomineral (schwertmannite) as well as biosolid-amended mesocosms soils, can be found elsewhere.[13,20]

Cryofixed samples. Cryofixation involves rapid freezing to prevent crystallization of water, and therefore, allows the shapes and dimensions of hydrated structures of the specimen intact for the TEM analysis. It also helps to reduce the radiation damages by as much as sixfold, compared to ambient temperatures.[10] Cryofixation technique is therefore very useful for both biological and colloidal samples, which range from living organisms, such as plants, viruses, microorganisms and their cells, to nanosized, inorganic and organic constructs (e.g. pharmaceutical formulations). Typically, two types of cryofixed TEM samples can be made, for example, a thin vitrified layer of aqueous specimens and a thin slice of vitrified bulk specimens. The former is simple and rapid, but is limited to liquid suspension samples having moderate viscosity with particles diameters of less than 200 nm. Whereas the latter can be done by cryosectioning using high-pressure freezing device after the bulk sample has been vitrified.[32,35]

Cryo-TEM sample preparation has three main step[31]: (1) a small aliquot (approximately 3 μL) of a liquid suspension containing the sample is loaded onto the surface of a supporting substrate (a holey or lacey carbon film) on a TEM grid. Then, the droplet is carefully blotted with filter paper until most of the supernatant liquid is removed, leaving only a thin layer of sample (approximately 100 nm) on the substrate. (2) The thin liquid layer is rapidly immersed into a suitable cryogen, typically, liquid ethane. Liquid nitrogen is also used to maintain the temperature of the ethane close to its melting point of −183 °C. Embedded within this layer of vitreous ice, the specimen is essentially preserved as it is, providing information of the fully native structure. This step requires a plunging device and an environmental vitrification system, with both humidity and temperature controlled. (3) The vitrified samples then need to be transferred into the TEM cryoholder of the microscope under liquid nitrogen and are examined at temperatures around 100 K (i.e. cryo-TEM). It is worthy to note that ice crystal can form due to humidity or temperature changes during the freezing process, which can generate artefacts on the TEM analysis.

4. CONCLUSIONS

Within the past several decades, transmission electron microscopes have been significantly improved. Analytical TEM is an exceptional tool for the materials characterization at the nanoscale, providing information of their size, morphology, aggregation state, chemical composition and crystal structure. It is by far the most powerful, single instrument for NMs characterization. By selecting different modes in (S)TEM, it is possible to obtain multiple, but complementary, perspectives of an element or chemical compound in laboratory samples, as well as in field-collected samples. As long as users understand the limitations of analytical TEM, it can be versatile and much more accessible for applications in various research areas.

ACKNOWLEDGEMENTS

Dr Kim is thankful for financial support from Temple University as a part of her start-up. The authors are also appreciative of support from the National Science Foundation (NSF) and the Environmental Protection Agency (EPA) under NSF Cooperative Agreement EF-0830093, entitled Center for the Environmental Implications of Nanotechnology (CEINT). Any opinions, findings, conclusions or recommendations expressed in this material are those of the author(s) and do not necessarily reflect the views of the NSF or the EPA. This work has not been subjected to EPA review and no official endorsement should be inferred.

REFERENCES

1. Anderson RM. *Material research society symposium proceedings*, vol. 254. Pittsburgh (PA, USA): Materials Research Society; 1992. p. 141–8.
2. Anderson RM, Klepeis SJ. *Material research society symposium proceedings*, vol. 480. Pittsburgh (PA, USA): Materials Research Society; 1997. p. 187.
3. Basile D, Boylan R, Baker B, Hayes K, Soza D. FIBXTEM-focused ion beammilling for TEM sample preparation. *Mater Res Sic Symp Proc* 1992;**254**:23–41.
4. Benninghoven A, Rüdenauer FG, Werner HW. *Secondary ion mass spectrometry: basic concepts, instrumental aspects, applications and trends*. New York (USA): John Wiley; 1987.
5. Borisevich AY, Lupini AR, Pennycook SJ. Depth sectioning with the aberration-corrected scanning transmission electron microscope. *PNAS* 2006;**103**:3044–8.
6. Browning ND, Wallis DJ, Nellist PD, Pennycook SJ. EELS in the STEM: determination of materials properties on the atomic scale. *Micron* 1997;**28**:333–48.
7. Chapman J, Steele J, Paterson J, Titchmarsh J. *The separation of characteristic signals from complex EELS spectra. Presented electron microscopy and analysis*. 1985 [Newcastle upon Tyne, UK].
8. Cowley JM. Applications of electron nanodiffraction. *Micron* 2004;**35**:345–60.
9. Daniels HR, Brydson R, Brown A, Rand B. Quantitative valence plasmon mapping in the TEM: viewing. *Ultramicroscopy* 2003;**96**:547–58.
10. Dubochet J, Adrian M, Chang JJ, Homo JC, Lepault J, McDowall AW, et al. Cryo electron microscopy of vitrified specimens. *Q Rev Biophys* 1988;**21**:129–228.
11. Echigo T, Monsegue N, Aruguete DM, Murayama M, Hochella Jr MF. Nanopores in hematite (α-Fe_2O_3) nanocrystals observed by electron tomography. *Am Mineral* 2013;**98**:154–62.
12. Egerton RF. Electron energy-loss spectroscopy in the TEM. *Rep Prog Phys* 2009;**72**: 1–25.
13. French RA, Caraballo M, Kim B, Rimstidt JD, Murayama M, Hochella Jr MF. The enigmatic iron oxyhydroxysulfate nanomineral schwertmannite: morphology, structure and composition. *Am Mineral* 2012;**97**:1469–82.
14. Ghaly M, Averback RS. Effect of viscous flow on ion damage near solid surfaces. *Phys Rev Lett* 1994;**72**:364.
15. Giannuzzi LA, Stevie FA. In: Giannuzzi S, editor. *Introduction to focused ion beams: instrumentation, theory, techniques, and practice*. USA: Springer; 2005.
16. Giannuzzi LA, Geurts R, Ringnalda J. *Microsc Microanal* 2005;**11**(Suppl. 2):828–9.

17. Haider M, Uhlermann S, Schwan E, Rose H, Kabius B, Urban K. Electron microscopy image enhanced. *Nature* 1998;**392**:768–9.
18. Janney DE, Cowley JM, Buseck PR. Transmission electron microscopy of synthetic 2- and 6- line ferrihydrite. *Clay Clay Miner* 2000;**48**:111–9.
19. Kim B, Park C-S, Murayama M, Hochella Jr MF. Discovery and characterization of silver sulfide nanoparticles in final sewage sludge products. *Environ Sci Technol* 2010;**44**: 7509–14.
20. Kim B, Murayama M, Colman BP, Hochella Jr MF. Characterization and environmental implications of nano- and larger TiO_2 particles in sewage sludge and soils amended with sewage sludge. *J Environ Monitor* 2012;**14**:1129–37.
21. Kim B, Levard C, Murayama M, Brown Jr GE, Hochella Jr MF. Integrated approaches of X-ray absorption spectroscopic and electron microscopic techniques on zinc speciation and characterization in a final sewage sludge product. *J Environ Qual* 2014;**43**: 908–16.
22. Krivanek OL, Dellby N, Lupini AR. Towards sub-angstrom electron beams. *Ultramicroscopy* 1999;**78**:1–11.
23. Krivanek OL, Nellist PD, Dellby N, Murfitt MF, Szilagyi Z. Towards sub-0.5 Å electron beams. *Ultramicroscopy* 2003;**96**:229–37.
24. Krivanek OL. An electron microscope for the aberration-corrected era. *Ultramicroscopy* 2007;**108**:179.
25. Kundmann M, Krivanek OL. Automated processing of parallel-detection EELS data. *Microsc Microanal M* 1991;**2**:245–56.
26. Lazar S, Botton GA, Wu M-Y, Tichelaar FD, Zandbergen HW. Materials science applications of HREELS in near edge structure analysis and low-energy loss spectroscopy. *Ultramicroscopy* 2003;**96**:535–46.
27. Lee MR, Bland PA, Graham G. Preparation of TEM samples by focused ion beam (FIB) techniques: applications to the study of clays and phyllosilicates in meteorites. *Mineral Mag* 2003;**67**(3):581–92.
28. Lee MR, Brown DJ, Smith CL, Hodson ME, Mackenzie M, Hellmann R. Characterization of mineral surfaces using FIB and TEM: a case study of naturally weathered alkali feldspars. *Am Mineral* 2007;**92**:1383–94.
29. Lee MR. Transmission electron microscopy (TEM) of earth and planetary materials: a review. *Mineral Mag* 2010;**74**(1):1–27.
30. Mayer J, Giannuzzi LA, Kamino T, Machael J. TEM sample preparation and FIB-induced damage. *MRS Bull* 2007;**32**:400–7.
31. Mittal V, Matsko NB. Visualization of organic-inorganic nanostructures in liquid. In: *Analytical imaging techniques for soft matter characterization, engineering materials*. Berlin (Heidelberg): Springer-Verlag; 2012. p. 13–29.
32. Moor H. Theory and practice of high-pressure freezing. In: Steinbrecht RA, Zierold K, editors. *Cryo-techniques in biological electron microscopy*. Berlin: Springer-Verlag; 1987. p. 175–91.
33. Moreau JW, Webb RI, Banfield JF. Ultrastructure, aggregation-state, and crystal growth of biogenic nanocrystalline sphalerite and wurtzite. *Am Mineral* 2004;**89**:950–60.
34. Mueller H, Uhlemann S, Hartel P, Haider M. Advancing the hexapole Cs-corrector for the scanning transmission electron microscope. *Microsc Microanal* 2006;**12**:442–55.

35. Mueller M, Moor H. Cryofixation of thick specimens by high pressure freezing. In: Mueller M, Becker RP, Boyde A, Wolosewick JJ, editors. *The science of biological specimen preparation*. Chicago: AMF O'Hare, SEM; 1985. p. 131–8.
36. Mugnaioli E, Capitani G, Nieto F, Mellini M. Accurate and precise lattice parameters by selected-area electron diffraction in the transmission electron microscope. *Am Mineral* 2009;**94**:793–800.
37. Muller DA. Atomic-scale chemical imaging of composition and bonding by aberration-corrected microscopy. *Science* 2008;**319**:1073–6.
38. Muller DA. Structure and bonding at the atomic scale by scanning transmission electron microscopy. *Nat Mater* 2009;**8**:263–70.
39. Nastasi M, Mayer JW, Hirvonen JK. *Ion-solid interactions: fundamentals and applications*. Great Britain: Cambridge University Press; 1996.
40. O'Keefe MA. Seeing atoms with aberration-corrected sub-Angstrom electron microscopy. *Ultramicroscopy* 2008;**108**:196–209.
41. Overwijk MHF, van den Heuvel FC, Bull-Lieuwma CWT. Novel scheme for the preparation of transmission electron-microscopy specimens with a focused ion beam. *J Vac Sci Technol B* 1993;**11**:2021–4.
42. Orloff J, Utlaut M, Swanson L. *High resolution focused ion beams: FIB and its applications*. NY (USA): Kluwer Academic/Plenum Publishers; 2003.
43. Schattschneider P, Rubino S, Stoeger-Pollach M, Hébert C, Rusz J, Calmels L, Snoeck E. Energy loss magnetic chiral dichroism: a new technique for the study of magnetic properties in the electron microscope. *J Appl Phys* 2008;**103**:07D931.
44. Scott MC, Chen C-C, Mecklenburg M, Zhu C, Xu R, Ercius P, et al. *Nature* 2012;**483**:444–7.
45. Shibata N, Findlay SD, Azuma S, Mizoguchi T, Yamamoto T, Ikuhara Y. Atomic-scale imaging of individual dopant atoms in a buried interface. *Nat Mater* 2009;**8**:654–8.
46. Sigmond P. Theory of sputtering. I. Sputtering yield of amorphous and polycrystalline targets. *Phys Rev* 1969;**184**:383–416.
47. Sigmond P. A mechanism of surface micro-roughening by ion bombardment. *J Mater Sci* 1973;**8**:1545–53.
48 Steeds JW, Morniroli J-P. Selected area electron diffraction (SAED) and convergent beam electron diffraction (CBED). In: *Minerals and reactions at the atomic scale: transmission electron microscopy*. Buseck PR, editor. *Reviews in Mineralogy*, vol. 27. Chantilly (VA, USA): Mineralogical Society of America; 1992. p. 37–84.
49. Tsuda K, Tanaka M. Refinement of crystal structural parameters using two-dimensional energy-filtered CBED patterns. *Acta Crystallogr* 1999;**A55**:939–54.
50. Urban K, Kabius B, Haider N, Rose H. A way to higher resolution: spherical-aberration correction in a 200 kV transmission electron microscope. *J Electron Microsc* 1999;**48**:821–6.
51. Urban KW. Studying atomic structures by aberration-corrected transmission electron microscopy. *Science* 2008;**321**:506–10.
52. Urban KW. Is science prepared for atomic resolution electron microscopy? *Nat Mater* 2009;**8**:260–2.
53. Verbeeck J, van Aert S. Model based quantification of EELS spectra. *Ultramicroscopy* 2004;**101**:207–24.
54. Volkert CA, Minor AM. Focused ion beam microscopy and micromachining. *MRS Bull* 2007;**32**:389–99.

55. Williams DB, Carter CB. *Transmission electron microscopy: a textbook for materials science*. 2nd ed. New York (USA): Springer; 2009.
56. Wirth R. Focused ion beam (FIB) combined with SEM and TEM: advanced analytical tools for studies of chemical composition, microstructure and crystal structure in geomaterials on a nanometre scale. *Chem Geol* 2009;**261**:217–29.
57. Browning ND, Arslan I, Bradley JP, Chi M, Dai Z, Erni R, et al. Aberration corrected and monochromated TEM/TEM for materials science. *Microsc Microanal* 2007;**13**(Suppl 2): 1180–1. CD.
58. Wirth R. Focused Ion Beam (FIB): A novel technology for advanced application of micro- and nanoanalysis in geosciences and applied mineralogy. *Eur J Mineral* 2004;**16**:863–76.
59. Wirth R. Focused Ion Beam (FIB): Applications in Micro- and Nanoanalysis in Geosciences and Applied Mineralogy. *Praktische Metallographie-Practical Metallography* 2005;**42**(4):188–205.

Chapter 5

Methods for Measuring Concentration (Mass, Surface Area and Number) of Nanomaterials

Soubantika Palchoudhury, Mohammed Baalousha and Jamie R. Lead[1]
Center for Environmental Nanoscience and Risk (CENR), Department of Environmental Health Sciences, Arnold School of Public Health, University of South Carolina, Columbia, SC, USA
[1]*Corresponding author: E-mail: jlead@mailbox.sc.edu*

Chapter Outline

Frontiers of Nanoscience, Vol. 8. http://dx.doi.org/10.1016/B978-0-08-099948-7.00005-1

1. INTRODUCTION

Widespread use of manufactured Nanomaterials (NMs) implies their inevitable release into the environment.[1] Therefore, accurate measurement of NM concentrations in complex media is significant for regulatory control and environmental protection.[2,3] Typically, NM concentration is measured in terms of mass, although surface area or number concentration have often been suggested but less frequently used. Mass concentration of NMs can be conveniently measured and is useful to interpret certain mass-driven phenomena e.g. sedimentation. However, the mass concentration cannot account for other, often surface- or number-related effects. In these cases, the surface area concentration might be more relevant to assess the environmental risk of NMs, including processes involved in biouptake along with pollutant and natural organic macromolecule (NOM) binding.[4,5] Other processes perhaps dominated by smaller NM fractions e.g. aggregation/disaggregation is best represented by number concentrations.[6] Compared to the mass concentration, there are fewer validated methods to measure the surface and number metrics till date. It should be noted that all the three NM metrics may be useful in the environment and in (eco)toxicology because sedimentation, surface sorption of contaminants and stability/breakage of NM aggregates are interrelated and important phenomena in fate and effects behaviours. Therefore, this chapter details the significance and measurement techniques for different NM concentrations in three broad sections corresponding to mass, surface area and number concentration.[7] The elaborate operating principles for these methods are explained in Chapter 2.

2. CONCENTRATION METRICS AND THEIR ENVIRONMENTAL AND TOXICOLOGICAL RELEVANCE

Mass concentration metric. Mass concentration (mass per unit volume e.g. ppb or $\mu g\ L^{-1}$) is the traditional metric to report the environmental concentration (exposure) and toxicity thresholds (dose) of conventional contaminants. Mass concentration is also the metric used when investigating the release and fate and transport of conventional contaminants in the environment. It is likely the best suitable metric to represent sedimentation of manufactured NMs after aggregation.[8,9] In addition, there are well-validated methods for NM mass concentration determination, making the production of accurate data easier. Therefore, mass concentration is the current practical choice for a rapid overall assessment of environmental processes related to the larger NM fractions (e.g. sedimentation).[10] For example, the mass concentration of fullerene C_{60} was considered critical to predict the fate of aqueous nanoscale fullerene (C_{60}) suspensions.[11]

Although mass concentration metrics have been widely adopted in the hazard and risk characterization of NMs, they are known to have limitations. Few studies have investigated the impact of dose metric (mass, number and surface area) on the dose–response relationship for NMs, but there is no consensus on the best dose metric to express NM toxicity. It is likely that the dose metric will be NM-dependent.[12,13] This is particularly true for polydisperse NM samples where mass concentration primarily reflects larger NM fractions.[14] A potential challenge is to quantify mass concentration of specific NMs in complex biological matrices.[15] The NMs can be altered via dynamic interaction with surrounding media. Finally, manufactured NMs are typically present at low concentrations (parts-per-trillion to ppb). Highly sensitive and specific analytical techniques are required to measure these toxicologically relevant mass concentrations. This metric is also not appropriate for specific toxicological processes, e.g. assessing inflammation under high aerosol concentration.[16] In these cases, the surface properties of NMs are found to play a dominant role.

Surface area concentration metric. Surface area is another important metric as the surface area per unit volume of NMs increases with the decrease in NM size and thus the surface of NMs becomes a dominant feature in controlling several environmental processes such as sorption of contaminants, NM dissolution and potentially toxicity.[17–19] NMs in the environment greatly influence speciation, transport and bioavailability of contaminants or nutrients, essentially because they possess high specific surface area (SSA) as compared to bulk materials or microparticles of the same composition.[17,18] The NMs contain a much higher proportion of atoms on the surface versus the particle interior than larger particles. These factors facilitate contaminant/NOM sorption on NM surfaces. The unique surface features are amplified with decreased NM sizes. Therefore, smaller iron oxide NMs showed significant binding capacity for trace metals.[19] This is particularly noticeable for the contaminant relocation process in porous media such as soils and aquifers. Surface features like the porosity and conformation of NMs dominate the sorption-desorption or retention of chemicals within the NM structure.[20] The NMs facilitate farther transport of trace metal and organic contaminants compared to noncolloidally bound components.[21] For instance, polychlorinated dibenzo-*p*-dioxins and dibenzofurans (polycyclic hydrocarbons or PAH) were relocated from soil to groundwater via NM binding.[22]

The attachment of pollutants to fractal-like NMs amplifies contaminant accessibility and toxicity to cells. For instance, arsenic adsorption on maghemite NMs at low doses disrupts the biotransformation mechanism of the contaminant and increases adverse effects.[23] Cd^{2+} accumulation in the viscera and gills of fish show steep increase in the presence of TiO_2 NMs.[24] This phenomenon called Trojan horse effect is better analyzed via surface area concentration of NMs. Additionally, the NM surface concentration is essential for analyzing inflammation responses related to airborne particles or particles from worn out biomaterial implants because the high NM surfaces could potentially induce reactive oxygen species production within the cell.[25,26] Another eco(toxicological) concern, particularly related to the high surface area and small size of metal-based NMs is the toxicity from dissolution of these metal from the NM surfaces. Therefore, several authors have suggested that the toxicity of NMs correlates best to the concentration of NMs when expressed as surface area concentration, particularly for the smaller NM fractions with high ratio of surface atoms. Most nanotoxicology-related interactions in the human body take place via surface contact with NMs and are strongly dependent on the deposition region (e.g. head, tracheobronchial and alveolar). Particularly, the surface area concentration is considered a better metrics to predict toxicity or inflammation for less soluble NMs.

Number concentration metric. Number concentration is highly useful to measure aggregation and disaggregation behaviour of NMs in the environment. Particularly, aggregation kinetics is best expressed as a function of NM number concentration.[27,28] The number-based metric gives a truer representation for the behaviour and fate of small NM fractions in environmental samples like aquatic colloids.

Several definitions are currently available for NMs and are being further improved by international organizations to regulate NMs for human and environmental safety.[29,30] Number size distribution and number particle concentration of NMs are significant metrics for rational regulation of NM-based ecotoxicity, according to the European Union (EU) definition.[31] The EU definition suggests that samples where 50% or more particles in the number size distribution have one or more dimensions in the range of 1–100 nm fall within the scope of nanotechnology for regulatory purposes. This keeps in consideration the potential toxicity from larger aggregates (>100 nm) of individual NMs. Therefore, number concentration has been suggested as a significant dose metric in nanotoxicological studies, though it is a relatively newer metric and few validated measurement methods are available.

Table 1 summarizes the significance of different NM concentration metrics to assess environmental processes and to interpret the dose–response relationships in nano(toxicology).

TABLE 1 Different Concentration Metrics for Nanomaterials

Concentration Metric	Unit	Relevance in Nano(Toxicology)	Relevance in Environmental Measurement
Mass concentration	$\mu g\ L^{-1}$	Analyzing the toxicity from larger NMs (high mass)	Sedimentation
Surface area concentration	$m^2\ cm^{-3}$	Better represents small NM-related toxicity effects like inflammation	Contaminant sorption
Number concentration	Particles L^{-1}	Relevant for small NM fraction	Aggregation/disaggregation

NM, nanomaterial.

3. MEASUREMENT OF MASS CONCENTRATION

Mass is the traditional and most commonly measured concentration metric of NMs. Several analytical techniques and methodologies are currently available to measure the total mass concentrations of NM samples such as inductively coupled plasma—mass spectrometry (ICP-MS), atomic absorption spectroscopy (AAS), UV-visible (UV-Vis) spectroscopy and quartz crystal microbalance (QCM). The basic principles behind the techniques are explained in Chapter 2. The following section discusses the pros and cons of the methods to measure NM mass concentrations in real systems.

3.1 Atomic Absorption Spectroscopy

AAS provides a robust analytical route to assess mass concentrations, potentially in the presence of NOM or heavy surfactant coverage.[32] Quantitative measurements in AAS are based on Beer's law (the concentration is proportional to absorbance, $C = KA$). AAS is well-known for the reliability in detecting a broad range of metal concentrations. Recently, it has been widely used to measure cellular uptake of Au and Ag NMs at trace levels.[33] Particularly, the electrothermal/graphite furnace (GF)-AAS could detect metal signal from Au NMs as low as 0.005 $\mu g\ L^{-1}$ with a NOM tolerance of up to $10^4\ \mu g\ L^{-1}$. In all cases, prior separation processes (e.g. cloud point extraction or flow field flow fractionation) and selective additives are required to distinguish NM-based metal signals from dissolved ions in these measurements. The GF-AAS was used to determine Ag NM-based metal signals from dissolved Ag in complex biological samples (fish) based on a higher thermal energy requirement for atomizing Ag NMs compared to Ag ions.[34] Compared to GF-AAS, a flame-AAS is more commonly used, with poorer limits of detection. However, a significant limitation of these AAS studies in measuring NM mass concentration is the use of digested metallic form of the sample. The NMs are not assessed in their original form, rather the metal content within the NM is measured, limiting true representation.

The sample digestion steps and the relatively high sample volume (2—5 mL) required for flame-AAS analysis adversely affect the detection limits for NM samples. Particularly, the nonaqueous samples have to be largely diluted to maintain the appropriate viscosity for the flame-AAS nebulizer. The chances of contamination are higher for solid samples because they require more sample preparation steps. The GF-AAS is sensitive and requires small sample volumes (μL), but is suitable for single element analysis in one run. To analyze multiple elements is time-consuming, therefore AAS measurement is typically limited to analysis of single element at a time. Additionally, the duplicate runs required in GF-AAS take longer analysis time. The necessity to use separate matrix modifiers for different samples to prevent chemical interference further limits the applicability of GF-AAS.

3.2 Inductively Coupled Plasma–Mass Spectrometry

ICP-MS is a leading analytical technique for assessing mass concentration of noncarbon NMs because it provides a sensitive detection limit ($\sim$0.1–0.05 μg L^{-1} of NMs) compared to other methods, has a large linear range and potential for isotopic measurement.[35] In ICP-MS, the ionized sample atoms travel through the electrically conductive rods of the quadrupole MS at a rate determined by their mass/charge ratio, allowing selected atoms to reach the detector. This arrangement makes the ICP-MS highly sensitive. In its standard operation mode, an ICP-MS reports total mass concentration of metals similar to AAS, but can be used to measure NM number concentration in the single particle mode (SP-ICP-MS). Instrument parameters, NM stability and the NM surroundings are important influences on ICP-MS results. The primary advantage of ICP-MS compared to other methods (e.g. AAS) is its ability to efficiently analyze versatile samples.

Typically, NM samples for ICP-MS are prepared via multiple acid digestions prior to measurement. This method presents potential limitations like contamination from digestion reagents, higher limit of detection and low sample throughput. Therefore, Fabricius et al. investigated the role of acid digestion and microwave-assisted sample preparation methods on the ICP-MS analysis of different inorganic NMs (Ag, Au, CeO_2, TiO_2 and ZnO).[36] A microwave-assisted digestion via two-step heating was recommended prior to ICP analysis. However, the faster acid digestion route was sufficient for some NMs. The digestion of NM samples is a major limitation of ICP-MS because it represents the total mass of metals rather than the mass of NMs in their true form. It essentially measures the mass concentration of metals within the NMs following sample digestion to dissolve the NMs. Therefore, it does not distinguish between a pure metal, metal oxide or composite NMs, limiting true representation of NM mass or composition. Recently, aqueous Au NMs were directly quantified via ICP-MS to avoid NM digestion.[37] The results showed a promising detection limit of 0.15 μg L^{-1} for spherical Au NMs ($\sim$15 nm), though a higher reliability is achieved with digested samples. This is a significant improvement because the de-ionized water will dissolve less contaminants compared to strong acids, minimizing interference from dissolved metals. Additionally, NMs in their original aqueous phase can be directly measured to give a better representation of the sample.

Based on currently available methods for nanotoxicological studies, ICP-MS served as a powerful tool to quantify uptake and biodistribution of metal-based NMs.[38] However, an important consideration in nanotoxicity analysis is to distinguish the metal NMs from the released metal ions and background interferences. The ICP-MS measures metal concentration rather than NM concentration and thus does not distinguish between dissolved metals and metals within NM. This limitation can be overcome by separating NMs from dissolved metals before ICP-MS analysis.[39] Therefore, separation routes like dialysis, ultracentrifugation, ultrafiltration, tangential flow filtration or more sophisticated field flow fractionation (FFF) methods are used for distinguishing dissolved fractions from NMs to enhance reliability. Among these techniques, the easy-to-use and cost-effective ultrafiltration method was found to be most easily implementable.[36] In the direct ICP-MS measurement of aqueous NMs reported recently, lab-made citrate-coated Au NMs were diluted using different solvents (de-ionized water, 1% v/v trisodium citrate and 1% v/v HCl), keeping the dilution factor (1:1000) same in all cases.[37] Each sample was spiked with Au(III) standard solution to maintain same final sample concentration. The accuracy of Au NM mass concentration measurement was evaluated on the basis of theoretically predicted values and the recovery of spikes. Results from aqueous Au NMs were comparable to the digested samples for highly stable and nonaggregated NM batches. This potential to measure NMs in their true form without digestion is promising, but the effects of dissolved ions were not considered in this study. Additionally, the method used noninteracting Au NMs for validation. This technique may not be applicable for soluble and reactive NM samples.

For quantitative elemental analysis of relatively higher NM concentrations (10^3 μg L^{-1}), an inductively coupled plasma–atomic emission spectrometry (ICP-AES) is commonly used. In an ICP-AES, the liquid sample is nebulized into small droplets (1–10 μm) prior to entering the argon plasma chamber. The sample atoms and ions are excited from their ground state inside the plasma chamber. The atoms emit characteristic emission spectra while returning to the ground state. An optical spectrometer detects the intensity of this fingerprint spectrum of the element to measure the sample concentration, based on linear calibration curves. The ICP-AES is useful for the multielement capacity, specificity and robust measurements. Both the ICP-MS and ICP-AES are widely applied for the analysis of real environmental or biological samples.[40]

3.3 UV-Vis Spectroscopy

UV-Vis spectroscopy is a simple and useful analytical technique to measure the mass concentration of NMs based on Beer's law (Figure 1).[41] Here, the sample solution in the UV cuvette is analyzed in comparison with the reference solvent. Light from two source lamps, hydrogen/deuterium for UV and tungsten/halogen for the visible region, is passed simultaneously through the sample and reference cuvettes. The source beam is separated into monochromatic components via a prism/diffraction grating and each wavelength splits into two equal-intensity beams with a half-mirror device prior to passage through the two cuvettes. The valence electrons of the sample absorb incident light for transition from ground state to higher energy (excited) orbitals. Ideally, the wavelength of absorbed light corresponds to the energy

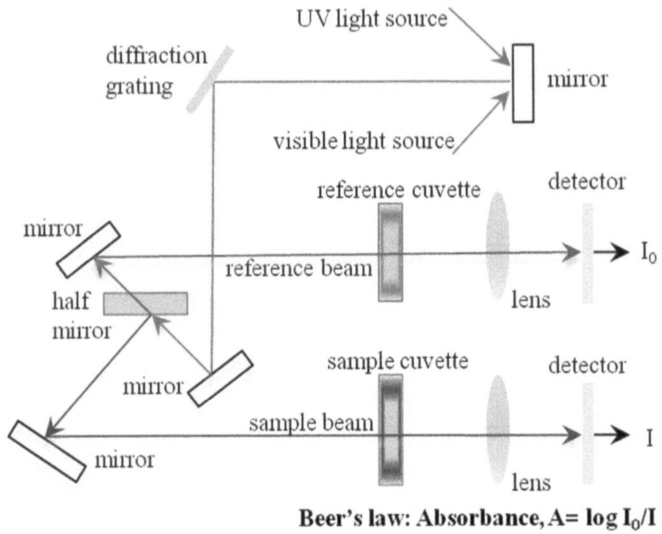

FIGURE 1 Schematic diagram of a UV-Vis spectroscope.

difference of the two states and is detected as a sharp characteristic sample peak. For low NM concentrations, these absorbance peaks show a linear correlation with the mass concentration (Beer's law), assuming no changes in NM behaviour. Some UV-Vis models require only microlitre sample volume.[42]

However, the UV-Vis is primarily suitable for a rough overall mass concentration measurement of unaggregated NM samples in simple media for lab-based studies. The method is mostly useful for monodisperse plasmonic NMs such as Au and Ag.[43] For aggregating samples such as aqueous phase fullerene NMs or charge-stabilized NMs in high ionic strength media, the UV-Vis data should be carefully analyzed because the spectroscopic properties of NMs vary considerably with size, surface properties and aggregation state of NMs (Figure 2).[28] To better characterize these samples with UV-Vis, a mass extinction coefficient was obtained via normalizing the UV-Vis spectra using fullerene concentration.[11] This coefficient was found to decrease with hydrodynamic diameter of the NMs. In contrast, a particle extinction coefficient based on the size distribution of fullerene NMs showed an increase with number weighted NM size. Therefore, effects of size, shape, aggregation and material-dependent extinction coefficients of NMs limit the use of UV-Vis to directly quantify mass concentration of polydisperse NMs. Recently, a UV-Vis-based titration method was designed with plasmonic NMs to determine the mass concentration of analyte NMs, regardless of the particle type.[44] In the above mentioned report, known concentrations of titrant Au NMs were self-assembled with different NMs via complementary

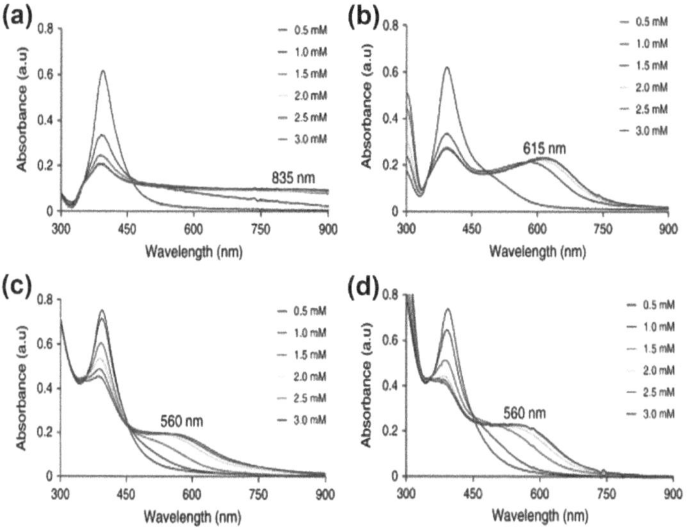

FIGURE 2 Effect of divalent/monovalent salts and Suwanee River fulvic acid (SRFA) on UV-Vis spectra of citrate-coated Au NMs. (a) $CaCl_2$, (b) $Ca(NO_3)_2$, (c) SRFA-$CaCl_2$ and (d) SRFA-$CaNO_3$. *Copyright: Sci. Total Environ.*

cDNA. However, these UV-Vis-based experiments were again limited to rough estimates in the lab scale. It is misleading to use UV-Vis measurements, particularly surface plasmon resonance (SPR) shifts for analysis of real environmental processes like sedimentation.[45,46] Several physical processes like aggregation, interaction with NOM and interaction with Cl, S ions influence SPR in NMs. SPR cannot be related to a single process. Particularly, mass concentration should not be assessed via SPR. It is also challenging to measure the mass concentrations for semiconductor NMs like Si via UV-Vis because particle size and surface play significant roles in the properties of these NMs.

In summary, it is difficult to obtain reliable quantitative results from the UV-Vis for aggregating NMs, NM mixtures or NMs in complex media. NM mixtures show complex peaks resulting in loss of selectivity. The absorbance peaks shift to higher wavelengths and decrease in intensity for aggregating systems, as mentioned above. Soluble NMs also show loss of absorbance signal from dissolution over time. This reduction in peak area may be useful to monitor NM dissolution rate, but for a simple system of unaggregated NMs.[47] These practical issues for measuring complex systems make the UV-Vis suitable for monitoring simplified lab-based systems such as a quick check of synthesis product. Availability of a wide range of cuvette path lengths (micro to several cm) has expanded the concentration ranges for applicability of the UV-Vis technique. However, the primary use of the method remains in lab studies.

3.4 Quartz Crystal Microbalance

A QCM can be successfully used to sense mass of NMs via a piezoelectric effect. QCM is a direct and sensitive technique to quantify mass concentration of nanocrystals without interference from size or surface of the NMs.[48] Typically, a liquid drop of the NM solution is deposited on one face of an oscillating crystal. The dry residue mass of NMs after solvent evaporation (Δm) is detected as a change in resonant frequency (Δf) of the quartz crystal, according to the Sauerbrey equation.

$$\Delta f = C_m \frac{\Delta m}{A}$$

where, C_m is a constant relative to the thickness and properties of the quartz crystal and A is the working area of the crystal. This technique has been extended to measure adsorption changes in liquid media for biological samples. Recently, QCM showed a detection limit of $5000-1.3 \times 10^6$ μg L^{-1} for Si NMs.[48] Compared to other methods, QCM has been applied in situ without additional sample preparation to measure small NMs (<10 nm).[49]

However, this protocol is unsuitable for true representation of NM mass concentration, as suggested in the following equation. Potential error in NM mass measurement may arise from residual salts and free capping agents since QCM measures the mass of all constituents within the dry residue after solvent evaporation. This will cause significant misrepresentation of the NM sample, particularly for in situ measurements.[49] Purifying the NM solution (e.g. dialysis) prior to QCM measurement could minimize the error in off-line systems.

Without purification, the measured mass = mass of all constituents in the suspension e.g. $M_{\text{NMs}} + M_{\text{dissolved metals}} + M_{\text{electrolytes}} + M_{\text{attached capping agent}} + M_{\text{free capping agent}}$.

With purification i.e. after removing electrolytes and free capping agent, the measured mass = $M_{\text{NMs}} + M_{\text{attached capping agent}}$.

Additionally, the physical property of the quartz crystal and the medium must be known for reliable concentration measurements in the QCM. The basic assumptions of the Sauerbrey equation such as a completely adsorbing system and a thin uniform mass distribution are other limitations of the technique, particularly for NMs in complex media.[50] Therefore, the QCM is best suitable to monitor lab studies, e.g. measuring the deposition rate of flame-synthesized iron oxide NMs.[49]

In summary, the mass concentration has been the traditional and widely used choice for NM measurement because it is comparable to common chemical concentration measurement techniques. The measurement methods for mass concentration are more common, compared to the newer surface area and number concentration measurements. However, these methods mostly measure metal concentration of acid digested NMs, limiting true representation of NM samples. Additionally, a single large particle can dominate the mass measurement results to misrepresent the exposure data for NMs. The different techniques to quantify mass concentration were discussed in the above section. Table 2 summarizes the advantages and disadvantages of the various analytical methods for determining mass concentrations. Among them, the ICP-MS recently showed great promise in terms of sensitivity and broad analysis range, but the method is unsuitable for carbon-based NMs (e.g. carbon nanotubes (CNTs) and fullerenes). Therefore, other NM concentration metrics like number and surface area are gaining significance in toxicity/exposure assessment as the mechanisms for biological and environmental interaction of NMs are largely unexplored.[51,52]

4. MEASUREMENT OF SURFACE AREA CONCENTRATION

To estimate the surface area concentration of the NMs, several analytical techniques such as the transmission electron microscope (TEM),[58] atomic force microscope (AFM),[59] small angle X-ray scattering,[60] nuclear magnetic resonance (NMR) and Brunauer–Emmett–Teller (BET) are currently available, but their suitability to measure NM surface has not been assessed. The principles behind these techniques are described in Chapter 2. This section compares the different techniques to measure NM surface area in terms of relevant evaluation criteria. For example, an important criterion is to directly measure the true surface area of NMs without relying on size, mass concentration or morphological assumptions. The measurement method should ideally account for surface details, aggregation and morphology of NMs because small changes in these parameters greatly influence the NM surface area concentrations. According to the European Union Commission Recommendation, a material should be considered nano where the SSA by volume of the material is $>60\ m^2\ cm^{-3}$.[31] Therefore, a suitable technique should be able to measure surface concentration of $>60\ m^2\ cm^{-3}$. Additionally, it is important to measure surface concentrations in solution for realistic studies of biointeraction and environmental toxicity. The measurement technique should ideally be able to isolate the NMs from dissolved compounds in complex media.

The following paragraphs list the few methods directly measuring NM surfaces without relying on size/shape-based assumptions. The main strengths of these direct techniques come from the capacity to detect all surface features for a truer representation of the NM surface area. The direct methods use adsorption of molecules on NM surfaces as probes to determine the surface area. This operating principle allows them to access the full surface area of the NMs with a high level of confidence. Among the direct methods, adsorption from solution and NMR showed most promise because they can measure particles in dispersion, as opposed to the dry samples required for BET. The techniques detailed in this section best fit the evaluation criteria for NM surface measurements, though they do not satisfy every requirement.

TABLE 2 Measurement Techniques for Mass Concentration of Nanomaterials

Technique	Basic Theory	Measured Parameter	Detection Limit	Selectivity	Common Applications	Advantages	Disadvantages
Atomic absorption spectroscopy	Beer's law	Total metal concentration	Element-dependent, varies with sample preparation, e.g. ~2 ng g^{-1} for Au NMs via graphite-AAS[53] and in the range of µg L^{-1} range for flame-AAS (Co, Fe)[54]	Cannot distinguish NMs from ions	Measurement of real environmental samples. E.g. water, soil, food or clinical analysis	Robust technique, can measure mass concentration of metals in presence of NOM or heavy surfactant coverage	Does not allow NM identification as it measures dissolved metals, multiple element analysis is time-consuming, high sample volume required in flame AAS, essential sample digestion steps limit measurement accuracy, multiple sample preparation steps required for solid samples, matrix modifiers required for GF-AAS
Inductively coupled plasma–mass spectrometry	Aerosol of the sample → sample ionization → extraction and separation of ions by mass detection of ions/concentration measurement	Total metal concentration	Depends on the size and chemical composition of the element, e.g. 0.80 µg L^{-1}[55] for Ag NMs, 0.0003 µg L^{-1} for TiO_2 NMs[56], and 0.16 µg L^{-1} for Zn NMs[57]	Selective for different metal ions, cannot separate NM from ions or background NOM	Industrial and real environmental sample analysis	High sensitivity and accuracy compared to other methods, can analyze a broad range of samples	Does not allow NM identification as it measures dissolved metals, careful sample preparation required, contamination from sample digestion lowers accuracy, expensive

UV-Vis	Beer's law	Detects NM mass concentration, under ideal conditions of uniform, nonaggregating sample and no interference from media	40–100 $\mu g\ L^{-1}$[115]	Detects NM aggregation, cannot distinguish NMs in complex media	Monitoring lab studies	Easy-to-use, rapid analysis and less expensive as compared to other techniques	Challenging to accurately analyze polydisperse, aggregating or nonplasmonic NMs, unreliable in complex media and for any NM mixtures because surface plasmon resonance depends on NM interaction with the surrounding environment
Quartz crystal microbalance	Piezoelectric effect, Sauerbrey equation	Total mass of constituents in a droplet after drying	~5000–1.3×10^6 $\mu g\ L^{-1}$ (based on Si and Ag NMs)[48]	Cannot detect NM mixtures or NMs in complex media	Monitoring lab studies	Minimum sample preparation, direct analysis, capable of measuring nonplasmonic NMs such as Si	Unsuitable for NMs in complex medium as it measures all constituents in the residue left after solvent evaporation, expensive, thickness and properties of the quartz crystal must be known, less sensitive than the ICP-MS, nonspecific (no compositional information)

NM, nanomaterial; AAS, atomic absorption spectroscopy; GF, graphite furnace; ICP-MS, inductively coupled plasma–mass spectrometry

For example, a potential disadvantage of the methods is the lack of selectivity to determine individual NMs from aggregates or other species (Table 3).

4.1 BET Gas Adsorption

This is the most common method to determine NM surface area.[20] The BET method was designed to directly measure the SSA and pore sizes of powdered samples under high-vacuum conditions.[61] In a typical BET analysis, NM surface is determined from the volume of N_2 gas adsorbed on the NMs. N_2 gas is assumed to have access to the entire NM surface. The surface area measurements are based on adsorption of gas molecules in infinite layers with no interlayer interaction. Under these conditions, the basic Langmuir theory can be applied to each layer to derive the BET surface area for NMs (S_{BET}).

$$S_{BET} = \frac{V_0 N_a s}{M_v}$$

where, V_0 is the volume of single monolayer of adsorbed gas, N_a is Avogadro's number, M_v is molar volume of gas adsorbate and s is the surface area of a single gas molecule adsorbed on the solid. The SSA for the known mass (M) of dry NM sample can be calculated from the BET surface area, as follows.

$$\text{SSA} = \frac{S_{BET}}{M}$$

This SSA for vacuum dried samples may differ from true surface concentrations of the NMs in solution, as suggested by recent reports on CeO_2 NMs.[61] Additionally, BET measurements are affected by loss of surface access at junctions due to interparticle interactions and sintering. These limitations can lead to an underestimation of the NM surface area with BET measurements. Additionally, BET does not provide an absolute measurement because the SSA value differs depending on the sorbent material (e.g. N_2). The key is to use the most applicable sorbent for a reliable surface area concentration with maximum surface details and minimum artifacts. Porosity and irregular morphology of the NMs further complicate the BET measurements. In particular, density of the NMs should be known to convert the BET reported surface area concentration ($m^2\,g^{-1}$) to volume SSA ($m^2\,cm^{-3}$) recommended by the EU.[62] The NM density may be significantly different from bulk density. The methods to measure NM density such as helium adsorption or pycnometry suffer similar limitations as the BET, reducing reliability of the surface area measurements. Therefore, measurement of NMs in dispersion is preferred, although BET is a powerful method for direct estimation of NM surface concentrations.

4.2 Adsorption from Solution

The method provides a simple route to quantify true geometric surface area of the NMs, without additional sample preparation or expensive instrumentation.

TABLE 3 Direct Measurement Techniques

Technique	Basic Theory	Detection Limit for Volume Specific Surface Area (SSA)	Selectivity to Represent True Surface	Common Applications	Advantages	Disadvantages
Brauner–Emmett–Teller	Measurement of SSA based on physical gas adsorption on the NM surface[20]	$>60\ m^2\ cm^{-3}$	Cannot isolate target NMs from aggregates or dissolved matter	Widely used to measure surface area of NMs in real systems	Provides direct measure of surface area, useful for measurement of nanoporous materials	High vacuum required, cannot measure NMs in dispersion, inaccessible surface areas for aggregated samples bias the measurements
Adsorption from solution	NM surface area measured from the decrease in concentration of the adsorbate solution after adsorption on NM surfaces	$>60\ m^2\ cm^{-3}$	Cannot isolate target NMs from aggregates or dissolved matter	Lab studies	Measures NMs in solution, inexpensive, negligible sample preparation required, directly represents the surface area	Selection of appropriate adsorbate is challenging, approximation of the ligand packing density required
Nuclear magnetic resonance	Difference in proton relaxation of adsorbed molecules compared to the free molecules provide a measure of the adsorbed surface area	$>60\ m^2\ cm^{-3}$	Cannot isolate target NMs from aggregates or dissolved matter	Diagnostics and lab studies	Measures NMs in solution, good reliability, direct measure of surface area	Expensive, requires reference material of known surface area for reliable measurement

NM, nanomaterial.

Here, the NM surface area is determined from the change in concentration of the molecular probe after adsorption on the NMs. The method has been widely used to assess NMs on solid substrates (e.g. catalysts).[63] Compared to BET, this method provides a more realistic measure of NM surface area because the NMs are measured in solution, not in vacuum. Unlike BET, dye molecules of choice can be used instead of N_2. For example, the surface area of Au NMs was successfully determined via adsorption of a strong binding ligand, 2-mercaptobenzimidazole (MBI).[64] The amount of MBI adsorbed (Γ_{MBI} nmol mL^{-1}) on Au NMs was easily quantified with a UV-Vis. The NM surface area (S) could be calculated from this value, using a known packing density of MBI ($P_{MBI} = 0.574$ nmol cm^{-2}) on Au NMs.

$$S = \frac{\Gamma_{MBI}}{P_{MBI}}$$

The small size, strong binding affinity and high packing density of MBI were suitable for accessing the Au NM surface. UV-Vis was selected for its simple operation. The organo-thiol, MBI was appropriate for this particular application because of its high stability in water and strong UV-activity. However, the stability and UV absorbance of MBI may be hindered in complex environmental media. A suitable adsorption probe will be required for environmentally relevant NM samples, in this case. Therefore, in solution adsorption technique, the choice of appropriate ligand and analytical instruments to match the NMs is potentially challenging. Another major limitation of the method is the approximation for ligand packing density on NM surfaces.

4.3 Nuclear Magnetic Resonance

NMR determines the surface area of NMs based on a difference in proton relaxation moments of the bulk liquid and liquid in contact with the NMs.[65] In a typical NMR set-up, the sample solution is placed in a coil between two permanent magnets. The protons in the sample initially align along the static magnetic field. A large magnetic field applied using a radio frequency pulse disrupts this magnetic orientation of the protons. On removal of this field, the sample protons realign with the initial static magnetic field to induce a measurable decaying voltage called 'free induction decay'. This relaxation process can be quantified through the longitudinal (T_1) and transverse relaxation (T_2) times to determine the relative ratio of bound and free liquid. The bound liquid provides a good representation of true NM surface.[66] Surface concentrations are measured in solution without additional sample preparation or assumptions on morphology. Compared to the available analytical techniques, NMR method is best suited to quantify NM surface area concentrations in terms of reliable measurement in solution (Table 3).

In conclusion, the surface area concentration of NMs is a significant dose metrics to assess pulmonary response and inflammation from insoluble NMs.[67] It is challenging to compare the surface concentrations reported in literature because different techniques focus on different types of surface area (e.g. geometric, superficial, Fuchs or active). Therefore, the requirements for an ideal surface area measurement technique were first identified. Direct measurement of surface area without relying on assumptions on morphology/size and the ability to measure NMs in dispersion were two primary evaluation criteria. The potential techniques for surface area measurements were identified through literature investigations and assessed against these requirements. However, direct measurements provided a more reliable representation of the NM surfaces. The solution phase adsorption and NMR were found to be the most suitable direct measurement techniques because they could measure NMs in solution.

5. MEASUREMENT OF NUMBER CONCENTRATION

The particle number concentration might be a suitable (eco)toxicology dose metric, as suggested in literature.[6,68] There is a need to accurately measure the NM number concentration and number size distribution for implementing NM regulations (e.g. EU recommendation), but the task is challenging.[69] There are several analytical techniques potentially capable of directly measuring NM number concentration such as microscopy techniques and single particle ICP-MS (SP-ICP-MS). These methods are discussed below together with their advantages and limitations, and recent developments in sample preparation and treatment for these analytical techniques. Other techniques, such as nanomaterial tracking analysis (NTA), dynamic light scattering and the FFF coupled with UV or ICP-MS can be used to indirectly estimate the number concentration of NMs following mathematical treatment of the measured data.[59] However, such mathematical treatments are error-prone because they indirectly calculate the particle number concentration from mass-/volume- or intensity-weighted measurements assuming a spherical, nonpermeable and noninteracting NM system.

5.1 Microscopy Techniques

Microscopy techniques have been widely used to measure NM size and size distribution, the structure and morphology of NMs and their aggregates or to investigate the interaction of NMs with other surfaces. However, microscopy techniques (AFM, TEM and NTA) have not been widely applied to measure number concentration of NMs in suspension due to the inherent sample preparation issues. Traditional sample preparation protocol based on drop deposition of the sample solution on microscopy substrates provides inaccurate representation of the NMs. This method introduces image artifacts and poses challenges in analyzing the NMs such as low sample recovery, NM aggregation during drying and precipitation of any salts present in the suspension. For standard sample preparation methods, observational biases for easier-to-see particles tend to limit the accuracy of the microscopy-based method with an overestimation of the NM size. Therefore, novel sample preparation routes are being explored to use the maximum potential of microscopy techniques in measuring NM number concentration. A sample preparation route with maximum sample recovery is targeted. Recently, the authors developed a fully quantitative sample preparation method to quantify the number concentration of NMs from stable Au NM suspension at environmentally relevant concentrations (ng to $\mu g\,L^{-1}$) using AFM.[70] This novel method is also applicable to TEM and is discussed in detail in the following section.

Microscopy techniques are indispensible in the nanoanalytical toolbox due to their high resolution and their ability to distinguish NM of different composition. This section discusses the different microscopy techniques to measure NM number concentration, the challenges and the potential to enhance measurement accuracy with careful sample preparation.

5.1.1 Atomic Force Microscopy

AFM is a powerful analytical technique to study the natural and manufactured NMs because it can potentially image samples in solution or in air at a high spatial resolution.[59,71] The sample preparation method is most significant in obtaining representative number concentration for NMs using AFM. For example, the adsorption method largely relies on the diffusion and interaction of NMs with the substrate and hence the property of the medium. Therefore, this method primarily detects the smallest fraction of NMs with high diffusion rates and the NMs tightly bound to the AFM substrate. In drop deposition method, the locally increased salt concentration during drying hinders the true representation of NMs via aggregation artifacts. Compared to these passive techniques, the ultracentrifugation method actively forces all NMs in suspension onto the AFM substrate. However, the thorough washing steps necessary to prevent image artifacts may cause loss of weakly bound NMs from the AFM substrate. Additionally, sample overloading was observed at high NM concentrations ($>1000\,\mu g\,L^{-1}$). Therefore, a combination of surface functionalization and ultracentrifugation was used as a sample preparation method to maximize NM recovery.[70] Before ultracentrifugation of the NM suspension on the substrate, the freshly cleaved mica substrates were immersed in 0.1% poly-L-lysine to functionalize the surfaces with this positively charged synthetic amino acid for enhanced sample-substrate adhesion. The particles on each image were counted manually and the number of NMs in suspension ($N_{suspension}$) can be calculated from counted number ($N_{counted}$) on each AFM image as follows:

$$N_{suspension} = \frac{N_{counted}}{V_{image}}$$

where, V_{image} is the volume of suspension on top of each AFM image, which can be calculated according to the following equation:

$$V_{image} = A_{image}h$$

where, A_{image} is the area of each AFM image (typically 1 × 1 μm to 5 × 5 μm) and h is the height of water column above the image, calculated as follows:

$$A_{image} = \frac{V_{centrifuged}}{\pi r^2}$$

Here, $V_{centrifuged}$ is the volume of the centrifuged NM suspension and r is the radius of the centrifuge tube ($r = 6.85$ mm, Figure 3).

This method gives a reliable measure of the NM number concentrations in terms of representing all size fractions of the sample and good sample recovery for pure and stable NM suspensions (Figure 4). However, the applicability of this quantitative sampling technique to measure number concentration of NMs in complex, natural media is likely to be a challenge. The method also requires further development to assess NM mixtures in the environment or to distinguish NMs from aggregates. Depending on the properties of NM samples (e.g. surface charge or hydrophobicity), materials other than positively charged poly-L-lysine may be required to suitably functionalize the mica substrate for good substrate-sample adhesion. Additionally, ultracentrifugation may not be sufficient to force low density NMs like polystyrene to the substrate. Ongoing investigations are being directed to address these potential issues for the quantitative sample preparation technique. This sampling method is also applicable for the TEM.

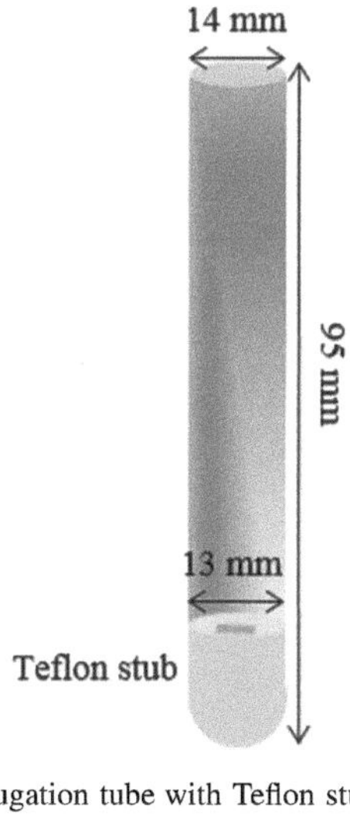

FIGURE 3 Beckman ultracentrifugation tube with Teflon stub to hold TEM/AFM grid.

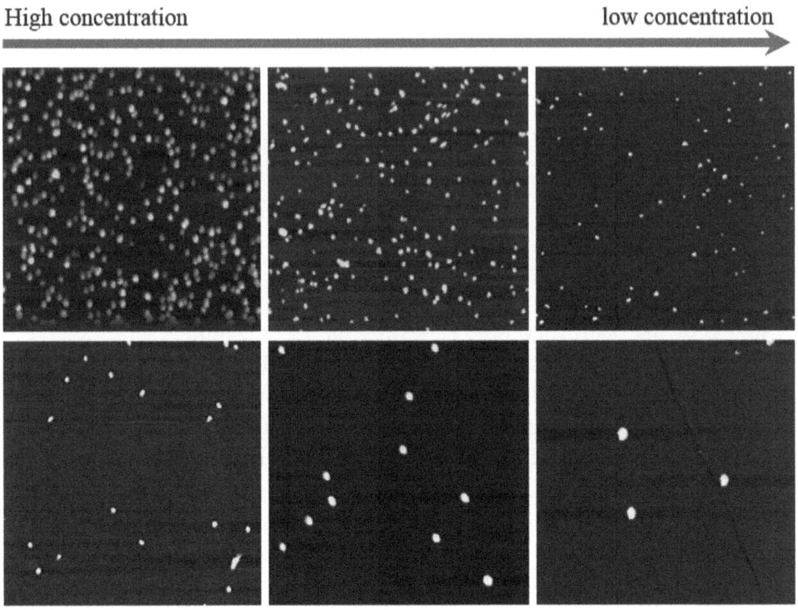

FIGURE 4 AFM micrographs (2 × 2 μm) of polyvinylpyrrolidone-coated Au NMs in $CaCl_2$ show decrease in number of NMs recovered on AFM substrate with decrease in NM mass concentrations, using the quantitative sampling technique.[70] *Reproduced by permission of the Royal Society of Chemistry. Copyright: Env. Sci. Process. Impact.*

AFM imaging of the NMs is based on the measurement of forces between the sharp AFM tip and the sample substrate. Typically, the NM samples can be imaged in ambient air or liquid environments. However, the NM samples are prone to perturbations in the liquid cells, and minimal differences in size and morphology are observed between the two modes for both natural and manufactured NMs. Therefore, the easy-to-use ambient mode is preferred for NM number size analysis.[72] Secondly, the AFM imaging mode plays a significant role in number concentration measurement. In the contact mode, the AFM tip scans the sample in contact at all times. The resulting frictional forces can cause displacement of NMs from the mica substrate or sample damage. Alternatively, the tapping or intermittent mode places the tip in contact with the sample for a high resolution, prior to oscillating off from the surface to prevent sample damage from friction. This mode is found to underestimate the height of soft materials like the DNA and polystyrene NMs, likely due to material deformation from high tip-sample interaction forces. In a noncontact mode, the cantilever oscillates slightly above its resonant frequency to probe the sample from a few nanometres above the surface. This nonperturbing mode is considered suitable for accurate measurements of NM number concentrations because it is less likely to induce sample changes.

However, the AFM is a nonelement specific measurement technique. The AFM can achieve a lower size limit of 0.1 nm only in the Z direction (image height). In comparison, the TEM can obtain this size resolution (0.1 nm) in the X and Y directions, but not Z. Therefore, the two techniques can be complementary in measuring the number concentration of different NMs.[62,73] The need to assess NMs with TEM is presented below.

5.1.2 Transmission Electron Microscopy

TEM is used to measure the number concentration of NM samples with a sufficient combination of size and electron density.[74] Typically, several TEM images are acquired from different sections of the NM sample for a good representation.[75] A large number of NMs (>100) from the images are analyzed to obtain a number average NM size. Similar to AFM, the NM sample preparation techniques affect the accuracy of TEM measurements. For example, air drying a drop of NM suspension on the TEM grid induces meniscus-based artifacts and changes in NM properties from the increased salt concentration. Additionally, concentration of the solution onto the TEM grid limits the ability to distinguish the NMs from impurity or salt contributions. Therefore, ultrathin sample preparation protocols were developed for detecting NM morphology in dense aggregates. Typically, a sample droplet is fast-frozen or added to a portion of Nanoplast resin, prior to sectioning and placing on the TEM grid.[76,77] However, these methods show inadequate surface coverage for a statistically representative result. Additionally, environmental samples such as aquatic colloids and biological specimens like NMs within cells require additional fixation steps to preserve the integrity of the samples.[74] Chemical fixation using glutaraldehyde or formaldehyde is required for some tissue specimens (e.g. alveolar). An alternate physical or cryofixation is available for larger cell samples. One approach is to freeze the water in the sample for arresting biological activity, but ice formation within the sample may lead to artifacts or sample damage. Therefore, a widely used sample preparation route is to dehydrate the specimen via freeze substitution or freeze drying. Based on a comparative study, fast-freezing of the sample droplet on the TEM grid followed by sublimation of water was found to give reproducible results.[7] However, the recent quantitative sample preparation technique reported for the AFM is likely best suitable to measure NM number concentrations on the TEM.[70]

The advantage of TEM over other methods is the high resolution achievable. Additionally, the crystal phase of the sample can be determined using electron diffraction (ED) patterns. Single-crystalline NM samples show well-defined points in ED, compared to concentric rings and diffused concentric rings for polycrystalline and amorphous NMs, respectively. The TEM also gives a rough chemical composition of the NM sample via energy dispersive spectroscopy (EDS) where the EDS plots the characteristic X-ray count versus

X-ray energy to show sample chemical composition of the NMs. Therefore, the TEM is a significant analytical tool for NM samples. However, the high expense, time and expertise required to operate TEM are potential limitations of the technique.

5.1.3 Nanomaterial Tracking Analysis

NTA is a recently developed method to indirectly measure NM number concentration in liquid dispersions.[30] It is a single particle technique for measuring the number concentrations of NMs because it tracks the Brownian movement of each individual NM in an ensemble to provide a real-time analysis. The laser beam is first refracted through a prism to obtain an incident beam with high power density and reduced profile. The NM solution in the sample flow cell scatter the incident beam in a manner such that it can be easily detected with an optical microscope. A charge-coupled device camera subsequently captures a video of the NMs in motion. The mean squared displacement is determined for each individual NM trajectory on a frame-by-frame basis with the NTA software. These values are used to calculate the diffusion coefficient and hydrodynamic diameter of the NMs, based on a modified Stokes–Einstein equation. The obtained size distribution provides a good measure for the NM number concentrations because the NTA technique relies on counting individual NMs. Additionally, the samples are prepared via simple dilutions. Fluorescent labelling of the NMs allows NTA to measure NM number concentrations in more complex media, but the method is limited to fluorophores that can be excited by the applied NTA laser.[78] Recent reports suggest the possibility to detect NM number concentrations in undiluted biological or environmental media with a lower wavelength laser source in the NTA.[79]

NTA is unsuitable for detecting small NM fractions. It also has a limited capacity to detect weakly scattering samples such as polymeric NMs. The upper size limit for NTA is typically governed by the diffusion/sedimentation of NMs. Therefore, the NM density greatly affects this value where the low density NMs show greater upper size limits, compared to the high density NMs. Additionally, the number size and concentration measurements are biased towards heavier or larger NMs. The detection limit for sample concentration is typically 10^{10}–10^{12} particles L^{-1} for a 30–60 s measurement span. Improving the optics could potentially enhance detection. The assumptions of a noninteracting system under infinite dilution and spherical NM shapes also limit the reliability of the method. It is challenging to distinguish different NMs in complex media with NTA because the method is not chemically specific.

5.2 Single Particle ICP-MS

The emerging SP-ICP-MS technology was designed to separately measure individual NMs from background including ionic components at realistic concentrations.[39] The principle is to detect one NM as a single pulse in each reading cycle. Sufficient NM dilution and data acquisition frequency is required for suitable vapourization of the sample in the ICP to prevent analysis of multiple NM events. Here, the frequency of pulses provides a direct measure of the number concentration of NMs. Compared to the dissolved ions with averaged pulse intensity, the pulse intensity associated with NMs is proportional to the atoms in the NM. The peak height is a measure of NM size. Several algorithms exist to separate the continuous signal of the dissolved analytes from the NMs. This method was first used to characterize Au NMs by Degueldre et al.[80] Currently, the SP-ICP-MS is a promising analytical tool to quantify number concentration of NMs at environmentally relevant concentrations.[81] The technique has a detection limit of 10^4 NMs L^{-1}. SP-ICP-MS can measure CNT concentrations by detecting the metal impurities in CNTs.[82] Normal mode ICP-MS cannot detect carbon-based NMs. SP-ICP-MS is being increasingly applied for wastewater analysis and assessing urban run-off. However, a potential limitation of SP-ICP-MS is the inability to measure smaller NMs. NMs smaller than the lower size limit are accounted as dissolved metals in SP-ICP-MS, inducing a bias in the number concentration and number size measurements. This could have serious implications for the EU definition of nanosized materials because the biased number concentration measurements may exclude several materials from being considered as nano. Additionally, the lower size limit of SP-ICP-MS is element-specific. Typically, a minimum NM size of $\sim$18 nm is required to generate measurable number of sample ions/atoms for plasmonic NMs, but it is higher for other NMs. Polydisperse NMs and heteroaggregates are highly challenging to measure in the SP-ICP-MS. Well-established protocols for different NM samples are still in progress. For example, the uncertainty in nebulization efficiency was recently minimized via analyte partitioning for better accuracy in NM number concentration measurement.[83]

The suitability of each method to measure NM number concentration is briefly summarized in Table 4.

In summary, number concentration of NMs is a relatively new but important metric to analyze eco(toxicological) phenomenon related to smaller NM fractions e.g. aggregation/disaggregation. The SP-ICP-MS is a promising technique to measure NM number concentration in terms of distinguishing individual NMs from background or ionic components, but is still in the research and developmental stage. Therefore, the well-established AFM and TEM techniques can serve as complementary methods to best represent NM number concentrations, given appropriate sample preparation routes are adopted. Ultracentrifugation of the NM samples onto a surface functionalized substrate was found to be the best suitable sample preparation technique for true representation of the NMs for the number concentration measurements.

TABLE 4 Measurement Techniques for Nanomaterial Number Concentration

Method	Lower Size Limit	Higher Size Limit	Lower Concentration Limit	Upper Concentration Limit	Comments
SP-ICP-MS	Element-dependent, e.g. ~20 nm for Ag NMs[20]	Element dependent (~80 nm for Ag NMs), relatively newer technique, well-established protocols and detection limits in progress	10^5 particles L^{-1}	10^8 particles L^{-1}	Multielement analysis at one time is challenging, accuracy limited by nebulizer efficiency and counting algorithms,[81] NMs lower than detection limit are measured as dissolved metal, limiting application
AFM	0.1 nm	2 μm	$\sim10^9$ particles L^{-1}	$\sim10^{15}$ particles L^{-1}	Recommended technique for measuring NM number concentration, quantitative sample preparation method enhances reliability[70]
TEM	0.1 nm	2 μm	$\sim10^9$ particles L^{-1}	$\sim10^{15}$ particles L^{-1}	Compared to the AFM, can provide information on NM composition, high vacuum system required, careful sample preparation critical to reduce measurement artifacts
NTA	10–20 nm for strongly scattering NMs like Au, element-dependent	1 μm, diffusion/ sedimentation dependent i.e. element-specific	10^{10} particles L^{-1}	10^{12} particles L^{-1}	Rapid and robust technique, less reliable because larger or highly scattering NMs skews measurement, cannot distinguish NMs in complex media

SP-ICP-MS, single particle–inductively coupled plasma–mass spectrometry; AFM, atomic force microscope; TEM, transmission electron microscope; NTA, nanomaterial tracking analysis; NM, nanomaterial.

6. CONCLUSION

Measuring the concentration of NMs using the three most commonly used dose metrics is a key to improve the fundamental understanding of NM toxicity, their risk assessment and the enforcement of NM regulation. However, best metric may be different when studying different processes. A definite answer of best dose metrics is still elusive and different metrics might be appropriate to describe different environmental processes. NM mass concentration is the traditional metric to assess risk. It is the best metric to represent sedimentation of NMs from heteroaggregation or homoaggregation in the environment. Relatively more established measurement techniques are used for mass concentration. Comparatively, the methods to measure NM number concentrations are less developed methods. The number concentration metric is best suitable to analyze NM aggregation. Even fewer developed methods are reported for measuring surface area concentration of NMs. This relatively new metric is considered best in representing processes related to small NM fractions like contaminant sorption. The three concentration metrics will be helpful in combination for an overall understanding and regulation of environmental and toxicity effects related to NMs.

REFERENCES

1. Wiesner MR, Plata DL. Environmental, health and safety issues: incinerator filters nanoparticles. *Nat Nanotechnol* 2012;**7**:487–8.
2. Nowack B, Bucheli TD. Occurrence, behavior and effects of nanoparticles in the environment. *Environ Pollut* 2007;**150**:5–22.
3. Hansen SF, Jensen KA, Baun A. NanoRiskCat: a conceptual tool for categorization and communication of exposure potentials and hazards of nanomaterials in consumer products. *J Nanopart Res* 2014;**16**(2195):1–25.
4. Lead JR, Wilkinson KJ. Aquatic colloids and nanoparticles: current knowledge and future trends. *Environ Chem* 2006;**3**:159–71.
5. Auffan M, Rose J, Bottero JY, Lowry GV, Jolivet JP, Wiesner MR. Towards a definition of inorganic nanoparticles from an environmental, health and safety perspective. *Nat Nanotechnol* 2009;**4**:634–41.
6. Baalousha M. Aggregation and disaggregation of iron oxide nanoparticles: influence of particle concentration, pH and natural organic matter. *Sci Total Environ* 2009;**407**:2093–101.
7. Domingos RF, Baalousha M, Ju-Nam Y, Reid MM, Tufenkji N, Lead JR, et al. Characterizing manufactured nanoparticles in the environment: multimethod determination of particle sizes. *Environ Sci Technol* 2009;**43**:7277–84.
8. Lead JR, Smith E. *Environmental and human health impacts of nanotechnology*. Wiley; 2009.
9. Quik JTK, Stuart MC, Wouterse M, Peijnenburg W, Hendriks AJ, van de Meent D. Natural colloids are the dominant factor in the sedimentation of nanoparticles. *Environ Toxicol Chem* 2012;**31**:1019–22.
10. AshaRani PV, Mun GLK, Hande MP, Valiyaveettil S. Cytotoxicity and genotoxicity of silver nanoparticles in human cells. *ACS Nano* 2009;**3**:279–90.

11. Chang X, Vikesland PJ. Uncontrolled variability in the extinction spectra of C-60 nanoparticle suspensions. *Langmuir* 2013;**29**:9685–93.
12. Ramachandran G, Paulsen D, Watts W, Kittelson D. Mass, surface area and number metrics in diesel occupational exposure assessment. *J Environ Monitor* 2005;**7**:728–35.
13. Teeguarden JG, Hinderliter PM, Orr G, Thrall BD, Pounds JG. Particokinetics in vitro: dosimetry considerations for in vitro nanoparticle toxicity assessments. *Toxicol Sci* 2007;**95**: 300–12.
14. Baalousha M, Lead JR. Nanoparticle dispersity in toxicology. *Nat Nanotechnol* 2013;**8**:308–9.
15. Bednar AJ, Poda AR, Mitrano DM, Kennedy AJ, Gray EP, Ranville JF, et al. Comparison of on-line detectors for field flow fractionation analysis of nanomaterials. *Talanta* 2013;**104**: 140–8.
16. Schwartz J, Marcus A. Mortality and air-pollution in London – a time-series analysis. *Am J Epidemiol* 1990;**131**:185–94.
17. Handy RD, von der Kammer F, Lead JR, Hassellöv M, Owen R, Crane M. The ecotoxicology and chemistry of manufactured nanoparticles. *Ecotoxicology* 2008;**17**:287–314.
18. Baun A, Hartmann NB, Grieger K, Kusk KO. Ecotoxicity of engineered nanoparticles to aquatic invertebrates: a brief review and recommendations for future toxicity testing. *Ecotoxicology* 2008;**17**:387–95.
19. Madden AS, Hochella MF, Luxton TP. Insights for size-dependent reactivity of hematite nanomineral surfaces through Cu^{2+} sorption. *Geochim Cosmochim Acta* 2006;**70**:4095–104.
20. Hassellöv M, Readman JW, Ranville JF, Tiede K. Nanoparticle analysis and characterization methodologies in environmental risk assessment of engineered nanoparticles. *Ecotoxicology* 2008;**17**:344–61.
21. Lead JR, Hamilton-Taylor J, Davison W, Harper M. Trace metal sorption by natural particles and coarse colloids. *Geochim Cosmochim Acta* 1999;**63**:1661–70.
22. Hofmann T, Wendelborn A. Colloid facilitated transport of polychlorinated dibenzo-p-dioxins and dibenzofurans (PCDD/Fs) to the groundwater at Ma Da area, Vietnam. *Environ Sci Pollut Res* 2007;**14**:223–4.
23. Auffan M, Rose J, Proux O, Masion A, Liu W, Benameur L, et al. Is there a Trojan-horse effect during magnetic nanoparticles and metalloid cocontamination of human dermal fibroblasts? *Environ Sci Technol* 2012;**46**:10789–96.
24. Zhang X, Sun H, Zhang Z, Niu Q, Chen Y, Crittenden JC. Enhanced bioaccumulation of cadmium in carp in the presence of titanium dioxide nanoparticles. *Chemosphere* 2007;**67**:160–6.
25. Wooley PH, Nasser S, Fitzgerald RH. The immune response to implant materials in humans. *Clin Orthop Relat Res* 1996;**326**:63–70.
26. Olivier V, Duval JL, Hindie M, Pouletaut P, Nagel MD. Comparative particle-induced cytotoxicity toward macrophages and fibroblasts. *Cell Biol Toxicol* 2003;**19**:145–59.
27. Chen KL, Elimelech M. Aggregation and deposition kinetics of fullerene (C-60) nanoparticles. *Langmuir* 2006;**22**:10994–1001.
28. Baalousha M, Nur Y, Roemer I, Tejamaya M, Lead JR. Effect of monovalent and divalent cations, anions and fulvic acid on aggregation of citrate-coated silver nanoparticles. *Sci Total Environ* 2013;**454**:119–31.
29. *BSI*. Terminology of nanomaterials. In: *PAS*, **136**; 2007.
30. ASTM. Active Standard ASTM E2834 Developed by Subcommittee: E56.02. *Standard guide for measurement of particle size distribution of nanomaterials in suspension by nanoparticle tracking analysis (NTA)*, vol. 14; 2012.

31. Commission Recommendation of October 18, 2011 on the Definition of Nanomaterial Text with EEA Relevance.
32. Hartmann G, Schuster M. Species selective preconcentration and quantification of gold nanoparticles using cloud point extraction and electrothermal atomic absorption spectrometry. *Anal Chim Acta* 2013;**761**:27–33.
33. Kulthong K, Srisung S, Boonpavanitchakul K, Kangwansupamonkon W, Maniratanachote R. Determination of silver nanoparticle release from antibacterial fabrics into artificial sweat. *Part Fibre Toxicol* 2010;**7**:1–9.
34. Gagne F, Turcotte P, Gagnon C. Screening test of silver nanoparticles in biological samples by graphite furnace-atomic absorption spectrometry. *Anal Bioanal Chem* 2012;**404**:2067–72.
35. Mitrano D, Ranville J, Neubauer K, Thomas R. Field-flow fractionation coupled with ICP-MS for the analysis of engineered nanoparticles in environmental samples. *Spectroscopy* 2012;**27**:36–44.
36. Fabricius A, Duester L, Meermann B, Ternes TA. ICP-MS-based characterization of inorganic nanoparticles—sample preparation and off-line fractionation strategies. *Anal Bioanal Chem* 2014;**406**:467–79.
37. Allabashi R, Stach W, de la Escosura-Muniz A, Liste-Calleja L, Merkoci A. ICP-MS: a powerful technique for quantitative determination of gold nanoparticles without previous dissolving. *J Nanopart Res* 2009;**11**:2003–11.
38. Unrine JM, Hunyadi SE, Tsyusko OV, Rao W, Shoults-Wilson WA, Bertsch PM. Evidence for bioavailability of Au nanoparticles from soil and biodistribution within earthworms (*Eisenia fetida*). *Environ Sci Technol* 2010;**44**:8308–13.
39. Laborda F, Jimenez-Lamana J, Bolea E, Castillo JR. Selective identification, characterization and determination of dissolved silver(I) and silver nanoparticles based on single particle detection by inductively coupled plasma mass spectrometry. *J Anal At Spectrom* 2011;**26**: 1362–71.
40. Chithrani BD, Ghazani AA, Chan WCW. Determining the size and shape dependence of gold nanoparticle uptake into mammalian cells. *Nano Lett* 2006;**6**:662–8.
41. Tejamaya M, Roemer I, Merrifield RC, Lead JR. Stability of citrate, PVP, and PEG coated silver nanoparticles in ecotoxicology media. *Environ Sci Technol* 2012;**46**:7011–7.
42. Hamner K, Maye MM, Ash DL, Page AF. Quantification of gold nanoparticles using the Thermo Scientific NanoDrop 2000 Spectrophotometer. *Thermo Sci* 2012;**1**.
43. Haiss W, Thanh NTK, Aveyard J, Fernig DG. Determination of size and concentration of gold nanoparticles from UV-Vis spectra. *Anal Chem* 2007;**79**:4215–21.
44. Park HG, Oh JH, Lee JS. Assembly-based titration for the determination of monodisperse plasmonic nanoparticle concentrations using DNA. *Anal Chem* 2011;**83**:4989–95.
45. Phenrat T, Saleh N, Sirk K, Tilton RD, Lowry GV. Aggregation and sedimentation of aqueous nanoscale zerovalent iron dispersions. *Environ Sci Technol* 2007;**41**:284–90.
46. Phenrat T, Saleh N, Sirk K, Kim HJ, Tilton RD, Lowry GV. Stabilization of aqueous nanoscale zerovalent iron dispersions by anionic polyelectrolytes: adsorbed anionic polyelectrolyte layer properties and their effect on aggregation and sedimentation. *J Nanopart Res* 2008;**10**:795–814.
47. Zook JM, Long SE, Cleveland D, Geronimo CLA, MacCuspie RI. Measuring silver nanoparticle dissolution in complex biological and environmental matrices using UV-visible absorbance. *Anal Bioanal Chem* 2011;**401**:1993–2002.
48. Reipa V, Purdum G, Choi J. Measurement of nanoparticle concentration using quartz crystal microgravimetry. *J Phys Chem B* 2010;**114**:16112–7.

49. Hevroni A, Golan H, Fialkov A, Rahinov I, Tsionsky V, Markovich G, et al. In situ measurement of the mass concentration of flame-synthesized nanoparticles using quartz-crystal microbalance. *Meas Sci Technol* 2011;**22**:115102.
50. Janshoff A, Galla HJ, Steinem C. Piezoelectric mass-sensing devices as biosensors – an alternative to optical biosensors? *Angew Chem Int Ed* 2000;**39**:4004–32.
51. Yildirimer L, Thanhb NTK, Loizidoua M, Seifalian AM. Toxicological considerations of clinically applicable nanoparticles. *Nano Today* 2011;**6**:585–607.
52. Oberdorster G, Oberdorster E, Oberdorster J. Nanotoxicology: an emerging discipline evolving from studies of ultrafine particles. *Environ Health Perspect* 2005;**113**:823–39.
53. Resano M, Mozas E, Crespo C, Briceno J, del Campo Menoyo J, Belarra MA. Solid sampling high-resolution continuum source graphite furnace atomic absorption spectrometry to monitor the biodistribution of gold nanoparticles in mice tissue after intravenous administration. *J Anal At Spectrom* 2010;**25**:1864–73.
54. Novak S, Drobne D, Goobic M, Zupanc J, Romih T, Gianoncelli A, et al. Cellular internalization of dissolved cobalt ions from ingested $CoFe_2O_4$ nanoparticles: in vivo experimental evidence. *Environ Sci Technol* 2013;**47**:5400–8.
55. Hoque ME, Khosravi K, Newman K, Metcalfe CD. Detection and characterization of silver nanoparticles in aqueous matrices using asymmetric-flow field flow fractionation with inductively coupled plasma mass spectrometry. *J Chromatogr A* 2012;**1233**:109–15.
56. Rashkow JT, Patel SC, Tappero R, Sitharaman B. Quantification of single-cell nanoparticle concentrations and the distribution of these concentrations in cell population. *J R Soc Interface* 2014:11.
57. Chalew TEA, Ajmani GS, Huang H, Schwab KJ. Evaluating nanoparticle breakthrough during drinking water treatment. *Environ Health Perspect* 2013;**121**:1161–6.
58. Van Doren EAF, De Temmerman P, Francisco MAD, Mast J. Determination of the volume-specific surface area by using transmission electron tomography for characterization and definition of nanomaterials. *J Nanobiotechnol* 2011;**9**:1–8.
59. Baalousha M, Lead JR. Rationalizing nanomaterial sizes measured by atomic force microscopy, flow field-flow fractionation, and dynamic light scattering: sample preparation, polydispersity, and particle structure. *Environ Sci Technol* 2012;**46**:6134–42.
60. Rasmussen K, Mast J, Temmerman P-JD, Verleysen E, Waegeneers N, Steen F, et al. *Titanium dioxide, NM-100, NM-101, NM-102, NM-103, NM-104, NM-105: characterization and physico-chemical properties*. Luxembourg. 2014.
61. Baalousha M, Ju-Nam Y, Cole PA, Gaiser B, Fernandes TF, Hriljac JA, et al. Characterization of cerium oxide nanoparticles-part 1: size measurements. *Environ Toxicol Chem* 2012;**31**:983–93.
62. Linsinger T, Roebben G, Gilliland D, Calzolai L, Rossi F, Gibson N, et al. *Requirements on measurements for the implementation of the European Commission definition of the term "nanomaterial"*. Luxemburg. 2012.
63. Janz A, Koeckritz A, Yao L, Martin A. Fundamental calculations on the surface area determination of supported gold nanoparticles by alkanethiol adsorption. *Langmuir* 2010;**26**:6783–9.
64. Gadogbe M, Ansar SM, He G, Collier WE, Rodriguez J, Liu D, et al. Determination of colloidal gold nanoparticle surface areas, concentrations, and sizes through quantitative ligand adsorption. *Anal Bioanal Chem* 2013;**405**:413–22.
65. Davis PJ, Gallegos DP, Smith DM. Rapid surface-area determination via NMR spin-lattice relaxation measurements. *Powder Technol* 1987;**53**:39–47.

66. Fairhurst D, Prescott S. The use of nuclear magnetic resonance as an analytical tool in the characterisation of dispersion behaviour. *Spectrosc Eur* 2011;**23**:13–6.
67. Park JY, Ramachandran G, Raynor PC, Olson Jr GM. Determination of particle concentration rankings by spatial mapping of particle surface area, number, and mass concentrations in a restaurant and a die casting plant. *J Occup Environ Hyg* 2010;**7**:466–76.
68. Vecchio G, Galeone A, Brunetti V, Maiorano G, Sabella S, Cingolani R, et al. Concentration-dependent, size-independent toxicity of citrate capped AuNPs in *Drosophila melanogaster. Plos One* 2012;**7**:e29980.
69. Commission E. Commission recommendation of October 18, 2011 on the definition of nanomaterial text with EEA relevance. *Off J Eur Union* 2011;**54**:L275/38.
70. Baalousha M, Prasad A, Lead JR. Quantitative measurement of nanoparticle size and number concentration from liquid suspensions by atomic force microscopy. *Environ Sci Process Impact* 2014;**16**:1338–47.
71. Lead JR, Muirhead D, Gibson CT. Characterization of freshwater natural aquatic colloids by atomic force microscopy (AFM). *Environ Sci Technol* 2005;**39**:6930–6.
72. Baalousha M, Lead JR. Size fractionation and characterization of natural aquatic colloids and nanoparticles. *Sci Total Environ* 2007;**386**:93–102.
73. Delvallee A, Feltin N, Ducourtieux S, Trabelsi M, Hochepied J-F. Comparison of nano-particle diameter measurements by atomic force microscopy and scanning electron microscopy. In: *16th International congress of metrology*, vol. 06007; 2013. p. 1–6.
74. Muhlfeld C, Rothen-Rutishauser B, Vanhecke D, Blank F, Gehr P, Ochs M. Visualization and quantitative analysis of nanoparticles in the respiratory tract by transmission electron microscopy. *Part Fibre Toxicol* 2007;**4**:1–17.
75. Williams DB, Carter CB. *Transmission electron microscopy.* New York and London: Plenum Press; 1996.
76. Leppard GG. Nanoparticles in the environment as revealed by transmission electron microscopy: detection, characterisation and activities. *Curr Nanosci* 2008;**4**:278–301.
77. Perret D, Leppard GG, Muller M, Belzile N, Devitre R, Buffle J. Electron-microscopy of aquatic colloids – non-perturbing preparation of specimens in the field. *Water Res* 1991;**25**:1333–43.
78. Dragovic RA, Gardiner C, Brooks AS, Tannetta DS, Ferguson DJP, Hole P, et al. Sizing and phenotyping of cellular vesicles using nanoparticle tracking analysis. *Nanomed-Nanotechnol* 2011;**7**:780–8.
79. Roding M, Deschout H, Braeckmans K, Rudemo M. Measuring absolute number concentrations of nanoparticles using single-particle tracking. *Phys Rev E* 2011;**84**:031920 (1–9).
80. Degueldre C, Favarger PY, Rosse R, Wold S. Uranium colloid analysis by single particle inductively coupled plasma-mass spectrometry. *Talanta* 2006;**68**:623–8.
81. Laborda F, Jimenez-Lamana J, Bolea E, Castillo JR. Critical considerations for the determination of nanoparticle number concentrations, size and number size distributions by single particle ICP-MS. *J Anal At Spectrom* 2013;**28**:1220–32.
82. Reed RB, Goodwin DG, Marsh KL, Capracotta SS, Higgins CP, Fairbrother DH, et al. Detection of single walled carbon nanotubes by monitoring embedded metals. *Environ Sci Process Impact* 2013;**15**:204–13.
83. Tuoriniemi J, Cornelis G, Hassellöv M. Improving the accuracy of single particle ICPMS for measurement of size distributions and number concentrations of nanoparticles by determining analyte partitioning during nebulisation. *J Anal At Spectrom* 2014;**29**:743–52.

Chapter 6

Nanomaterials: Dispersion, Dissolution and Dose

Nicole Hondow, Andy Brown and Rik Brydson[1]
Institute for Materials Research, School of Chemical and Process Engineering, University of Leeds, Leeds, West Yorkshire, UK
[1]*Corresponding author: E-mail: r.m.drummond-brydson@leeds.ac.uk*

Chapter Outline

1. INTRODUCTION

Engineered nanomaterials (NM) have been the focus of much recent research with a wide range of potential applications proposed in biomedicine[1] (as imaging probes for cells and tissues,[2] drug delivery[3] and sensing of target molecules[4]), environmental remediation,[5] catalysis,[6] data storage[7] and lighting technology.[8] Commercial exploitation is however still limited, in part because uncertainty remains regarding the risk to human health and the environment.[9] To address this uncertainty we need to identify and quantify the hazard that engineered NM present in a particular application or exposure scenario, also importantly we need to understand how a NM type may be transformed by dispersion in biological or environmental media. The latter consideration has significant implications for the ecotoxicity of engineered NM.[10,11]

Frontiers of Nanoscience, Vol. 8. http://dx.doi.org/10.1016/B978-0-08-099948-7.00006-3

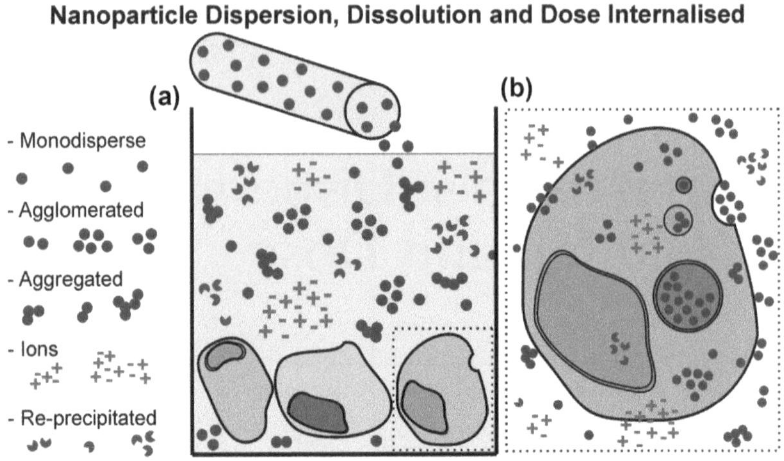

FIGURE 1 Schematic of the nanomaterial (NM) Dispersion-Dissolution-Dose problem for a simple case of NM exposure to an in vitro cell line. (a) Stock NM may be monodispersed in a simple fluid but once placed in a complex (ionic) medium required for cell culture, we need to readdress the NM agglomeration and aggregation alongside any NM dissolution and reprecipitation. (b) We then need to consider how these different material forms are internalized and at what number or concentration, plus whether they then remain stable in the particular cellular environment in which they are encapsulated.

When placed in biological or environmental media, engineered NM can disperse individually, agglomerate, aggregate, sediment, dissolve (totally or only partially) and even reprecipitate or complex as a second phase (Figure 1(a)). There is a sensitive interrelationship between the transformation of NM as they enter the environment, the transport, final fate and any toxicity resulting from the particles. The exact nature of the dispersion of engineered NM in environmental media (i.e. the chemical and structural form and any agglomeration or aggregation) combined with any resultant dissolution or transformation in the medium will strongly influence the final dose (or perhaps more appropriately the final number), which is internalized, including its exact location, by a cell or organism (Figure 1(b)). We will refer to this concept as the dispersion – dissolution – dose (DDD) relationship. Ultimately it is the dose internalized, i.e. the number not simply the concentration of NM internalized including the chemical and structural form that is of critical interest since, in terms of toxicology, it is this which acts within a biochemical context.[12]

This chapter discusses each aspect of the DDD relationship in turn and provides examples highlighting how the individual quantities and the interrelationship may be faithfully determined and potentially modelled. Only relatively few types of engineered NM are currently used in high volume or in significant numbers of products and therefore these present the largest environmental concern: they include silver, titanium dioxide, zinc oxide, silica and carbon-based particles or nanomaterials.[13] We will not however restrict our discussion to just these NM types because we wish to highlight the emerging techniques that can be used to quantify the DDD relationship. We also recognize that the field of environmental geochemistry has been using similar methods to characterize NM that occur naturally in the environment for many decades and has already made a significant contribution to our understanding of how engineered NM will disperse in environmental media.[14] Similarly, there is a wealth of experience that already exists in detecting NM in cells and organisms; for example, the use of NM composed of atomically heavy materials as labels in electron microscopy for the localization of specific proteins in eukaryotic cells is well established and has made a significant contribution to our understanding of cellular function.[15]

One of the key messages of this chapter is that there are many characterization techniques, which can provide quantitative measures of NM dispersion, dissolution and dose provided that the user is aware of the chemistries of the NM, of any coatings or dispersants and of the media they are dispersed in and crucially, also understands the limitations of the characterization techniques to be applied. It is this understanding of the analytical limitations that has led the field of nanotoxicology to take on ex situ methods that capture or freeze the composition and spatial distribution of NM in a given matrix or that separate and analyze individual components.[16] There is an obvious push for reliable in situ characterization, however, we will see that this is challenging given the complex environments and relatively low concentrations that engineered NM can be distributed.[13] Thus at the risk of repetition, we believe the most pressing point to stress in this chapter is the DDD relationship and how quantification of the appropriate metric of NM number will lead to a more complete understanding of the toxicological response of a cell or organism to engineered NM exposure; there are many examples in the literature of toxicology studies, which do not take account of this concept rendering their studies virtually useless. Complete understanding can only be achieved with appropriate characterization and expertise already discussed by Rivera Gil et al.[17] and presented in some detail here.

The ultimate goal of establishing the DDD relationship is to link to biological endpoints in order to provide a mechanistic understanding of the outcomes of exposure to NM thus for example, avoiding misinterpreting artefactual responses.[18] Measures of NM number distributions at given time points, when dispersed and/or transformed in media and similarly for particular locations within cells and organisms can be used to also establish the statistics behind NM and NM agglomerates traversing biological barriers and undergoing intracellular NM agglomerate merging events. From these statistics, mathematical transfer functions can be developed to provide predictive modelling of NM toxicity/pharmacology. We believe that establishing in this way, the exposure pathways that control the hazard presented by a given NM type will enable a more complete risk-benefit analysis of NM exposure than is currently available.

2. DETECTION

To begin investigating and quantifying the DDD relationship, one must first be able to detect (identify and confirm) engineered NM when dispersed in a particular medium. For environmental media, this is no simple task given the low-predicted exposure concentrations of engineered NM (ng−μg/L).[13] In fact this presents an immediate concern regarding the sensitivity and accuracy of a particular NM concentration measurement, and we will take care in the following discussion to consider these factors. For example, natural NM and colloids (i.e. minerals, clays and products of bacteria) are a consistent feature of low-temperature surface waters and ground waters and distinguishing these from engineered NM is considered a grand challenge in econanotoxicity.[13] This can in principle be achieved by isotope tracer studies[19,20] if the NM contain elements that are uncommon in the environment, e.g. semiconductor quantum dots or Ag, by the use of acid digestion combined with atomic spectrometries such as inductively coupled plasma-mass spectroscopy (ICP-MS) or atomic absorption spectroscopy (AAS).[21] While these techniques have very sensitive elemental detection limits, they cannot identify the size and form of the NM, essential information if number distribution of NM is required. We also note that in the field of geochemistry, the use of trace element ratios is an established technique for identifying the provenance of minerals.[22] We envisage such an approach could also be used to highlight the presence of engineered NM in complex media.

Over the years there have been examples of detection and relatively detailed geochemical analyses of NM characteristics in natural environments, with analytical electron microscopy emerging as a powerful tool for the identification and characterization of trace metal contaminants on and within nanoparticulate phases.[23,24] In general however, electron microscopy and specifically transmission electron microscopy (TEM) analysis is not used as a stand-alone characterization technique but more often it provides nanoscale information to support bulk analytical data at the single NM level with a chemical sensitivity of 0.1 atom %. For example, the identification of Cr^{3+} substitution in synthetic haematite NM within samples from the Ajka red mud spill site, Hungary, used X-ray absorption spectroscopy (XAS) to identify the signature and valence state of Cr contamination but required a combination of TEM-selected area electron diffraction and energy dispersive X-ray (EDX) analysis to colocalize the Cr to the haematite NM (Figure 2).[25]

These multitechnique approaches are equally applicable to a biological setting. For example, evidence for a Si-specific intracellular mechanism for Al detoxification in aquatic snails, involving regulation of orthosilicic acid [$Si(OH)_4$] was supported with intracellular analysis by TEM to confirm the formation of hydroxyaluminosilicate NM agglomerates in exposed snails.[26] Finally, as already highlighted, Ag NM are now extensively produced and used in consumer products, and sulphidation is thought to be the dominant

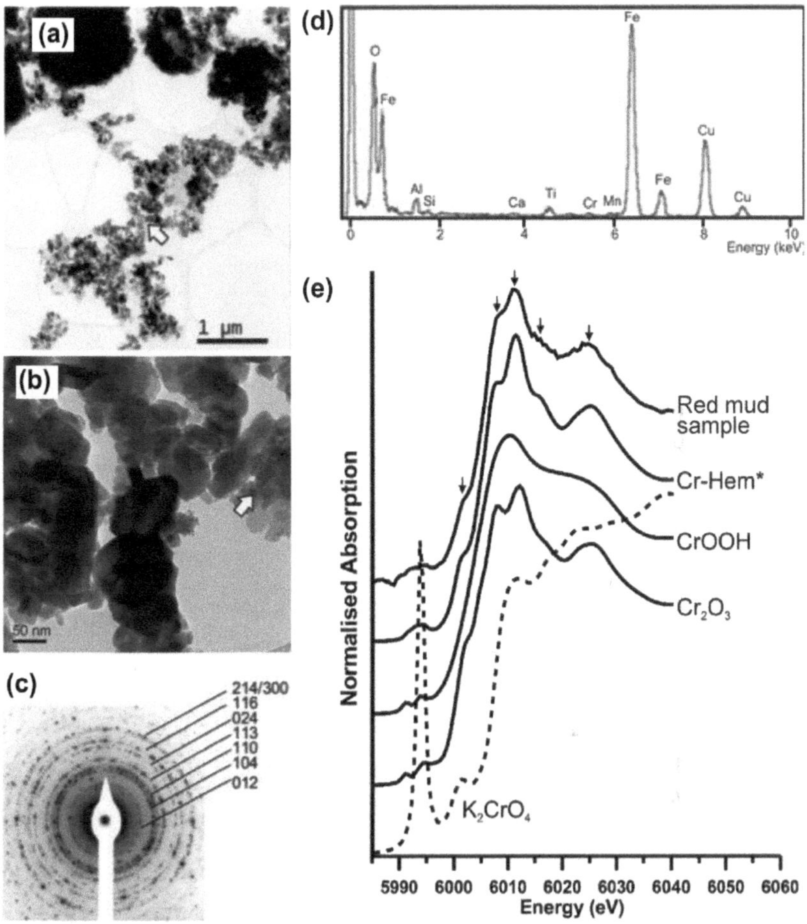

FIGURE 2 Identification of Cr^{3+} substitution in haematite within samples from the Ajka red mud spill site, Hungary.[25] (a) Low-resolution transmission electron microscopy (TEM) image of the red mud particles from the spill site. (b) Higher magnification TEM image of area highlighted by white arrow in (a). (c) Polycrystalline selected area electron diffraction pattern indexed to haematite. (d) Energy dispersive X-ray spectrum collected from the point shown with a white arrow in (b), confirming the additional presence of Ti and Cr in the haematite phase. (e) Normalized Cr *K*-edge X-ray absorption near edge structure spectra collected from the red mud sample and standards (three Cr^{3+} and one Cr^{6+}) showing the red mud fraction is consistent with the Cr^{3+}-substituted haematite and Cr_2O_3 standards. *Adapted from Burke et al.*[25]

degradation product as they are released into the environment[27]; silver sulphide (α-Ag_2S) NM have been identified in the final stage sewage sludge materials of a full-scale municipal wastewater treatment plant using analytical TEM.[28]

Having established the actual presence and form of engineered NM when suspended in a given media, the goal must then be to quantify both the dispersion (i.e. agglomeration and sedimentation) and any dissolution which may be achieved by analysis in the complex media directly or by inference from dispersion measurements in more simple media.[29]

3. DISPERSION

The distribution of engineered NM in a simple aqueous suspension is controlled by factors such as ionic strength of the solution, pH, size and surface chemistry of the particles.[30] In biological media additional factors need to be considered; typically NM for cellular delivery are coated in a functional polymer and delivered in a liquid-containing biomacromolecules including simple amino acids, vitamins and serum proteins in addition to being buffered with ionic salts, all of which can interact to affect the hydrodynamic behaviour and bio-interaction of the particles.[31,32] The size including the degree of agglomeration or aggregation (which can be difficult to distinguish even after ultrasonication to break up agglomerates) and the surface charge plus chemistry are important factors to measure in order to understand the biological response of, for example, in vitro cells exposed to dispersed NM.[33] In this respect, it is clear that different cell types respond in different ways when exposed to aggregated NM.[34] How these findings then transfer to whole organisms is still to be fully elucidated, despite an ever growing number of in vivo studies.[35]

Dispersion of engineered NM in environmental media is potentially more complex than the biological case. The exposure concentrations of the NM are predicted to be low in, for example, surface water and/or wastewater treatment plant effluent and sludge.[36] Here there is also the presence of natural organic matter (such as humic and fulvic acids) and dissolved molecules (such as phosphates and sulphates) combined with the presence of natural mineral colloids to contend.[37] Identifying which media-NM interactions dominate under different conditions is essential for understanding and predicting the fate of these NM in the environments. As a result of this complexity, one has to be careful to probe the distribution of the engineered NM and not that of the insoluble components of the medium.

Here, we have separated measures of particle dispersion into techniques which estimate (bulk) and those which directly count (by microscopy) NM number so that we can identify the strengths and weaknesses of each approach for establishing the NM number distribution we are targeting with the DDD relationship.

3.1 Bulk Techniques

NM dispersion in liquid is most commonly measured by dynamic light scattering (DLS) as this assesses large numbers of particles in solution and provides a robust and quantitative measure for narrow particle size distributions.[38,39] DLS is not, however, able to distinguish between different types of particles in solution and importantly becomes significantly less reliable for the measurement of large particle size dispersity (because of the sixth power dependence of light scattering intensity on the sizes of the scattering particles). Accuracy is also reliant on knowing the refractive index of the NM and the viscosity and refractive index of the dispersant.[40,41] These confounding factors can render DLS ineffective for measuring accurately the dispersity of low concentrations of NM in environmental media, and it has been shown beneficial to couple it to a size fractionation method as discussed below.[16]

In addition to DLS, other techniques such as nanoparticle tracking analysis (NTA) and field flow fractionation (FFF) can be used in the measurement of primary and agglomerate or aggregated NM size distribution. Similar to DLS, NTA and FFF measure diffusion coefficients of suspended particles (and hence assume a potentially incorrect spherical particle shape) and have limited applicability at low concentrations ($\sim$1 mg/L).[42] Both techniques are regularly compared to microscopy such as TEM, often with close agreement and with any discrepancies assumed to be due to the measurement of hydrodynamic diameter or surface coatings not imaged by TEM.[43–46] In the investigation of the dispersion of ceria NM in water and fish medium, NTA has been shown to have difficulties in the measurement of large agglomerates more than 1000 nm in diameter.[41,47] The size range is further limited at the lower end,[41] with effectiveness defined by the sensitivity of the camera.[43] Farkas et al. have shown that a high sensitivity camera can both increase the size range detected (down to 30 nm from 40 nm), but also enables the differentiation of 125 nm silver NM agglomerates in media and 40 nm agglomerates in water (the low sensitivity camera identified 60 nm particle sizes in both cases).[43] NTA has been described as not as reproducible as DLS,[41] however, it does directly provide a number concentration of particles.[47]

FFF can be coupled with detectors to obtain complementary data, with numerous types of FFF separation (e.g. flow, sedimentation, etc.) combined with various detectors (UV-visible (UV-Vis) spectroscopy, ICP-MS, ICP-atomic emission spectroscopy (ICP-AES), etc.) in order to provide analysis opportunities.[48–50] The variety of combinations have limitations that depend upon the exact set-up; there is the potential for interaction between the analyte and membrane in flow-FFF, there is a lower size detection of 30 nm for sedimentation-FFF, and certain elements are not detectable by ICP-MS and ICP-AES.[49,50] A potential limitation that reduces application of both NTA and FFF to representative characterization of NM under relevant conditions is the concentration requirement, with a dilution to 10^7-10^9 particles/mL needed for NTA,[45] and dilution or preconcentration commonly required for FFF.[49,50] Domingoes et al. comment on the comparison of numerous techniques (including DLS, NTA, FFF and TEM) that there is no one ideal technique as all provide both advantages and limitations, however, the importance of representative sample preparation for electron microscopy is emphasized (e.g. prevention of drying artefacts, more of which is discussed later).[42]

Small-angle X-ray scattering (SAXS) and potentially small-angle neutron scattering can provide further liquid-based analysis on a statistically relevant number (10^5) of particles.[51,52] SAXS has been used to measure the size of Au NM, with a larger size measured in comparison to TEM as SAXS is again sensitive to the 6th power of the particle diameter (d^6).[53] Furthermore, SAXS has been used to infer the aggregation state of Ag NM synthesized using reducing and capping agents.[51] SAXS is however limited by the need to fit and model scattering data[52,53] and is often conducted at a synchrotron.[51,52] Also, scattering of NM dispersions in biological or environmental media can be dominated by the liquid (and other components in the medium), creating issues for measurement of small NM sizes and dilute concentrations.[52] Other more specialist techniques for particle sizing are disc centrifuge[54] and Coulter counters,[55] although their general applicability to analysis of NM dispersion in complex media remains to be determined.

3.2 Microscopy Techniques

As may already be evident, microscopy and in particular TEM, is a technique that can complement bulk dispersion measures such as that by DLS, and is principally used to measure the primary size of NM.[56–59] In simple monodisperse systems, there is often excellent agreement between DLS and TEM,[60] and despite the general requirements and conditions for electron microscopy (ultrahigh vacuum, thin samples) there are studies that have used TEM to confirm qualitatively the degree of NM agglomeration suggested by DLS.[61] The aggregates or agglomerates formed by NM can have complicated 3-D shapes, which are not easy to assess by DLS. For example, magnetic polymer coated iron oxide NM form linear aggregates in the presence of a magnetic field, the structure of which have been examined by TEM imaging of sections of the aggregates captured in gel and resin-embedded for imaging.[62] Furthermore, scanning electron microscopy (SEM) has been used to investigate the dispersion of TiO_2 NM agglomerates in the presence of bacteria.[63]

Analysis by TEM however must avoid using sample preparation involving dropping and drying a solution on a grid (the drop-cast method), since it induces NM agglomeration, negating quantitative sizing.[64] Imaging of NM-bearing solutions in an environmental/liquid cell within a TEM's vacuum system is being developed but is costly and currently a relatively low resolution approach.[65] Nonetheless, the prospect of real-time analysis of NM agglomeration in suspensions viewed by electron microscopy (both SEM and TEM) of liquid cells is plausible and exciting[65–68]; however, the effect of the electron beam on the sample, the membrane and the liquid must be understood before robust comparison to complex samples can be made. Alternatively, it is possible to prepare samples by plunge freezing a blotted solution on a TEM grid with sufficient speed to ensure the aqueous phase vitrifies with no significant redistribution of suspended material. This technique has been used to measure and quantify the dispersion of silica,[69] ZnO,[70,71] and polymer-coated CdTe/ZnS NM in cell delivery media.[72]

3.3 Case Studies

Mu et al. have shown by DLS and plunge freeze TEM that amorphous silica NM agglomerate in cell culture media (that is not supplemented with serum proteins) and have identified physical interactions between plasma lipid membrane models and the agglomerated NM (Figure 3).[69] They demonstrate passive cellular uptake in vitro and measure a cyto- and genotoxic response consistent with this, suggesting that the response is a result of the adhesive interaction between the cellular membrane and the agglomerated NM. In parallel, Zhang et al. have shown that the amorphous silica NM interaction with a lipid membrane is related to NM size where small NM (18 nm) cause a 'freezing effect' of otherwise fluid phospholipid bilayers that can result in membrane fracture while large particles (>78 nm) promote membrane wrapping and internalization.[73]

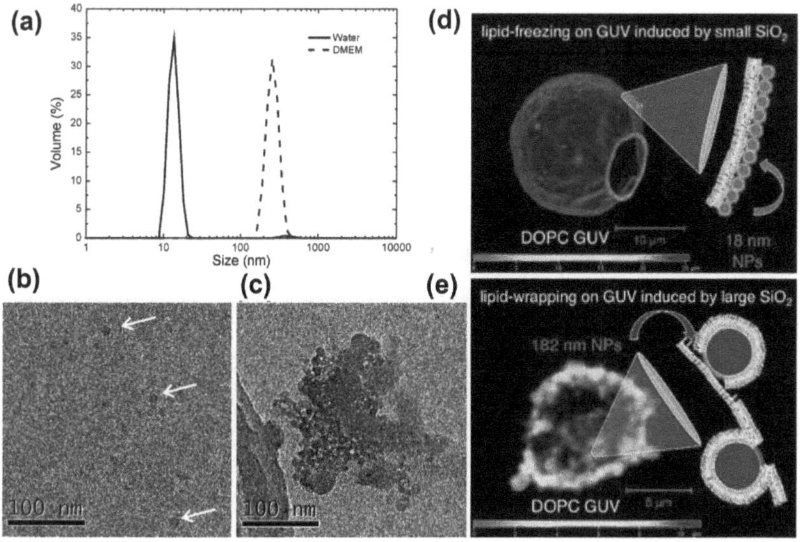

FIGURE 3 The dispersion of silica nanomaterials (NM) in water and media and the interactions with model membranes.[69,73] (a) Dynamic light scattering (DLS) of silica NM (100 μg/mL; 37 °C) indicates a monodispersion in MilliQ water (solid line) and agglomeration in Dulbecco's modified eagle medium (DMEM; dashed line). (b, c) Transmission electron microscopy images from NM in water (b) and DMEM (c) plunge frozen to retain sample dispersion correlate with DLS, arrows indicating monodisperse NM in (c). (d, e) Confocal laser scanning microscopy of giant unilamellar vesicles (GUV) annotated with a schematic of the interaction mechanism of small (d; 18 nm) silica NM causing a 'freeze effect' and large (e; >78 nm) causing membrane wrapping. *Adapted from Mu et al.*[69] *and Zhang et al.*[73]

The dispersion stability of ZnO NM in water and cell culture media has been shown to be dependent on particle shape and coating,[74] and on the additional presence of serum proteins.[70] As might be expected, specific polymer coatings designed to improve the dispersion of ZnO NM such as coating with an aliphatic polyether do produce a stable dispersion by DLS in both water and cell culture media, while uncoated ZnO NM are significantly more agglomerated and less stable regardless of size.[74] A characterization protocol for ZnO NM has been described which identifies the presence of zinc carbonate and hydroxide impurities on the surface of uncoated ZnO.[75] These impurities probably impact on the dispersion stability and may form from atmospheric contamination[75] or because of the partial solubility and reprecipitation of ZnO in water[76] or certain cell culture media such as Dulbecco's Modified Eagle Medium (DMEM) (Figure 6). Comparison between DLS and plunge freeze TEM of uncoated ZnO NM dispersed in water and serum protein supplemented water are in broad agreement; serum proteins produce a dispersion of significantly finer agglomerates by providing a size-stabilizing coating or corona on the NM (Figure 4). Uncoated ZnO NM dispersed in water have been shown to interact with phospholipid monolayers[77] and the addition of serum proteins shown to enhance synthetic membrane activity of the ZnO. The in vitro toxicity of ZnO NM dispersed in cell culture media without the addition of serum proteins is discussed later (in Section 4), however, understanding the full impact of serum proteins on toxicology assays remains an important challenge for ZnO and nanotoxicology in general.[78]

Hondow et al. demonstrate the advantage of plunge freeze TEM over DLS and NTA when assessing the distribution of a low concentration of quantum dots (10 nM) in water and a cell culture medium with 10% supplementation of serum proteins (Figure 5).[72] The TEM particle size distribution in the media that cells are exposed to can then be correlated to the endocytic uptake of the quantum dots by U-2 OS cells and specifically to the overdispersed Poisson distribution of quantum dot loaded vesicles per cell for a cell population of 10,000 measured by imaging flow cell cytometry.[79] Going further, the individual quantum dot distribution within cellular endosomes measured by TEM has been used to provide a calibration factor to transform the flow cytometry metric and obtain an internalized NM dose distribution (Figure 7).[12]

Thus, by a combination of bulk and microscopic techniques and by emerging ex situ freezing or in situ capture techniques, it is possible to capture and quantify NM dispersion in complex media, however, if the engineered NM are not chemically stable in the dispersion media then one must also consider dissolution.

4. DISSOLUTION

Determining NM dissolution and dissolution kinetics in biological and environmental media is essential when the solution chemistry of the dispersion media

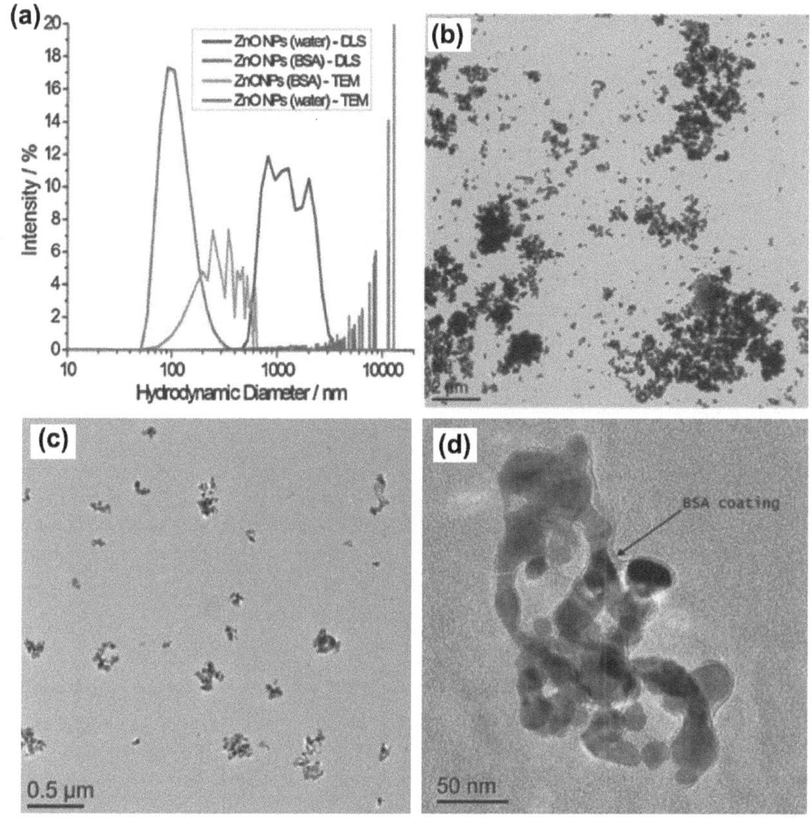

FIGURE 4 Dispersion of ZnO nanomaterials (NM) in water with and without serum (BSA; bovine serum albumin). (a) Particle size distributions for ZnO NM with and without the presence of serum obtained by dynamic light scattering (DLS) and transmission electron microscopy (TEM) (with plunge freezing of samples). (b, c) Larger agglomerates are imaged by TEM when ZnO NM are dispersed in water alone (b) than when in the presence of serum (c). (d) Higher magnification TEM image of ZnO NM suggesting the presence of a serum (BSA) coating, as indicated by the arrow. *Adapted from Churchman et al.*[70]

can significantly affect the thermodynamic equilibrium and equilibrium dynamics of the NM. Factors such as ionic strength, NM size, agglomeration state, temperature and NM surface coating are all critical. Recent research has demonstrated that some NM types such as Ag and ZnO are likely to undergo dissolution and or phase transformations in either aqueous environments or cell culture media.[80–82] These phase transformations are driven by differences in the equilibrium solubility of a NM type when dispersed in different media and these typically can result in the partial reprecipitation to nanoparticulate compounds of lower solubility, for example, the case of the transformation of Ag to Ag_2S NM in typical environmental media.[27] Dissolution of ZnO NM in water[83] and reprecipitation as zinc carbonates and phosphates in different cell culture media has

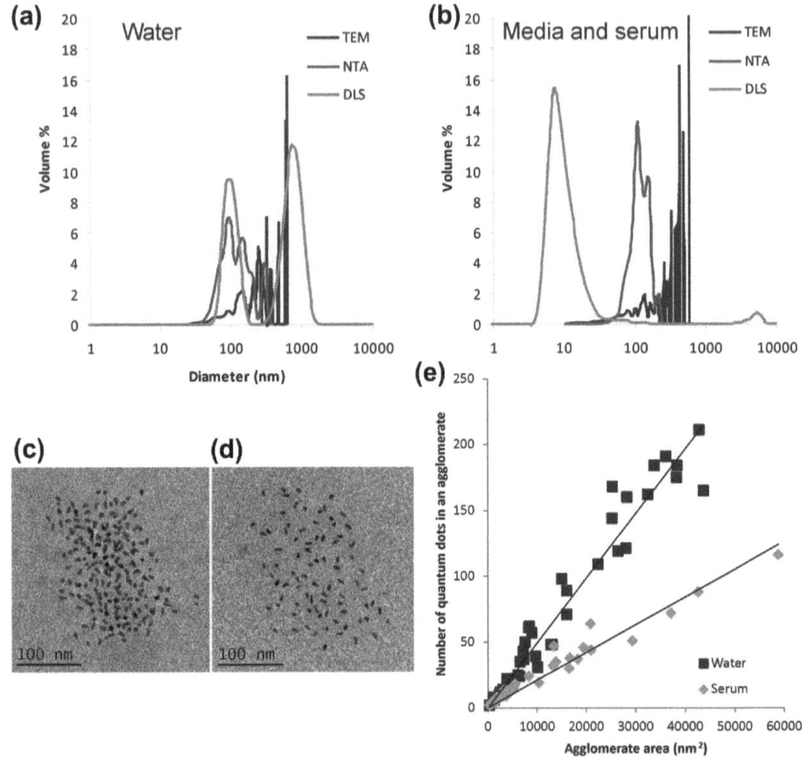

FIGURE 5 Comparison of quantum dot dispersion in water to dispersion in media and serum. (a) Dynamic light scattering (DLS), nanoparticle tracking analysis (NTA) and transmission electron microscopy (TEM) analysis of quantum dot dispersion in water, showing good agreement between the methods. (b) Disagreement between the three analysis methods when analyzing the quantum dots dispersion in media and serum. While DLS has only probed the distribution of the serum ($\sim$8 nm, as compared to the 10 nm primary particle size of the quantum dots) and the NTA did not identify the larger agglomerates, TEM of samples prepared by the plunge-freezing route enabled the measurement of just the nanomaterials (NM) dispersion state. (c, d) TEM images from samples prepared by the plunge-freezing route when dispersed in water (c) and media and serum (d), indicating not only agglomerate size, but also the number of particles per agglomerate and therefore the agglomerate packing fraction, which may be a more relevant metric for cellular uptake (Figure 7). (e) Examination of the packing fraction shows a reduction in the number of particles per agglomerate when dispersed in media and serum, which may have implications for the number of NM internalized by a cell or organism; this analysis is only possible with TEM. *Adapted from Hondow et al.*[72]

been established; with the phosphate concentration of the media having a major effect on the equilibrium Zn^{2+} levels retained in solution,[76,84] and more recently atmospheric buffering has been shown to influence that equilibrium.[74]

Knowledge of the dispersion state is also vital when examining the dissolution of NM, for example, the magnitude of dissolution of α-FeOOH (goethite) nanorods is severely quenched (by orders of magnitude) when aggregated,[85] and aggregated PbS (galena) NM (240 nm) having dissolution characteristics closer to microparticles (3 μm) than monodispersed NM (14 nm).[86]

NM dissolution also impacts on cellular uptake, and dissolution can continue once within a cell. While specific uptake mechanisms of cells and higher organisms will not be discussed here, it is important to be aware that the uptake pathway may have an effect on the final intracellular location and form of the NM. Different entry mechanisms lead to different transport pathways in a cell and different NM endpoints; for example, caveolae-mediated endosomes contain fewer enzymes than clathrin-mediated endosomes and will therefore degrade inorganic materials less.[88]

Photoluminescent quantum dots, for example, highlight that even intracellular dissolution assays must reflect the conditions of exposure since the factors that affect solubility (pH, ionic strength, concentration) vary between different cellular components (e.g. early endosome, lysosome, cytosol),[89] where the intracellular pH can range between 7 and 5.[90] Investigations using fluorescence measurements to study the time- and pH-dependent release of Cd^{2+} from Cd-containing quantum dots have been conducted in buffers to match the pH of the cytoplasm, endosome and lysosome.[91] This raises the question of what is an effective simulated biofluid, as it is not only pH but also the protein, enzyme, reactive oxygen species, etc. content that potentially can have an effect. Various different assays are used; Mancini et al. use hypochlorous acid and hydrogen peroxide as reactive oxygen species that are produced during cellular metabolism,[92] while Metz et al. have developed a biomimetic assay to simulate oxidative environmental conditions, which mimic the reactive oxygen species produced by fungi.[93]

Once again, we will discuss the strengths and weaknesses of various techniques for measurement of dissolution, and ultimately we will conclude that multiple and potentially correlative measures of dissolution for a given NM dispersion are the most effective way to address this complex issue.

4.1 Atomic Spectroscopies

Dissolution is commonly measured by atomic spectroscopic techniques such as ICP-MS, ICP-AES (also known as ICP-optical emission spectroscopy, ICP-OES) and AAS, the details of which have been discussed in previous chapters. In general, these techniques are very sensitive to elemental content, however, they do not determine chemical speciation and therefore require some prior separation of dissolved species from NM (or from cells for in vitro, or tissue for in vivo investigations), which can confound the analysis:

- NM are commonly separated from any dissolved component using techniques such as dialysis (for example, across porous membranes with a molecular weight cut off on the order of 12 kDa), centrifugation (generally with a relative centrifugal force maximum value between 10,000 and

200,000 g) and filtration (typically with filters down to 3 kDa or 20 nm equivalent pore sizes),[27,87] however, numerous sample preparation issues have been noted.
 - Instances have been reported both where filters have not adequately removed smaller NM (as determined by DLS and TEM of a centrifuged and/or filtered supernatant),[94] and where a known metal content has been lost or not detected (e.g. in the calibration of a soluble control).[93]
 - Issues have been reported with centrifugation, with even harsh conditions (150,000 g) resulting in incomplete separation of dissolved and ultrafine nanoparticulate ZnO (as determined by the TEM imaging of NM in the supernatant).[95]
- Most common atomic spectroscopic techniques cannot distinguish between different oxidation states present in solution, compounding the difficulty in distinguishing between dissolved and pristine NM elements. Although as already discussed ICP-MS can be coupled to FFF to analyze the elemental composition of different size fractions of agglomerated NM.[50]
 - Single particle ICP-MS can potentially distinguish between particulate and dissolved fractions, but requires dilution to the point that, as the name suggests, only a single particle is analyzed per measurement. This has been used in the differentiation of Ag NM and Ag^+ in solution.[96]
 - ICP-MS has been useful in the tracking of CdTe-containing quantum dots in vivo, with acid digested tissue analyzed at different time points.[29,97] While ICP-MS cannot distinguish between Cd originally bound in a quantum dot and the biologically reactive 'free' Cd^{2+}, it is possible to compare the Cd:Te ratio, something which is not possible with many other NM systems (e.g. metal and metal oxides).

Dissolution is rarely measured by atomic spectroscopic techniques alone, with other techniques (including sizing, imaging and electrochemical techniques) used to investigate unaltered particles,[98,99] and to quantify the limits in any applied separation techniques.[100]

4.2 Particle-Based Spectroscopies

The stability and dissolution of NM can also be probed by techniques related to the intrinsic properties of the NM of interest, such as UV-Vis and fluorescence spectroscopies. While useful for in vitro investigations, these techniques are limited for in vivo investigations, where general background levels can be both high and variable, limiting the use of, for example, fluorescence spectroscopy in the examination of quantum dots in mice.[29] NM dissolution is inevitably a dynamic process and under biological conditions dispersion and exposure time become important factors.[89] Dissolution dynamics under environmental conditions are perhaps less critical when there is sufficient time for the NM to reach thermodynamic equilibrium, however, for some NM types dissolution is considerably slowed, for example, by the presence of coatings.[98,101] Nonetheless, the importance of dispersion media and control of exposure conditions is critical because subtle changes in the choice of medium or dissolution assay can significantly impact upon the amount and rate of some NM dissolution, for example, ZnO dissolution is sensitive to the CO_2 buffering of cell culture media.[74]

When attempting to replicate environmental or biological exposure in the laboratory, it is common to use model NM. However, care needs to be taken as it has been shown that the NM synthesis method can have an effect on dissolution. For example, the dissolution of silica NM in biological fluids was examined using fluorescence spectroscopy by monitoring a fluorescent dye incorporated during synthesis into the core of the NM, however, the dissolution was subsequently found to be synthesis method dependent.[102]

Fluorescence and UV-Vis spectroscopies of model CdSe/ZnS quantum dots have been used to indicate the stabilizing effect on the NM dispersion of serum proteins that supplement cell culture media. This temporal stability can be demonstrated by comparison to quantum dots dispersed in serum-free suspensions in water and different media and can be attributed to the formation of protein corona on the NM.[103] The observed changes in fluorescence and UV-Vis intensities of the serum-free suspensions indicate aggregation or surface degradation of the quantum dots.

Flow cytometry is commonly used for the in vitro investigations of fluorescent NM exposures and has highlighted the importance of cell type used, with a loss of fluorescence intensity shown to be both experiment and cell-type specific.[104] This is particularly evident in a study of quantum dot uptake in two types of mouse embryonic cells (stem cells and fibroblasts), where only one cell type (stem cells) showed a large decrease in quantum dot fluorescence over 72 h.[105] After inhibiting cell proliferation (to exclude dilution of fluorescence due to mitosis or excretion) similar results were obtained by flow cytometry, along with supporting TEM imaging, confirming the selective dissolution of NM in the stem cells.

Translating these conditions to in vivo exposure becomes even more challenging because the previously mentioned factors that affect solubility are even more variable.[106] As such, some recent reports have shown that in vitro results have not matched those in vivo, emphasizing the need for identification of the relevant media and assays.[29] In particular, artefactual responses to in vitro assays need to be identified and eliminated, for example, artefactual activation of the inflammasome platform in response to particle gorging by cells in vitro that is unlikely to occur in vivo,[18] or false positives in colourimetric or fluorescence-based toxicity assays resulting from interaction of the NM with the dyes or fluorophores.[78]

4.3 Electrochemical Analysis

Provided the precise NM being investigated dissolve to ionic components, electrochemical analysis can provide another avenue for dissolution studies. An advantage of electrochemical techniques, which include the use of ion selective electrodes (ISE) and absence of gradients and Nernstian equilibrium stripping (AGNES), is direct measurement with potentially no need for separation of the dispersed and dissolved components.[87,107] ISE have been used in the study of Ag NM in water and bacterial growth media (ferric citrate) showing that a greater percentage of the NM dissolve in growth media than in water (28 compared to 13%) and that the majority of the dissolved silver is present as ammonia–silver complexes.[100] This type of quantitative analysis of dissolution has also been applied to ZnO NM using AGNES and will be discussed in more detail later in this chapter (Section 4.5 and Figure 6).[74] These techniques are not universal because first, there are only a limited number of ISE (and interferences) and secondly, AGNES is only relevant for elements that alloy with mercury electrodes i.e. has a limited ability to detect metals in the context of toxicity.[108]

4.4 Microscopy

Various imaging techniques can be used to examine NM exposed to biological or environmental media, dissolution assays or even after in vitro cellular uptake. All offer the advantage of being able to resolve (partially) transformed NM (or at least NM agglomeration), however, analysis of individual NM requires electron microscopy (or scanning probe microscopy) and imaging, in general, in high vacuum conditions that require specific sample preparation, which limits the temporal resolution of the sampling. The effects of intracellular chemistry on ZnO nanowires have been examined using confocal microscopy, with the cells loaded with an intracellular Zn^{2+} marker, which fluoresces green upon binding with Zn^{2+}.[109] Due to the resolution limitations of light microscopy, SEM and TEM have both been used to study the morphology of ZnO after exposure to biofluids or inside cells after in vitro exposure.[74,109,110] Furthermore, the structural (e.g. atomic lattice imaging and electron diffraction) and analytical capabilities (e.g. EDX spectroscopy) of the TEM have been utilized in the analysis of NM retained after CdSe/ZnS quantum dots were exposed to dissolution tests, with transformation to amorphous selenium-containing particles being identified.[93]

Electron microscopy is commonly used in the structural and chemical analysis of pristine NM, and there is further potential for high resolution TEM analysis of NM subjected to dissolution assays. Scanning TEM combined with parallel electron energy loss spectroscopy (STEM-EELS) has previously shown the nonuniform ZnS shell distribution on 4.3 nm CdSe/ZnS quantum dots[111] and has been used to image and analyze NM containing both inorganic

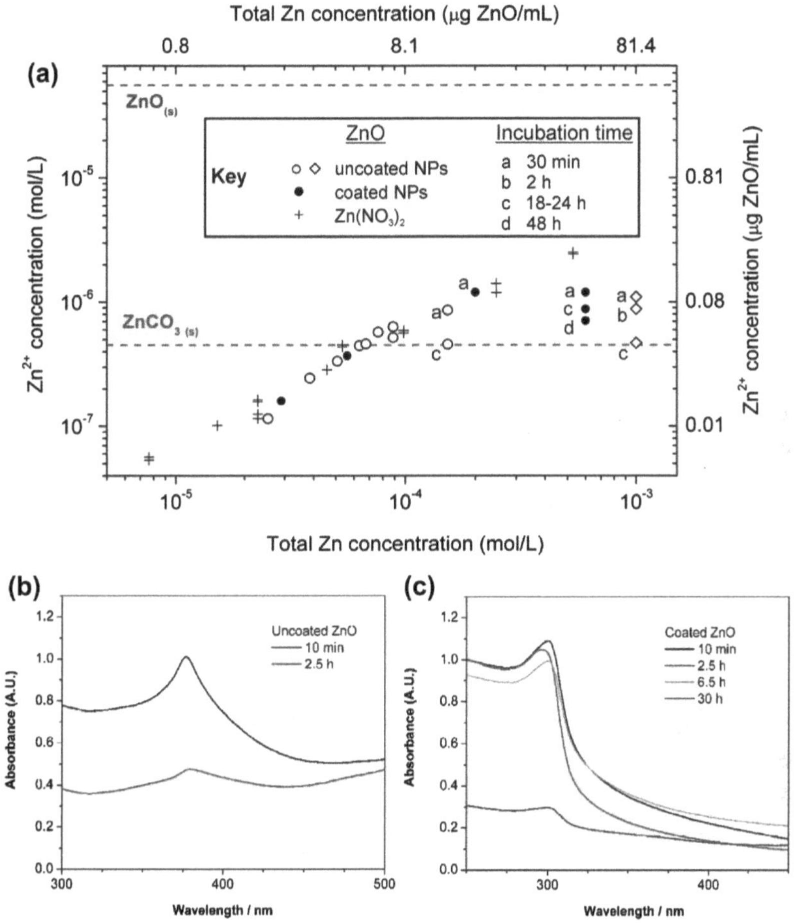

FIGURE 6 Investigation of coated and uncoated ZnO nanomaterials (NM) in Dulbecco's modified eagle medium (DMEM) at 37.0 °C under 5% CO_2/95% N_2 atmosphere with pH = 7.68. (a) Free Zn^{2+} concentration (in mol/L – left and bottom, and its equivalence in μg ZnO/mL – right and top) measured by absence of gradients and Nernstian equilibrium stripping. $Zn(NO_3)_2$ remained constant for at least 6 h. The red (dark grey in print versions) and blue (grey in print versions) dashed lines denote the free Zn^{2+} concentration in equilibrium with solid $ZnCO_3$ (smithsonite) and ZnO (zincite) bulk phases, respectively. All the samples approached the zinc carbonate equilibrium solubility level over time and this is equivalent to the complete dissolution of 5.5 μg/mL of ZnO in this medium. (b) UV-Vis spectrum of uncoated ZnO NM (from experiment shown in (a) at 50–80 μg-ZnO/mL) in DMEM showing the near complete dissolution within 2.5 h. (c) UV-Vis spectrum of coated ZnO NM (from experiment shown in (a) at 50–80 μg-ZnO/mL) in DMEM showing significantly slower dissolution. *Adapted from Mu et al.*[74]

and organic components at sub-nanometre resolution.[112] STEM-EELS does not have the sensitivity associated with techniques such as atomic spectroscopy, and there are potential sample related issues (e.g. the requirement for preparation of a thin specimen and the possibility of alteration under vacuum and upon electron irradiation), however, it can provide nano and even atomic scale spatial analysis of location, structure and oxidation state.[113] Specialist imaging and spectroscopy can also be performed on thicker samples (50 nm–20 μm) by synchrotron-based scanning transmission X-ray microscopy (STXM), with the associated XAS providing lower spatial resolution but improved element sensitivity than STEM-EELS.[114,115]

Clearly, there is no one simple technique to answer all questions regarding dissolution. All of the above techniques have limitations; atomic spectroscopy cannot (in general) distinguish between dissolved and solid NM forms, analysis with fluorescence spectroscopy and electrochemical techniques is only applicable for certain NM types (and adding fluorescent probes may change the NM protein corona, therefore changing the dispersion and dissolution properties) and imaging techniques are resolution (light) or volume and time (TEM) limited. Thus, applying multiple, and potentially correlative, techniques is often the answer, especially given the complexities added by experimental design including the appropriate choice of medium, buffer or dissolution assay plus in vitro and in vivo models.

4.5 Case Study

An example of this multitechnique approach to assessing dissolution is given by Mu et al. who established ZnO NM solubility in media (DMEM without the addition of serum proteins) held under identical conditions to those employed for cell culture (37 °C, 5% CO_2 and pH 7.68).[74] They showed by ICP-MS and AGNES that for NM at concentrations up to 5.5 μg ZnO/mL, dissolution is complete (with the majority of the soluble zinc complexed to dissolved ligands in the medium), taking *ca.* 1 h for uncoated and *ca.* 6 h for polymer-coated particles to dissolve (Figure 6). AGNES and thermodynamic modelling show that above 5.5 μg/mL results are consistent with the reprecipitation of zinc carbonate, keeping the solubilized zinc fixed to 67 μM of which only 0.45 μM existed as free Zn^{2+}, i.e. not complexed to dissolved ligands. At these relatively high concentrations, NM with an aliphatic polyether-coating showed slower dissolution (i.e. slower free Zn^{2+} release) and reprecipitation kinetics compared to uncoated NM, that required more than 48 h to approach thermodynamic equilibrium.

The work of Mu et al. suggested that cytotoxicity and DNA damage of three epithelial cell lines exposed to uncoated ZnO NM presented in serum-free cell culture media are dominated by ZnO dissolution and reprecipitation as zinc carbonate. They also showed by TEM combined with monitoring of intracellular Zn^{2+} concentrations and ZnO-NM interactions with model lipid membranes, that an aliphatic polyether-coat on ZnO NM increases cellular uptake, enhancing toxicity by enabling intracellular dissolution and release of Zn^{2+}. Meanwhile, needle-like NM morphologies were shown to enhance toxicity because of incomplete cellular uptake resulting from an inferred lower rate of dissolution combined with the size and aspect ratio of the needles. Therefore, to limit toxicity, ZnO NM with nonacicular morphologies and coatings that only weakly interact with cellular membranes were recommended, for example, Hsiao and Huang have used a relatively inert titanium oxide coating on ZnO to reduce its cytotoxicity.[116]

Ultimately, establishing and quantifying the form and number in which NM are presented to cells or organisms is a prerequisite to fully assessing and understanding any biological interaction with NM. Relating NM exposure levels to internalized NM dose is an important step for the development of the field, and while the former can be obtained with the techniques described above or in the preceding chapters, the latter will be discussed next.

5. DOSE

Obtaining quantitative measures of NM uptake is an absolute requirement of nanotoxicology. There has been much debate over the appropriate dose metric (particle number, mass, surface area, etc.),[117–119] with specific examples where surface area[120] or particle number[121] have been deemed more appropriate. It is, however, perhaps more important to relate the exposure level to the internalized NM dose, and while mass concentrations are often convenient for exposure levels and most interactions are proportional to surface area, in order to quantify the internalized dose only the particle number truly reflects NM uptake.[122] Furthermore, it is not only the number of internalized NM in a cell, but the distribution of NM internalized per cell that must be quantified. Measurement of NM dose can be achieved indirectly using atomic spectroscopy of cells or tissues following acid digestion/lysis of the cells or tissue and an assumption regarding NM size, or via other indirect methods such as magnetic measurements, radiolabeling or more directly by microscopy.

5.1 Atomic Spectroscopy

The atomic spectroscopy techniques used for the study of dissolution (ICP-MS, ICP-AES and AAS) can provide bulk measures of NM dose after a known amount of cells (e.g. measured using a haemocytometer) are digested or lysed, commonly through the addition of acids. An intracellular molar concentration of 0.1 nM of Au NM was measured after the addition of 6 nM of cell-penetrating peptide modified gold NM into HeLa cells,[123] however, as a bulk measurement this assumes a homogenous NM distribution across the cell population. ICP-AES has been used in the quantification of positively and negatively charged surface modified iron oxide NM in A549 cells, with a higher uptake noted for the positively charged particles.[124] Imaging techniques however showed that the positively charged particles agglomerate and many were only attached to the surface of cells.[124] Similarly, ICP-MS was used to quantify the uptake of gold NM and nanorods by HeLa cells with conversion of measured numbers of atoms to estimated numbers of NM to give an NM dose,[125] assuming all the Au atoms/particle were internalized (not surface attached) and also assuming a homogenous NM size and cellular uptake.

Atomic spectroscopy techniques have the highest elemental detection sensitivity (parts per billion) but lack spatial resolution, even including the techniques with 2-D mapping capability such as laser ablation ICP-MS, which has a resolution of 5–20 μm,[126] and nano-secondary ion mass spectrometry with ~100 nm resolution.[127] Atomic spectroscopy has been used in the bulk measurement of Cd levels in the kidneys and liver of mice exposed to Cd-containing quantum dots,[128] but there can be difficulties when attempting to quantify NM dose above that of background elements, for example, the detection of superparamagnetic iron oxide NM above iron levels already present in tissue.[129]

5.2 Particle Spectroscopy

Light microscopy and associated fluorescent NM detection techniques offer an avenue for studying live cells, but there is often not sufficient resolution to identify individual NM. Confocal laser scanning microscopy (CLSM) has been used to visualize and quantify the uptake of 50 nm and 1 μm fluorescent polystyrene NM into A549 cells, although the 50 nm NM were rarely observed and a lack of uptake had to be supported by TEM.[130] CLSM has also been used to confirm the presence of positively charged iron oxide NM agglomerates attached to the surface of A549 cells, adding uncertainty to the uptake quantified by ICP-AES.[124] Fluorescent-based microscopy (and new opportunities opened up by super resolution microscopy) is limited by the requirement of a fluorescing sample, with the possibility of photobleaching or destabilization of the fluorophore and possible interference effects from the biological background.[122,126]

The emission of fluorescent dye-modified polymer NM has also been utilized in the quantification of NM dose.[131] Here, a microplate reader was used to measure fluorescence per in vitro well and having determined the number of cells per well (using a Coulter counter), Ferrari et al. estimated the number of NM per cell using the known number of dye molecules per particle. Confocal microscopy was used to confirm internalization, and a homogeneous NM dose was assumed.

A useful technique in the examination and quantification of dose distribution in a cell population is flow cytometry. Intercell dose variability has been noted with both quantum dots and fluorescent dye-modified polymer NM,[79,131] but often the resolution is limited to analysis of NM-loaded vesicles rather than individual NM.[79] Using flow cytometry, it is also possible to both identify (and therefore remove from analysis) dead cells (through addition of a molecular probe)[131] and to study the inheritance of NM through differing cell generations.[132]

5.3 Microscopy

Electron microscopy allows direct imaging and analysis on the scale of both cells and NM. A relative or absolute quantification of NM distribution can be achieved through a stereological approach,[133] and random sampling in defined compartments,[89,90] despite the challenge of collecting sufficient results to have statistical significance, as discussed in Mayhew et al.[133] and Elsaesser et al.[122] The analytical capabilities of the TEM (e.g. EDX and EELS) can be utilized to confirm the identity of NM from similarly sized cellular features.[134]

A recent report demonstrates the quantification of gold NM distributions in whole cells by STEM tomography (achieved by lowering the level of membrane staining of the cells and limiting the tilt range of image acquisition). The method was originally developed for the analysis of low density lipoprotein uptake by macrophage cells but would be equally applicable to the quantification of NM uptake, although the resolution of cellular ultrastructure is limited.[135] A more conventional approach is to extract quantifiable information by TEM from many, single thin-sections using a sampling technique to make representative measurements from a specimen such as a pellet of cultured cells. This stereological approach has been used to identify the cellular location and number of individual gold particles taken up by fibroblast cells.[136,137] There is however debate over the relative merits of the sampling method.[138,139] Others have used optical dark field microscopy in conjunction with TEM to quantify gold NM uptake into epithelial cells, measuring uptake as a function of NM shape, stabilizing agent and surface charge.[140]

The above approaches are challenging because of the time required for processing or analysis and the high number of cells required for meaningful statistics. Time-consuming NM counting from electron micrographs can be reduced if they are used to calibrate a lower resolution high-throughput technique such as cytometry. Indeed precise quantification of NM internalization has been demonstrated by correlation of flow cell cytometry to confocal light microscopy.[141]

5.4 Correlative Analysis

Cytometry correlation to electron microscopy has also been achieved by monitoring the distribution of semiconductor quantum dots internalized in U-2 OS cells by endocytosis (Figure 7).[12] Probability distribution functions of NM-loaded vesicles per cell and NM per vesicle were obtained by correlating

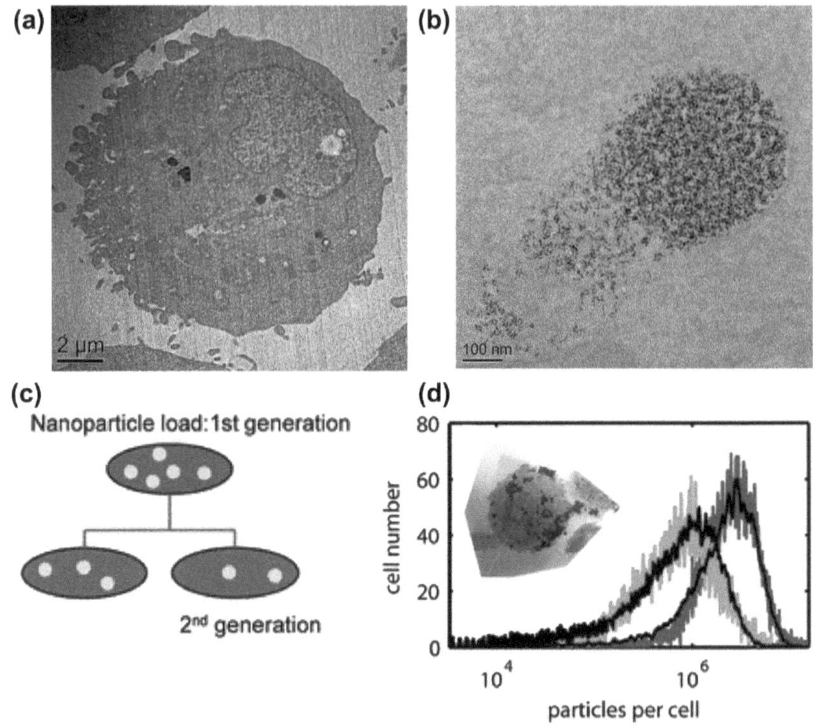

FIGURE 7 The use of correlative transmission electron microscopy (TEM) and flow cytometry to measure an internalized dose of quantum dots. (a, b) TEM images from a single (~100 nm thick) section showing a cell (a) and a vesicle from within this cell containing quantum dots. (c) Schematic showing the partitioning of quantum dots during cell division, where vesicle (yellow (white in print versions)) inheritance is near symmetric but highly variable vesicle loading leads to a highly asymmetric particle dose in daughter cells. (d) Measured (coloured) and predicted (black) frequency distributions of internalized quantum dots per cell at 24 h (red (dark grey in print versions) and 48 h (green (light grey in print versions)), agreement is achieved by predicting a slight asymmetry of inheritance of vesicles in the model (with the known, inhomogeneous quantum dot loading), rather than an equal split of quantum dot numbers. Inset is a 2-D representation of a 3-D reconstruction from serial block face scanning electron microscope of a cell with the internalized quantum dots visible. *Adapted from Summers et al.*[12]

vesicle counts in optical (fluorescence) images of a large population of cells (~10,000) to NM count in TEM images of thin sections of a sub population of the cells (~100). TEM is limited by the volume which can be analyzed,[122] with the thickness of a thin section being one to two orders of magnitude smaller than the average diameter of a cell. It is however possible to subject the resin-embedded sample prepared for TEM analysis to serial sectioning in the SEM and collect full 3-D information regarding the NM intracellular distribution in a subset of cells. Using serial block face SEM to measure the distribution of NM-loaded vesicles through whole cells, the calibration factor required to transform the cytometry metric to total particle dose per U-2 OS cell was estimated (mean dose 2.4×10^6).[12] This type of correlative approach to generate NM number or dose taken up by a cell population, including the identification of the form of its vesicular packaging and the (re-)distribution upon cell division will enable the accurate prediction of dose kinetics for specific NM systems.

5.5 Other Methods

The flow cytometry to TEM link can only be established by the use of fluorescent NM, and other NM types must be quantified using their intrinsic properties. For example, the quantification of superparamagnetic iron oxide NM uptake for which the dose has been quantified by a number of means, including:

- By 'magnetophoresis', where the motion of a cell under a magnetic field is quantitatively related to the physicochemical properties of the cell-label complex, and is therefore directly related to the number of phagocytized NM.[142]
- By the magnetic spectroscopy of Brownian motion which has been used for the monitoring of NM binding to cell surface proteins and uptake by cancer cells.[143]

Noninvasive MRI has been used to quantify the dose of quantum dot and superparamagnetic iron oxide NM in lipoproteins (micelles) delivered to live animals (mice),[144] however, there are reports detailing the limitation of in vivo quantification of iron as most tissue contains iron and will therefore give a variable background signal.[129] To overcome this issue, Freund et al. have developed a new preparation method for the radiolabeling (^{59}Fe) of superparamagnetic iron oxide NM, with ^{59}Fe monitored inside mice using a large volume Hamburg whole body radioactivity counter.[129] Radiolabeling has sub-ppb (1–5 pg) detection, but thick tissue autoradiography is limited to a spatial resolution of approximately 1 μm.[126]

Synchrotron methods offer alternative routes for quantification of internalized NM dose, with STXM or X-ray fluorescence microscopy (XFM) and X-ray absorption near edge structure spectroscopy offering routes to obtain quantitative maps and local chemistry information.[115,126] XFM has been used in the study of Co-doped ZnO (25 nm primary particle size, 1.4 μm agglomerate size) in THP monocyte-derived macrophages.[145] Synchrotron XFM allows analysis of 1000 cells/h and can produce statistically representative histograms of cellular NM loading. The overall dose can be determined by ICP-MS (1.5×10^4 NM), but location determined in parallel by a combination of focused ion beam SEM to ablate part of the cell and XRF to quantify NM levels at the plasma membrane/in early endosomes.[145]

If only bulk techniques or techniques related to NM properties (e.g. fluorescence, magnetics) are to be used in the quantification of internalized dose, there needs to be confidence that all measured particles are internalized (i.e. those surface bound NM or those in the cell suspension are efficiently removed). Density gradient centrifugation and flow cytometry are two possible approaches that can be used to remove extracellular NM.[146] Furthermore, there needs to be a measure of the fluorescence after dispersion, as fluorescent NM may present a change in quantum yield or optical properties (e.g. quenching or enhancement) due to agglomeration/aggregation.[146]

6. FINAL SUMMARY

To understand and quantify the hazard that engineered NM suspended in biological and environmental media present to cells and organisms, we need to characterize the transformation, transport and final fate of the NM i.e. the exposure pathways. We have indicated there is a sensitive relationship between these parameters, which we have defined as the DDD relationship. In the case studies provided here, we have demonstrated that even when this relationship is only partially quantified it can be linked to biological endpoints such as toxicity in order to obtain a mechanistic understanding of the outcomes of in vitro exposure to NM.[12,69,72,74]

To characterize and quantify the relationship, one must begin by understanding the intrinsic chemistry of the NM in combination with that of the dispersion media. Then, we can employ appropriate strategies to detect the NM and measure the dispersion and dissolution relationship when suspended in media (Figure 8(a)). This is best achieved by using a multitechnique approach with bulk and microscopic characterization both required and often destructive, ex situ specimen preparation procedures are needed. Ultimately, we can establish the primary NM dispersion (size distribution of individual and agglomerated NM), the secondary NM dispersion (size distribution of those aggregated to natural colloids and those reprecipitated in a second phase) and the dissolved fraction (the concentration bound to dissolved ligands and that free in solution) (Figure 8(b)). We must then establish similar NM dispersions and dissolved ion concentrations internalized by cells or organisms to obtain the dose distribution (also Figure 8(b)). We should however note that cellular material may not differentiate between primary and secondary NM during uptake and that there will be a probability distribution of internalized NM dose across a cell population or tissue. The dose distribution will be time dependent because of either, further uptake or, dilution due to intracellular dissolution, mitosis and excretion.

Having established size, number and concentration distributions of dispersed and dissolved NM in media plus the probability distribution of numbers of NM internalized by cells we can:

1. link to measured biological endpoints as discussed above and;
2. model the DDD relationship using a series of mathematical transformations (Figure 8(c)).[147,148]

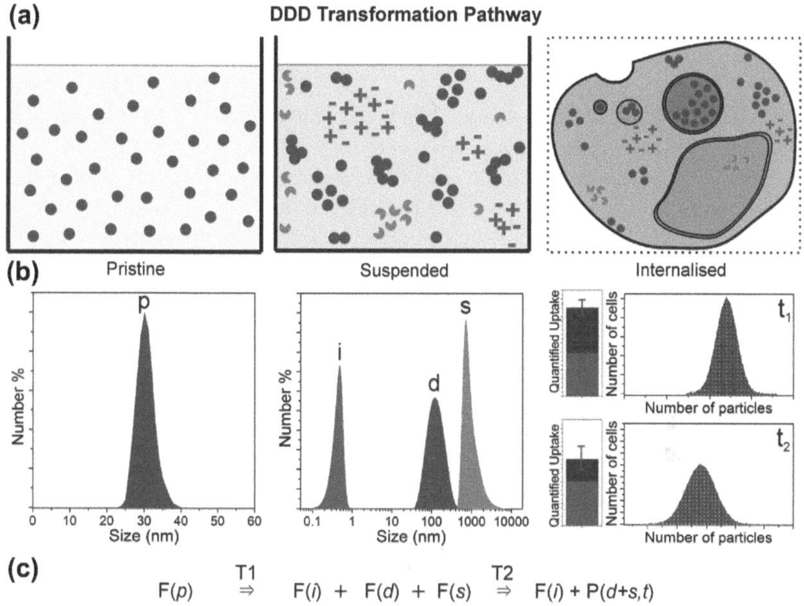

FIGURE 8 Schematic of the three stages to quantifying the dispersion, dissolution and dose (DDD) relationship. (a) Recognition that engineered nanomaterials (NM) may exhibit a narrow or well-characterized size distribution in the intended application that once dispersed in a complex medium NM agglomeration, aggregation, dissolution and reprecipitation may take place and that when internalized by a cell or organism the number distribution of the transformed NM will change. (b) Quantitative measures of: the original NM size distribution; the dispersed primary (p) and secondary (s) size distribution with the concentration of the dissolved (i) fraction (free and complexed in solution) in a complex media; and the time (t) dependent probability distribution of internalized primary and secondary NM with the concentration of the dissolved fraction (free and complexed) within a cell population or tissue. (c) Mathematical fits of the quantitative characterization profiles in B can be used to extract transformation functions (T1 and T2) between the original and dispersed NM in media and the NM dose distribution within a cell population. The resulting data driven model could enable prediction of NM uptake for different pristine size distributions and potentially at different exposure concentrations.

Step 1 will enable us to understand the key steps in the exposure pathway that may enhance a hazard presented by an NM type, for example, excessive particle uptake or gorging.[18] Step 2 will enable us to predict NM dose internalized by cells or organisms exposed to NM of a known primary particle size distribution in a given media. Key to this quantification process however is an appropriate understanding of the limitations of the characterization techniques to be applied and therefore an awareness of how robust the above data driven models can be. Fundamentally, we believe most of the techniques described above would be appropriate if analyzed with sufficient expertise; however, multiple approaches are recommended for a truly robust and general conclusion to be reached.

REFERENCES

1. Shi J, Votruba AR, Farokhzad OC, Langer R. Nanotechnology in drug delivery and tissue engineering: from discovery to applications. *Nano Lett* 2010;**10**:3223–30.
2. Lee D-E, Koo H, Sun I-C, Ryu JH, Kim K, Kwon IC. Multifunctional nanoparticles for multimodal imaging and theragnosis. *Chem Soc Rev* 2012;**41**:2656–72.
3. Doane TL, Burda C. The unique role of nanoparticles in nanomedicine: imaging, drug delivery and therapy. *Chem Soc Rev* 2012;**41**:2885–911.
4. Saha K, Agasti SS, Kim C, Li X, Rotello VM. Gold nanoparticles in chemical and biological sensing. *Chem Rev* 2012;**112**:2739–79.
5. Khin MM, Nair AS, Babu VJ, Murugan R, Ramakrishna S. A review on nanomaterials for environmental remediation. *Energ Environ Sci* 2012;**5**:8075–109.
6. Zaera F. Nanostructured materials for applications in heterogeneous catalysis. *Chem Soc Rev* 2013;**42**:2746–62.
7. Zeng H, Sun S. Syntheses, properties, and potential applications of multicomponent magnetic nanoparticles. *Adv Funct Mater* 2008;**18**:391–400.
8. Shirasaki Y, Supran GJ, Bawendi MG, Bulovic V. Emergence of colloidal quantum-dot light-emitting technologies. *Nat Photon* 2013;**7**:13–23.
9. Van Broekhuizen P, Van Veelen W, Streekstra WH, Schulte P, Reijnders L. Exposure limits for nanoparticles: report of an international workshop on nano reference values. *Ann Occup Hyg* 2012;**56**:515–24.
10. Batley GE, Kirby JK, McLaughlin MJ. Fate and risks of nanomaterials in aquatic and terrestrial environments. *Acc Chem Res* 2012;**46**:854–62.
11. Levard C, Hotze EM, Lowry GV, Brown GE. Environmental transformations of silver nanoparticles: impact on stability and toxicity. *Environ Sci Technol* 2012;**46**:6900–14.
12. Summers HD, Brown MR, Holton MD, Tonkin JA, Hondow N, Brown AP, et al. Quantification of nanoparticle dose and vesicular inheritance in proliferating cells. *ACS Nano* 2013;**7**:6129–37.
13. Maurer-Jones MA, Gunsolus IL, Murphy CJ, Haynes CL. Toxicity of engineered nanoparticles in the environment. *Anal Chem* 2013;**85**:3036–49.
14. Kretzschmar R, Schafer T. Metal retention and transport on colloidal particles in the environment. *Elements* 2005;**1**:205–10.
15. Allen TD. Introduction to electron microscopy for biologists. In: *Methods in cell biology*, vol. 88. Academic Press; 2008.
16. Hassellöv M, Readman J, Ranville J, Tiede K. Nanoparticle analysis and characterization methodologies in environmental risk assessment of engineered nanoparticles. *Ecotoxicology* 2008;**17**:344–61.
17. Rivera Gil P, Oberdörster G, Elder A, Puntes V, Parak WJ. Correlating physico-chemical with toxicological properties of nanoparticles: the present and the future. *ACS Nano* 2010;**4**:5527–31.
18. Pele L, Haas CT, Hewitt R, Faria N, Brown A, Powell J. Artefactual nanoparticle activation of the inflammasome platform: in vitro evidence with a nano-formed calcium phosphate. *Nanomedicine* 2014. http://dx.doi.org/10.2217/NNM.14.58.
19. Handler RM, Beard BL, Johnson CM, Scherer MM. Atom exchange between aqueous Fe(II) and Goethite: an Fe isotope tracer study. *Environ Sci Technol* 2009;**43**:1102–7.
20. Zhang L, Petersen EJ, Huang Q. Phase distribution of 14C-labeled multiwalled carbon nanotubes in aqueous systems containing model solids: peat. *Environ Sci Technol* 2011;**45**:1356–62.

21. Holbrook RD, Galyean A, Gorham J, Herzing A, Pettibone J. Chapter 2-Overview of nanomaterial characterization and metrology. In: Baalousha MA, Saleh N, Lead JR, editors. *Fronteirs of nanoscience − characterization of nanomaterials in complex environmental and biological media*. Elsevier; 2014.
22. Rollinson HR. *Using geochemical data: evaluation, presentation, interpretation*. New York: J. Wiley & Sons; 1993.
23. Utsunomiya S, Ewing RC. Application of high-angle annular dark field scanning transmission electron microscopy, scanning transmission electron microscopy-energy dispersive X-ray spectrometry, and energy-filtered transmission electron microscopy to the characterization of nanoparticles in the environment. *Environ Sci Technol* 2003; **37**:786−91.
24. Utsunomiya S, Kersting AB, Ewing RC. Groundwater nanoparticles in the far-field at the Nevada Test Site: mechanism for radionuclide transport. *Environ Sci Technol* 2009;**43**:1293−8.
25. Burke IT, Mayes WM, Peacock CL, Brown AP, Jarvis AP, Gruiz K. Speciation of arsenic, chromium, and vanadium in red mud samples from the Ajka spill site, Hungary. *Environ Sci Technol* 2012;**46**:3085−92.
26. White KN, Ejim AI, Walton RC, Brown AP, Jugdaohsingh R, Powell JJ, et al. Avoidance of aluminum toxicity in freshwater snails involves intracellular silicon−aluminum biointeraction. *Environ Sci Technol* 2008;**42**:2189−94.
27. Levard C, Reinsch BC, Michel FM, Oumahi C, Lowry GV, Brown GE. Sulfidation processes of PVP-coated silver nanoparticles in aqueous solution: impact on dissolution rate. *Environ Sci Technol* 2011;**45**:5260−6.
28. Kim B, Park C-S, Murayama M, Hochella MF. Discovery and characterization of silver sulfide nanoparticles in final sewage sludge products. *Environ Sci Technol* 2010;**44**:7509−14.
29. Liu N, Mu Y, Chen Y, Sun H, Han S, Wang M, et al. Degradation of aqueous synthesized CdTe/ZnS quantum dots in mice: differential blood kinetics and biodistribution of cadmium and tellurium. *Part Fibre Toxicol* 2013;**10**:37.
30. Jiang J, Oberdörster G, Biswas P. Characterization of size, surface charge, and agglomeration state of nanoparticle dispersions for toxicological studies. *J. Nanopart Res* 2009;**11**:77−89.
31. Lynch I, Dawson KA. Protein-nanoparticle interactions. *Nano Today* 2008;**3**:40−7.
32. Nel AE, Mädler L, Velegol D, Xia T, Hoek EMV, Somasundaran P, et al. Understanding biophysicochemical interactions at the nano-bio interface. *Nat Mater* 2009;**8**:543−57.
33. Fröhlich E. The role of surface charge in cellular uptake and cytotoxicity of medical nanoparticles. *Int J Nanomed* 2012;**7**:5577−91.
34. Albanese A, Chan WCW. Effect of gold nanoparticle aggregation on cell uptake and toxicity. *ACS Nano* 2011;**5**:5478−89.
35. Joris F, Manshian BB, Peynshaert K, De Smedt SC, Braeckmans K, Soenen SJ. Assessing nanoparticle toxicity in cell-based assays: influence of cell culture parameters and optimized models for bridging the in vitro-in vivo gap. *Chem Soc Rev* 2013;**42**:8339−59.
36. Maurer-Jones MA, Bantz KC, Love SA, Marquis BJ, Haynes CL. Toxicity of therapeutic nanoparticles. *Nanomedicine* 2009;**4**:219−41.
37. Petosa AR, Jaisi DP, Quevedo IR, Elimelech M, Tufenkji N. Aggregation and deposition of engineered nanomaterials in aquatic environments: role of physicochemical interactions. *Environ Sci Technol* 2010;**44**:6532−49.

38. Powers KW, Brown SC, Krishna VB, Wasdo SC, Moudgil BM, Roberts SM. Research strategies for safety evaluation of nanomaterials. Part VI. Characterization of nanoscale particles for toxicological evaluation. *Toxicol Sci* 2006;**90**:296–303.
39. Powers KW, Palazuelos M, Moudgil BM, Roberts SM. Characterization of the size, shape, and state of dispersion of nanoparticles for toxicological studies. *Nanotoxicology* 2007;**1**:42–51.
40. Berne BJ, Pecora R. *Dynamic light scattering with applications to chemistry, biology, and physics*. New York: Dover Publications Inc; 2000.
41. Filipe V, Hawe A, Jiskoot W. Critical evaluation of nanoparticle tracking analysis (NTA) by nanosight for the measurement of nanoparticles and protein aggregates. *Pharm Res* 2010;**27**:796–810.
42. Domingos RF, Baalousha MA, Ju-Nam Y, Reid MM, Tufenkji N, Lead JR, et al. Characterizing manufactured nanoparticles in the environment: multimethod determination of particle sizes. *Environ Sci Technol* 2009;**43**:7277–84.
43. Farkas J, Christian P, Urrea JAG, Roos N, Hassellöv M, Tollefsen KE, et al. Effects of silver and gold nanoparticles on rainbow trout (*Oncorhynchus mykiss*) hepatocytes. *Aquat Toxicol* 2010;**96**:44–52.
44. M-M P, Somchue W, Shiowatana J, Siripinyanond A. Flow field-flow fractionation for particle size characterization of selenium nanoparticles incubated in gastrointestinal conditions. *Food Res Int* 2014;**57**:203–9.
45. Montes-Burgos I, Walczyk D, Hole P, Smith J, Lynch I, Dawson K. Characterisation of nanoparticle size and state prior to nanotoxicological studies. *J. Nanopart Res* 2010;**12**:47–53.
46. Poda AR, Bednar AJ, Kennedy AJ, Harmon A, Hull M, Mitrano DM, et al. Characterization of silver nanoparticles using flow-field flow fractionation interfaced to inductively coupled plasma mass spectrometry. *J Chromatogr A* 2011;**1218**:4219–25.
47. Tantra R, Jing S, Pichaimuthu S, Walker N, Noble J, Hackley V. Dispersion stability of nanoparticles in ecotoxicological investigations: the need for adequate measurement tools. *J Nanopart Res* 2011;**13**:3765–80.
48. Baalousha M, Lead JR. Nanoparticle dispersity in toxicology. *Nat Nanotechnol* 2013;**8**:308–9.
49. Bednar AJ, Poda AR, Mitrano DM, Kennedy AJ, Gray EP, Ranville JF, et al. Comparison of on-line detectors for field flow fractionation analysis of nanomaterials. *Talanta* 2013;**104**:140–8.
50. Dubascoux S, Le Hecho I, Hassellov M, Von Der Kammer F, Potin Gautier M, Lespes G. Field-flow fractionation and inductively coupled plasma mass spectrometer coupling: history, development and applications. *J Anal Atom Spectrom* 2010;**25**:613–23.
51. de Oliveira JFA, Cardoso MB. Partial aggregation of silver nanoparticles induced by capping and reducing agents competition. *Langmuir* 2013;**30**:4879–86.
52. McKenzie LC, Haben PM, Kevan SD, Hutchison JE. Determining nanoparticle size in real time by small-angle X-ray scattering in a microscale flow system. *J Phys Chem C* 2010;**114**:22055–63.
53. Mori Y, Furukawa M, Hayashi T, Nakamura K. Size distribution of gold nanoparticles used by small angle X-ray scattering. *Particul Sci Technol* 2006;**24**:97–103.
54. Mahl D, Diendorf J, Meyer-Zaika W, Epple M. Possibilities and limitations of different analytical methods for the size determination of a bimodal dispersion of metallic nanoparticles. *Colloid Surf A* 2011;**377**:386–92.

55. Henriquez RR, Ito T, Sun L, Crooks RM. The resurgence of Coulter counting for analyzing nanoscale objects. *Analyst* 2004;**129**:478–82.
56. Borchert H, Shevchenko EV, Robert A, Mekis I, Kornowski A, Grübel G, et al. Determination of nanocrystal sizes: A comparison of TEM, SAXS, and XRD studies of highly monodisperse $CoPt_3$ particles. *Langmuir* 2005;**21**:1931–6.
57. Ji Z, Jin X, George S, Xia T, Meng H, Wang X, et al. Dispersion and stability optimization of TiO_2 nanoparticles in cell culture media. *Environ Sci Technol* 2010;**44**:7309–14.
58. Murdock RC, Braydich-Stolle L, Schrand AM, Schlager JJ, Hussain SM. Characterization of nanomaterial dispersion in solution prior to in vitro exposure using dynamic light scattering technique. *Toxicol Sci* 2008;**101**:239–53.
59. Orts-Gil G, Natte K, Drescher D, Bresch H, Mantion A, Kneipp J, et al. Characterisation of silica nanoparticles prior to in vitro studies: from primary particles to agglomerates. *J Nanopart Res* 2011;**13**:1593–604.
60. Zheng J, Stevenson MS, Hikida RS, Van Patten PG. Influence of pH on dendrimer-protected nanoparticles. *J Phys Chem B* 2002;**106**:1252–5.
61. Allouni ZE, Cimpan MR, Høl PJ, Skodvin T, Gjerdet NR. Agglomeration and sedimentation of TiO_2 nanoparticles in cell culture medium. *Colloid Surf B* 2009;**68**:83–7.
62. Saville SL, Woodward RC, House MJ, Tokarev A, Hammers J, Qi B, et al. The effect of magnetically induced linear aggregates on proton transverse relaxation rates of aqueous suspensions of polymer coated magnetic nanoparticles. *Nanoscale* 2013;**5**:2152–63.
63. Horst AM, Neal AC, Mielke RE, Sislian PR, Suh WH, Mädler L, et al. Dispersion of TiO_2 nanoparticle agglomerates by *Pseudomonas aeruginosa*. *Appl Environ Microbiol* 2010;**76**:7292–8.
64. Dutta D, Sundaram SK, Teeguarden JG, Riley BJ, Fifield LS, Jacobs JM, et al. Adsorbed proteins influence the biological activity and molecular targeting of nanomaterials. *Toxicol Sci* 2007;**100**:303–15.
65. Klein KL, Anderson IM, De Jonge N. Transmission electron microscopy with a liquid flow cell. *J Microsc* 2011;**242**:117–23.
66. Grogan JM, Rotkina L, Bau HH. In situ liquid-cell electron microscopy of colloid aggregation and growth dynamics. *Phys Rev E* 2011;**83**:061405.
67. Tiede K, Tear SP, David H, Boxall ABA. Imaging of engineered nanoparticles and their aggregates under fully liquid conditions in environmental matrices. *Water Res* 2009;**43**:3335–43.
68. Yuk JM, Park J, Ercius P, Kim K, Hellebusch DJ, Crommie MF, et al. High-resolution EM of colloidal nanocrystal growth using graphene liquid cells. *Science* 2012;**336**:61–4.
69. Mu Q, Hondow NS, Krzemiński Ł, Brown AP, Jeuken LJC, Routledge MN. Mechanism of cellular uptake of genotoxic silica nanoparticles. *Part Fibre Toxicol* 2012;**9**:29.
70. Churchman AH, Wallace R, Milne SJ, Brown AP, Brydson R, Beales PA. Serum albumin enhances the membrane activity of ZnO nanoparticles. *Chem Commun* 2013;**49**:4172–4.
71. Wallace R, Brown AP, Brydson R, Milne SJ, Hondow N, Wang P. Characterisation of ZnO nanoparticle suspensions for toxicological applications. *J Phys Conf Ser* 2012;**371**:012080.
72. Hondow N, Brydson R, Wang P, Holton MD, Brown MR, Rees P, et al. Quantitative characterization of nanoparticle agglomeration within biological media. *J Nanopart Res* 2012;**14**:977.
73. Zhang S, Nelson A, Beales PA. Freezing or wrapping: the role of particle size in the mechanism of nanoparticle–biomembrane interaction. *Langmuir* 2012;**28**:12831–7.

74. Mu Q, David CA, Galceran J, Rey-Castro C, Krzemiński Ł, Wallace R, et al. Systematic investigation of the physicochemical factors that contribute to the toxicity of ZnO nanoparticles. *Chem Res Toxicol* 2014;**27**:558–67.
75. Wallace R, Brown AP, Brydson R, Wegner K, Milne SJ. Synthesis of ZnO nanoparticles by flame spray pyrolysis and characterisation protocol. *J Mater Sci* 2013;**48**:6393–403.
76. Reed RB, Ladner DA, Higgins CP, Westerhoff P, Ranville JF. Solubility of nano-zinc oxide in environmentally and biologically important matrices. *Environ Toxicol Chem* 2012;**31**:93–9.
77. Vakurov A, Guillermo M, Drummond-Brydson R, Wallace R, Svendsen C, Nelson A. ZnO nanoparticle interactions with phospholipid monolayers. *J Colloid Interf Sci* 2013;**404**:161–8.
78. Doak SH, Griffiths SM, Manshian B, Singh N, Williams PM, Brown AP, et al. Confounding experimental considerations in nanogenotoxicology. *Mutagenesis* 2009;**24**:285–93.
79. Summers HD, Rees P, Holton MD, Rowan Brown M, Chappell SC, Smith PJ, et al. Statistical analysis of nanoparticle dosing in a dynamic cellular system. *Nat Nanotechnol* 2011;**6**:170–4.
80. Lombi E, Donner E, Tavakkoli E, Turney TW, Naidu R, Miller BW, et al. Fate of zinc oxide nanoparticles during anaerobic digestion of wastewater and post-treatment processing of sewage sludge. *Environ Sci Technol* 2012;**46**:9089–96.
81. Lv J, Zhang S, Luo L, Han W, Zhang J, Yang K, et al. Dissolution and microstructural transformation of ZnO nanoparticles under the influence of phosphate. *Environ Sci Technol* 2012;**46**:7215–21.
82. Ma R, Levard C, Michel FM, Brown GE, Lowry GV. Sulfidation mechanism for zinc oxide nanoparticles and the effect of sulfidation on their solubility. *Environ Sci Technol* 2013;**47**:2527–34.
83. Franklin NM, Rogers NJ, Apte SC, Batley GE, Gadd GE, Casey PS. Comparative toxicity of nanoparticulate ZnO, bulk ZnO, and $ZnCl_2$ to a freshwater microalga (*Pseudokirchneriella subcapitata*): the importance of particle solubility. *Environ Sci Technol* 2007;**41**:8484–90.
84. Turney TW, Duriska MB, Jayaratne V, Elbaz A, O'Keefe SJ, Hastings AS, et al. Formation of zinc-containing nanoparticles from Zn^{2+} ions in cell culture media: implications for the nanotoxicology of ZnO. *Chem Res Toxicol* 2012;**25**:2057–66.
85. Rubasinghege G, Lentz RW, Park H, Scherer MM, Grassian VH. Nanorod dissolution quenched in the aggregated state. *Langmuir* 2009;**26**:1524–7.
86. Liu J, Aruguete DM, Murayama M, Hochella MF. Influence of size and aggregation on the reactivity of an environmentally and industrially relevant nanomaterial (PbS). *Environ Sci Technol* 2009;**43**:8178–83.
87. Misra SK, Dybowska A, Berhanu D, Luoma SN, Valsami-Jones E. The complexity of nanoparticle dissolution and its importance in nanotoxicological studies. *Sci Total Environ* 2012;**438**:225–32.
88. Hillaireau H, Couvreur P. Nanocarriers' entry into the cell: relevance to drug delivery. *Cell Mol Life Sci* 2009;**66**:2873–96.
89. Zhu M, Nie G, Meng H, Xia T, Nel A, Zhao Y. Physicochemical properties determine nanomaterial cellular uptake, transport, and fate. *Acc Chem Res* 2012;**46**:622–31.
90. Kneipp J, Kneipp H, Wittig B, Kneipp K. Following the dynamics of pH in endosomes of live cells with SERS nanosensors. *J Phys Chem C* 2010;**114**:7421–6.
91. Soenen SJ, Demeester J, De Smedt SC, Braeckmans K. The cytotoxic effects of polymer-coated quantum dots and restrictions for live cell applications. *Biomaterials* 2012;**33**:4882–8.

92. Mancini MC, Kairdolf BA, Smith AM, Nie S. Oxidative quenching and degradation of polymer-encapsulated quantum dots: new insights into the long-term fate and toxicity of nanocrystals in vivo. *J Am Chem Soc* 2008;**130**:10836–7.
93. Metz KM, Mangham AN, Bierman MJ, Jin S, Hamers RJ, Pedersen JA. Engineered nanomaterial transformation under oxidative environmental conditions: development of an in vitro biomimetic assay. *Environ Sci Technol* 2009;**43**:1598–604.
94. Mudunkotuwa IA, Rupasinghe T, Wu C-M, Grassian VH. Dissolution of ZnO nanoparticles at circumneutral pH: a study of size effects in the presence and absence of citric acid. *Langmuir* 2011;**28**:396–403.
95. Xu MS, Li J, Hanagata N, Su HX, Chen HZ, Fujita D. Challenge to assess the toxic contribution of metal cation released from nanomaterials for nanotoxicology – the case of ZnO nanoparticles. *Nanoscale* 2013;**5**:4763–9.
96. Hadioui M, Leclerc S, Wilkinson KJ. Multimethod quantification of Ag^{+} release from nanosilver. *Talanta* 2013;**105**:15–9.
97. Lin C-H, Chang LW, Chang H, Yang M-H, Yang C-S, Lai W-H, et al. The chemical fate of the Cd/Se/Te-based quantum dot 705 in the biological system: toxicity implications. *Nanotechnology* 2009;**20**:215101.
98. Kittler S, Greulich C, Diendorf J, Köller M, Epple M. Toxicity of silver nanoparticles increases during storage because of slow dissolution under release of silver ions. *Chem Mater* 2010;**22**:4548–54.
99. Miao A-J, Zhang X-Y, Luo Z, Chen C-S, Chin W-C, Santschi PH, et al. Zinc oxide–engineered nanoparticles: dissolution and toxicity to marine phytoplankton. *Environ Toxicol Chem* 2010;**29**:2814–22.
100. Maurer-Jones MA, Mousavi MPS, Chen LD, Buhlmann P, Haynes CL. Characterization of silver ion dissolution from silver nanoparticles using fluorous-phase ion-selective electrodes and assessment of resultant toxicity to Shewanella oneidensis. *Chem Sci* 2013;**4**:2564–72.
101. Martin MN, Allen AJ, MacCuspie RI, Hackley VA. Dissolution, agglomerate morphology, and stability limits of protein-coated silver nanoparticles. *Langmuir* 2014;**30**:11442–52.
102. Mahon E, Hristov DR, Dawson KA. Stabilising fluorescent silica nanoparticles against dissolution effects for biological studies. *Chem Commun* 2012;**48**:7970–2.
103. Ojea-Jiménez I, Piella J, Nguyen T-L, Bestetti A, Ryan AD, Puntes V. Stability of polymer encapsulated quantum dots in cell culture media. *J Phys Conf Ser* 2013;**429**:012009.
104. Summers HD, Holton MD, Rees P, Williams PM, Thornton CA. Analysis of quantum dot fluorescence stability in primary blood mononuclear cells. *Cytom Part A* 2010;**77A**:933–9.
105. Pi Q, Zhang W, Zhou G, Liu W, Cao Y. Degradation or excretion of quantum dots in mouse embryonic stem cells. *BMC Biotechnol* 2010;**10**:36.
106. Borm P, Klaessig FC, Landry TD, Moudgil B, Pauluhn J, Thomas K, et al. Research strategies for safety evaluation of nanomaterials, part V: role of dissolution in biological fate and effects of nanoscale particles. *Toxicol Sci* 2006;**90**:23–32.
107. Galceran J, Companys E, Puy J, Cecilia J, Garces JL. AGNES: a new electroanalytical technique for measuring free metal ion concentration. *J Electroanal Chem* 2004;**566**:95–109.
108. Chito D, Weng L, Galceran J, Companys E, Puy J, van Riemsdijk WH, et al. Determination of free Zn^{2+} concentration in synthetic and natural samples with AGNES (Absence of Gradients and Nernstian Equilibrium Stripping) and DMT (Donnan Membrane Technique). *Sci Total Environ* 2012;**421–422**:238–44.
109. Müller KH, Kulkarni J, Motskin M, Goode A, Winship P, Skepper JN, et al. pH-dependent toxicity of high aspect ratio ZnO nanowires in macrophages due to intracellular dissolution. *ACS Nano* 2010;**4**:6767–79.

110. Zhou J, Xu NS, Wang ZL. Dissolving behavior and stability of ZnO wires in biofluids: a study on biodegradability and biocompatibility of ZnO nanostructures. *Adv Mater* 2006;**18**:2432–5.
111. Yu Z, Guo L, Du H, Krauss T, Silcox J. Shell distribution on colloidal CdSe/ZnS quantum dots. *Nano Lett* 2005;**5**:565–70.
112. van Schooneveld MM, Gloter A, Stephan O, Zagonel LF, Koole R, Meijerink A, et al. Imaging and quantifying the morphology of an organic-inorganic nanoparticle at the sub-nanometre level. *Nat Nano* 2010;**5**:538–44.
113. Ward MB, Hondow N, Brown AP, Brydson R. Electron energy loss spectroscopy and energy dispersive X-ray analysis. In: Kirkland AI, Haigh SJ, editors. *Nanocharacterisation*. 2nd ed. Cambridge: RSC Publishing; 2014.
114. de Groot FMF, de Smit E, van Schooneveld MM, Aramburo LR, Weckhuysen BM. In-situ scanning transmission X-ray microscopy of catalytic solids and related nanomaterials. *ChemPhysChem* 2010;**11**:951–62.
115. Goode AE, Perkins JM, Sandison A, Karunakaran C, Cheng H, Wall D, et al. Chemical speciation of nanoparticles surrounding metal-on-metal hips. *Chem Commun* 2012;**48**:8335–7.
116. Hsiao IL, Huang Y-J. Titanium oxide shell coatings decrease the cytotoxicity of ZnO nanoparticles. *Chem Res Toxicol* 2011;**24**:303–13.
117. Oberdorster G, Oberdorster E, Oberdorster J. Concepts of nanoparticle dose metric and response metric. *Environ Health Persp* 2007;**115**:A290.
118. Stoeger T, Schmid O, Takenaka S, Schulz H. Inflammatory response to TiO_2 and carbonaceous particles scales best with BET surface area. *Environ Health Perspect* 2007;**115**:A290–1.
119. Wittmaack K. Dose and response metrics in nanotoxicology: Wittmaack responds to Oberdorster et al. and Stoeger et al. *Environ Health Persp* 2007;**115**:A291–2.
120. Oberdorster G, Oberdorster E, Oberdorster J. Nanotoxicology: an emerging discipline evolving from studies of ultrafine particles. *Environ Health Persp* 2005;**113**:823–39.
121. Wittmaack K. In search of the most relevant parameter for quantifying lung inflammatory response to nanoparticle exposure: particle number, surface area, or what? *Environ Health Persp* 2007;**115**:187–94.
122. Elsaesser A, Taylor A, de Yanés GS, McKerr G, Kim E-M, O'Hare E, et al. Quantification of nanoparticle uptake by cells using microscopical and analytical techniques. *Nanomedicine* 2010;**5**:1447–57.
123. Krpetić Ž, Saleemi S, Prior IA, Sée V, Qureshi R, Brust M. Negotiation of intracellular membrane barriers by TAT-modified gold nanoparticles. *ACS Nano* 2011;**5**:5195–201.
124. Schweiger C, Hartmann R, Zhang F, Parak W, Kissel T, Rivera Gil P. Quantification of the internalization patterns of superparamagnetic iron oxide nanoparticles with opposite charge. *J Nanobiotechnol* 2012;**10**:28.
125. Chithrani BD, Ghazani AA, Chan WCW. Determining the size and shape dependence of gold nanoparticle uptake into mammalian cells. *Nano Lett* 2006;**6**:662–8.
126. He X, Ma Y, Li M, Zhang P, Li Y, Zhang Z. Quantifying and imaging engineered nanomaterials in vivo: challenges and techniques. *Small* 2013;**9**:1482–91.
127. Audinot J-N, Georgantzopoulou A, Piret J-P, Gutleb AC, Dowsett D, Migeon HN, et al. Identification and localization of nanoparticles in tissues by mass spectrometry. *Surf Interface Anal* 2013;**45**:230–3.
128. Yang RSH, Chang LW, Wu J-P, Tsai M-H, Wang H-J, Kuo Y-C, et al. Persistent tissue kinetics and redistribution of nanoparticles, quantum dot 705, in mice: ICP-MS quantitative assessment. *Environ Health Persp* 2007;**115**:1339–43.

129. Freund B, Tromsdorf UI, Bruns OT, Heine M, Giemsa A, Bartelt A, et al. A simple and widely applicable method to ^{59}Fe-radiolabel monodisperse superparamagnetic iron oxide nanoparticles for in vivo quantification studies. *ACS Nano* 2012;**6**:7318–25.
130. Brandenberger C, Rothen-Rutishauser B, Blank F, Gehr P, Muhlfeld C. Particles induce apical plasma membrane enlargement in epithelial lung cell line depending on particle surface area dose. *Resp Res* 2009;**10**.
131. Ferrari R, Lupi M, Falcetta F, Bigini P, Paolella K, Fiordaliso F, et al. Integrated multiplatform method for in vitro quantitative assessment of cellular uptake for fluorescent polymer nanoparticles. *Nanotechnology* 2014;**25**:045102.
132. Rees P, Brown MR, Summers H, Holton M, Errington R, Chappell S, et al. A transfer function approach to measuring cell inheritance. *BMC Syst Biol* 2011;**5**:31.
133. Mayhew TM, Mühlfeld C, Vanhecke D, Ochs M. A review of recent methods for efficiently quantifying immunogold and other nanoparticles using TEM sections through cells, tissues and organs. *Ann Anat* 2009;**191**:153–70.
134. Muhlfeld C, Rothen-Rutishauser B, Vanhecke D, Blank F, Gehr P, Ochs M. Visualization and quantitative analysis of nanoparticles in the respiratory tract by transmission electron microscopy. *Part Fibre Toxicol* 2007;**4**:11.
135. Baudoin J-P, Jerome WG, Kübel C, de Jonge N. Whole-cell analysis of low-density lipoprotein uptake by macrophages using STEM tomography. *PLoS One* 2013;**8**:e55022.
136. Brandenberger C, Mühlfeld C, Ali Z, Lenz A-G, Schmid O, Parak WJ, et al. Quantitative evaluation of cellular uptake and trafficking of plain and polyethylene glycol-coated gold nanoparticles. *Small* 2010;**6**:1669–78.
137. Elsaesser A, Barnes CA, McKerr G, Salvati A, Lynch I, Dawson KA, et al. Quantification of nanoparticle uptake by cells using an unbiased sampling method and electron microscopy. *Nanomedicine* 2011;**6**:1189–98.
138. Howard CV, Elsaesser A. Letter to the editor: quantification of nanoparticle 'dose'. *Nanomedicine* 2011;**6**:1153–4.
139. Mühlfeld C, Brandenberger C. Letter to the editor: uptake of nanoparticles by cells: do you know their number? *Nanomedicine* 2011;**6**:1149–51.
140. Rosman C, Pierrat S, Henkel A, Tarantola M, Schneider D, Sunnick E, et al. A new approach to assess gold nanoparticle uptake by mammalian cells: combining optical dark-field and transmission electron microscopy. *Small* 2012;**8**:3683–90.
141. Gottstein C, Wu G, Wong BJ, Zasadzinski JA. Precise quantification of nanoparticle internalization. *ACS Nano* 2013;**7**:4933–45.
142. Jing Y, Mal N, Williams PS, Mayorga M, Penn MS, Chalmers JJ, et al. Quantitative intracellular magnetic nanoparticle uptake measured by live cell magnetophoresis. *FASEB J* 2008;**22**:4239–47.
143. Giustini AJ, Perreard I, Rauwerdink AM, Hoopes PJ, Weaver JB. Noninvasive assessment of magnetic nanoparticle-cancer cell interactions. *Integr Biol* 2012;**4**:1283–8.
144. Bruns OT, Ittrich H, Peldschus K, Kaul MG, Tromsdorf UI, Lauterwasser J, et al. Real-time magnetic resonance imaging and quantification of lipoprotein metabolism in vivo using nanocrystals. *Nat Nanotechnol* 2009;**4**:193–201.
145. James SA, Feltis BN, de Jonge MD, Sridhar M, Kimpton JA, Altissimo M, et al. Quantification of ZnO nanoparticle uptake, distribution, and dissolution within individual human macrophages. *ACS Nano* 2013;**7**:10621–35.
146. Taylor A, Wilson KM, Murray P, Fernig DG, Levy R. Long-term tracking of cells using inorganic nanoparticles as contrast agents: are we there yet? *Chem Soc Rev* 2012;**41**:2707–17.

147. Brown MR, Hondow N, Brydson R, Rees P, Brown AP, Summers HD. Statistical prediction of nanoparticle delivery: from cell culture media to cell. *Nanotechnology* 2015;**26**:155101.
148. Hondow N, Brown MR, Starborg T, Monteith AG, Brydson R, Summers HD, et al. Quantifying the cellular uptake of semiconductor quantum dot nanoparticles by analytical electron microscopy. *J Microsc* 2015. http://dx.doi.org/10.1111/jmi.12239.

Chapter 7

Surface Properties (Physical and Chemical) and Related Reactions: Characterization via a Multi-Technique Approach

Jerome Rose
CNRS-Aix Marseille University, IRD, UM34, UMR 7330, Europole de l'arbois, Aix en Provence, France
E-mail: rose@cerege.fr

Chapter Outline

Frontiers of Nanoscience, Vol. 8. http://dx.doi.org/10.1016/B978-0-08-099948-7.00007-5

1. INTRODUCTION: RELATIONSHIP BETWEEN SURFACE REACTIVITY, SURFACE ENERGY AND SURFACE ATOMS

Even if the definition of nanomaterials (NMs) remains a question of debate, most of the craze of nanotechnology is related to the fact that NMs exhibit new or strongly modified properties compared to their bulk counterparts. However the 'nano' prefix is used too many times with only a reference to particles size, i.e. smaller than 100 nm without any specific 'new' or 'enhanced' properties.[1] Gold NMs are one of the most famous cited examples to highlight the so-called 'nano' effect i.e. new size-dependent properties below a certain particle size. Gold metal has been known for tens of thousands of years for its inalterability over ageing effects (atmospheric alteration) and more specifically gold is known as nonliable to rust.[2] Gold does not exchange electron with its surrounding environment. However, metallic gold NMs smaller than 5 nm are known to be excellent catalysts,[3,4] that is to say that metallic gold NMs can exchange electrons with molecules at their surface. Other examples of 'nano' effects can be cited like the quantum dots optical fluorescence property with modification of the fluorescence energy as function of the size of NMs[5–7] or the decrease of melting temperature for gold and many other NMs (e.g. Refs 8–10). One of the objective of the chapter is to highlight to what extend 'nano' specific properties are related to or impact surface atoms.

The high surface to mass ratio is one of the certainties which is conferred by a very small size. Indeed a surface of 1200 m^2 is calculated for 1 cm^3 of 5-nm cubic particles. NM surfaces represent the contact zone between particles and their surrounding environment and therefore corresponds to the interface, where all chemical (anions and cations) and electrochemical (electron, protons) exchanges will occur. Even if it is not specific to NMs, the high surface to mass ratio makes the surface a region of intense activity in the case of NMs.

With this in mind, and before a deeper investigation of the surface of NMs, it is worth remembering why atoms at the surface differ from atoms in the bulk. In the case of crystallized inorganic particles, surfaces need to be considered as a zone of thermodynamic nonequilibrium. The surface's location is very well defined between solid and liquid or solid and gas. This specificity is at the origin of their reactivity. In the case of metallic oxides NMs (in aqueous media), surfaces consist of oxygen atoms with a lower coordination number due to the disruption of the crystal periodicity. This disrupted coordination leads to a violation of the electro-neutrality of minerals between anions and cations and therefore to surface charge. These oxygen atoms will preserve partial charge leading to a surface of acidic—basic properties more or less pronounced governed by the protonation constants (in solution) as follow:

$$M_n - O^{(nv-2)} + H^+ \Leftrightarrow M_n - OH^{(nv-1)} \quad K_{n,1} \tag{1}$$

$$M_n - OH^{(nv-1)} + H^+ \Leftrightarrow M_n - OH_2^{(nv)} \quad K_{n,2} \tag{2}$$

with M represents, metals from the particle surface; n represents the number of metallic cations linked to oxygen atoms and v is the charge over coordination number (z/N), which is the formal bond valence introduced by Pauling[11] (see further the example of Fe oxyhydroxides mineral surfaces, Eqns (3) and (4)). The latter equilibria are at the origin of surface charge in water but also reveal their high complexity because the surface oxygen atoms' protonation constants are related to the cation intrinsic properties (size, formal charge) and their coordination mode (apex atoms or bridges).

Energy balance is another example of the thermodynamic nonequilibrium of surfaces. If we consider the energy requirement to generate a new surface after crystal cleavage, its contribution will depend on the size of the surface domain that is created. Calculation based on NaCl crystals revealed that this energy is negligible for micronic crystals (<1 J/g) while it becomes predominant for nanometric particles (560 J/g).[12] The free surface energy contribution is therefore not in favour to the thermodynamic stabilization of very small objects that can therefore be considered as meta-stable systems.

From the surface free energy $\Delta G_0(surface) = \gamma A$, where γ is the surface tension and A the interface area, two scenarios can be involved to minimize this energetic contribution. First, the NMs can grow through Ostwald ripening or through aggregation to decrease the area (A) of the solid—solution interface. The second possible case consists in decreasing the interfacial tension (γ) through specific surface adsorption.[13]

Surface energy is therefore the driving force of NM fate in solution. It is then crucial to consider NMs as meta-stable products that are prone to be transformed as a function of their surrounding environment.

2. SURFACE PROPERTIES AND SURFACE REACTIONS

2.1 Introduction

Generally speaking, we can consider that surface properties of NMs are closely related to the chemical properties of the atoms at the very near surface of the NMs. Acid—base (Lewis or Brönsted) as well as redox surface properties can confer to the particles the ability to form hydrogen bonds especially in the case of OH or NH surface groups. Hydrogen bonds are at the origin of surface solvation. The latter properties can also enable the formation of surface complexes with molecules that can mediate the hydrophobia/hydrophilia of particles. Chemical properties can also provide the possibilities of NMs serving as central sites for electron and ionic transfer from the surface to the core. It can also promote the adsorption of cations that can nucleate the formation of a core/shell structure. In addition to chemical properties, the structure of NM surface can also alter surface reactions and reactivities. It can be the case for surface defects which enhance the adsorption capacity of NMs. Both aspects are detailed below.

Even if most mechanisms detailed further are not nano-specific, we will highlight how the decrease of particle size can affect the surface properties of NMs.

2.2 Proton and Surface Charge

The protonation–deprotonation equilibria of the oxygenated surface sites (Eqns (1) and (2)) contribute to the electrostatic surface charge of the particles. Acid–base surface reactions contribute also to change the surface energy as a function of the crystal faces exposed to water. For isotropic particles for which all crystal faces possess almost the same energy, the decrease of interfacial tension through the increase of surface charge can be a way to control the size of particles.

Protonation–deprotonation can be considered as proton adsorption–desorption at the surface of particles whatever their size. However, it is a very specific adsorption case since it is at the origin of the pH-dependent charge of most of the NMs in polar solvent and of course in water. As indicated by Eqns (1) and (2) all particles with –OH functional group will exhibit positive surface charges at a pH lower than the point of zero charge (PZC) and negative surface charges at a pH higher than the PZC. It can be illustrated by the specific case of an iron oxyhydroxide surface

$$\begin{aligned} &\equiv \mathrm{FeOH_2^+} \Leftrightarrow \equiv \mathrm{FeOH^0} + \mathrm{H^+} \\ &\mathrm{K_1^{int}} = [\mathrm{H^+}][\equiv \mathrm{FeOH^0}][\equiv \mathrm{FeOH_2^+}]^{-1} \exp(-F\psi/RT) \end{aligned} \tag{3}$$

$$\begin{aligned} &\equiv \mathrm{FeOH^0} \Leftrightarrow \equiv \mathrm{FeO^-} + \mathrm{H^+} \\ &\mathrm{K_2^{int}} = [\mathrm{H^+}][\equiv \mathrm{FeO^-}][\equiv \mathrm{FeOH^0}]^{-1} \exp(-F\psi/RT) \end{aligned} \tag{4}$$

with $\equiv FeX$ all Fe surface functional groups, F is the Faraday constant, ψ is the surface potential, R is the gas constant, and T is the absolute temperature.

Figure 1 indicates that the PZC for ferrihydrite, for instance, is at pH = 7.9.

For some minerals (like clays), fixed charges exist that are added with the pH-dependent charges to give the total surface charge. Such permanent charges are generally based on metal substitutions inside the NM structure. The density of charge can be approximated by a potential $\tilde{\psi}$.

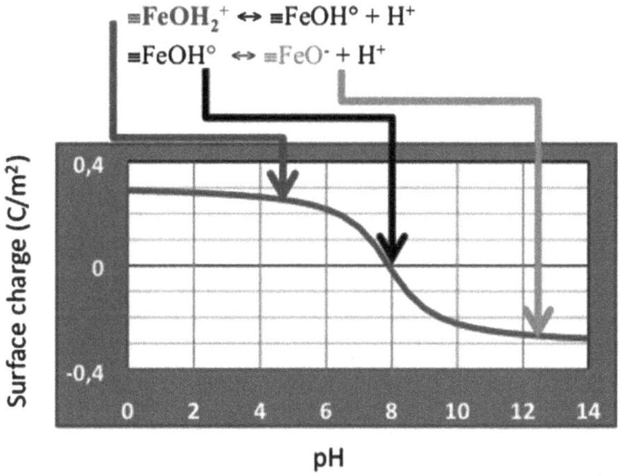

FIGURE 1 Surface charge of FeOOH (2L ferrihydrite nanomaterials) as function of pH.

When particle size decreases down to the nanometre range it has been proved in many cases that the number of functional groups, i.e. the surface charge density is much larger than for micrometric particles.[14–16] Theoretical calculations[15] suggested that the decrease of surface charge density reached an asymptotic limit at particle diameters >20 nm (Figure 2).

Moreover in some cases the acid–base constant can change with size[14,15,17] and therefore modify the PZC. An increase of the pH_{PZC} is observed for TiO_2, α-FeOOH (goethite) and γ-Fe_2O_3 (maghemite) particles smaller than 5 nm.[14,15,17] In the case of γ-Fe_2O_3, the pH_{PZC} increased from 6.3 to 8.4 for particle size of 3.5 and 12 nm respectively.[17] The increase of the pH_{PZC} with size is supposed to depend on the curvature of the solid–liquid

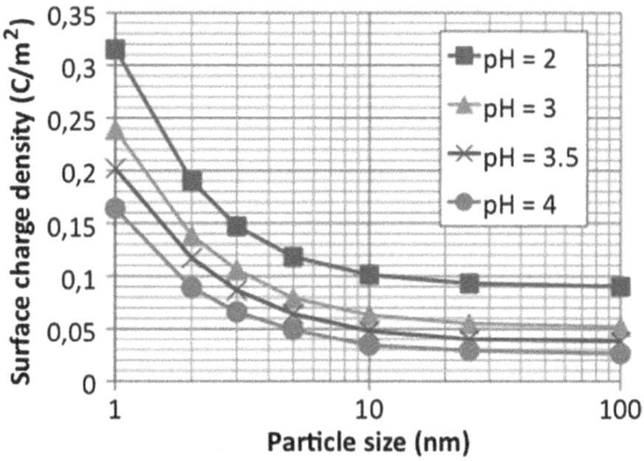

FIGURE 2 Modification of the surface charge density as a function of the particle size. *From Ref. 15.*

interface. As for negatively charged system, as the curvature increases, the surfaces will desorb more protons. Consequently the basicity increases.[17]

2.3 Adsorption Reactions

2.3.1 Introduction

Various mechanisms can be at the origin of molecules adsorption at the surface of particles. van der Waals forces, hydrogen bonds, Lewis acid forces, electrostatic forces and also entropic forces such as hydrophobic forces can all lead to adsorption of organic or inorganic molecules at the surface of any particles.

van der Waals forces and more importantly electrostatic forces due to surface charges are at the origin of reversible and nonspecific adsorptions of anionic or cationic molecules that are generally referred to 'physical' adsorption. With more details, in an aqueous medium, a diffuse electrical double layer will form at the solid—liquid interface as ionic species are attracted to the charged surface. Ions with a same charge compared to the NM surface are repelled from the surface, while those ions having the opposite charge are attracted to it. (Quasi) nonreversible adsorptions are more often the consequence of chemical reactions between molecules and surface functional groups ('chemisorption'). Hydrophobic interactions may also lead to irreversible adsorption.

Even if adsorption occurs at the surface of any particle, adsorption mechanisms may strongly be altered as the size of the adsorbent decrease to the nanometric range.

2.3.2 Inorganic Molecules

One characteristic of inorganic adsorption is that most of the adsorbed molecules are small compared to the size of the NMs. As discussed above reversible and nonreversible adsorption can occur. To further illustrate the adsorption mechanism, let us take the example of the adsorption of divalent metal at the surface of metal oxyhydroxide NM like FeOOH or AlOOH for instance. All cations (and anions) are solvated in water. At the surface of NMs, two main mechanisms can occur. If the ion is attracted to the surface via long-range coulombic forces, it can retain its water of hydration. The presence of hydration sphere maintains the adsorbed inorganic molecule at a certain distance from the surface atoms of the NM. Such surface structure or surface complexes are called 'outer sphere complexes' (Figure 3). In our specific example 'outer sphere complexes' can form at pH higher than the PZC of the NM. Indeed at pHs higher than the PZC, metal cations and NM surfaces are oppositely charged, which can lead to reversible and nonspecific bonds.

If the ion loses some of its hydration water molecules, it can become chemically bound to the NM surface via short-range forces. This reaction can be considered as a ligand exchange between the surface functional group and the hydration sphere of the adsorbed metal ion. In that case the interatomic

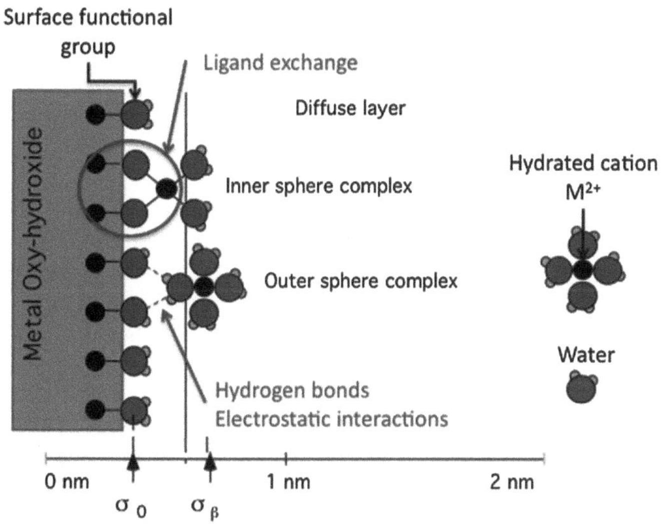

$$\equiv S - OH + M^{2+}(H_2O)_x \Leftrightarrow \equiv S - OM^{+}(H_2O)_{x-1} + H^{+} + H_2O$$

FIGURE 3 Modes of surface coordination to the oxide surface.

distance between the Fe-surface atom and the adsorbed divalent cations M^{2+} is short (<3.6–3.8 Å). Such surface structures or surface complexes are called 'inner sphere complexes'.

2.3.3 Organic Molecules

Adsorption of natural organic matter or organic (macro) molecules from biologic media (proteins, enzymes, etc.) to the surface of NMs will follow previous discussed principles. These principles lead to strong and irreversible adsorption (ligand exchange, cation bridges) or weaker surface bonds (anion exchange, cation exchange, van der Waals interactions). Hydrophobic bonding may be more difficult to predict in terms of reversibility even if most of the time quasi-irreversible adsorption occurred. The specific case of protein adsorption discussed in Chapter 11 will not be detailed in the current chapter. Briefly, recent papers[18] have shown that the quantity of proteins bound to the NM surfaces can be divided into two layers, with a first strongly bound layer up to the saturation of the surface (monolayer) and a second weekly bound protein layer.

2.4 Photocatalytic Reactions

Surface reactivity can take various forms. One unique reaction is the photocatalytic effect. This effect is not specific to NMs but it is clearly modified as the size of particles decrease. Photocatalytic effect is generally described as any catalytic chemical reaction induced by light irradiation and that occurs at the solid surface.[19–21] The photocatalytic reaction is initiated by the transition of one electron from the valence band (VB) to the conduction band (CB). This transition only occurs if the energy of the irradiating light (hν) is equal or higher than the energy difference between the VB and the CB. In the case of TiO_2 (anatase), the transition energy is 3.2 eV on average. When the transition occurs a hole is created in the VB while an electron is added to the CB. Even if this transition can occur in many materials, the hole–electron pair will be affected by mutual electrostatic attractions that are inversely proportional to the dielectric constant of the solid. Therefore solids with very high dielectric constant will enable the pair to reach the surface before they recombine. Then at the surface, the hole and electron will react with an electron donor or an electron acceptor species leading to oxidation and reduction of adsorbed surface molecules (water, organic or inorganic adsorbed molecules) as illustrated in Figure 4.

One characteristic of the smallest NMs is that the transition energy is size-dependent. Indeed, in the specific case of TiO_2 anatase particles, transition energy increases as size decreases[22–25] (Figure 5). In fact, many semi-conductors that are described as 'quantum particles' or 'quantum dots' exhibit changes in their light absorption spectra with a shift to the blue as particle size decreases. This phenomenon has been experimentally observed for anatase

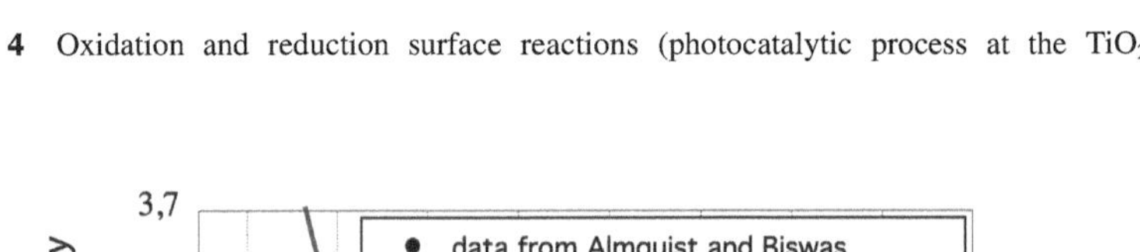

FIGURE 4 Oxidation and reduction surface reactions (photocatalytic process at the TiO_2 surface).

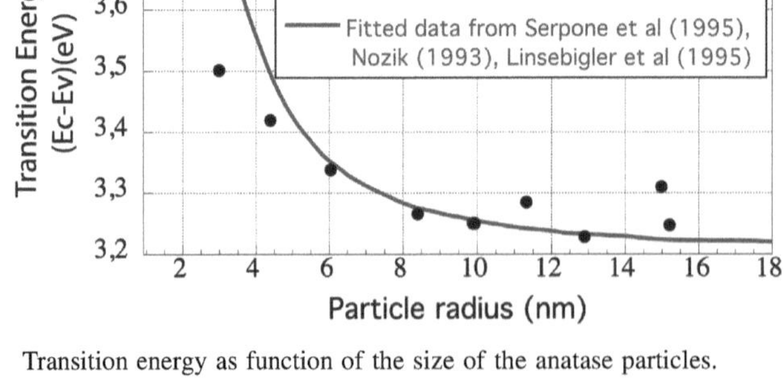

FIGURE 5 Transition energy as function of the size of the anatase particles.

(i.e. Ref. 22) with an increase from 3.2 to 3.5 eV for particles smaller than 12–16 nm (radius). The increase of the transition energy will lead to an increase in the photocatalytic surface reactivity through the increase of the redox potential. It has been recently shown that in addition to the effect of size, the photocatalytic yield can be influenced by the crystallographic defects at the surface.[26]

2.5 Surface Redox Reactions

In the specific case of metal NMs whether under metallic, sulphide or oxides forms, Eh and pH conditions will control their redox state. Indeed, most of metals can exist under various possible redox forms that can be modified at the NM surface due to modification of the external conditions. Metals like Fe, Cu, Ag or rare earths like Ce can illustrate the redox changes: Fe^{2+}, Fe^{3+}; Cu^{2+}, Cu^{1+}, Cu^{0}; Ag^{1+}, Ag^{0}; Ce^{3+}, Ce^{4+}. The modifications of the surrounding environment can be due to physicochemical changes when NMs are transferred from oxic to anoxic waters for example, but also it can be due to interactions with living organisms. Living organisms can control or even modify the Eh through oxidative phosphorylation in mitochondria or by thiol/disulphide redox systems and with mainly the glutathione and cysteine systems. Then many surface atoms to metal NMs may be instable. The case of cerium NMs is interesting. Various studies have shown that the interactions between nano-ceria and bacteria lead to Ce(IV) reduction at the surface of the particles and this phenomenon could be the reason for their toxicity (e.g. Refs 27–29).

2.6 Surface Structure: Facets and Defects

Crystallized particle shapes depend most of the time on crystal growth conditions. However when particles reach thermodynamic equilibrium, the final shape and morphology depend mainly (1) on the crystalline structure of the material (e.g. Ref. 16) (crystallographic space group) and (2) on the total surface energy minimization. As early as 1901, the Wulff theorem[30]; in Ref. 31 enables to determine the equilibrium shape of crystals for large particles. However it is not applicable for nanometric scale. Atomistic calculations are therefore the only alternative for shape prediction. For the early stage growth of *cfc* structure metal NMs for instance, it was calculated that the icosahedron (20 facets) is the more stable shape. Theoretical calculations only work for a limited number of atoms within the particles and failed to predict the particle shape as size is higher than few nanometres.

There is a large literature on the experimental determination of the shape of NMs as function of synthesis conditions. Most of the work is devoted to control the shape through the adsorption of specific molecules, oriented attachment, etc. Indeed, the fine control of the shape of NMs, i.e. the competing growth of each crystal facet requires to control face energies. The synthesis of NMs with control shape remains challenging due to the difficulty to deal with very reactive objects.

A consequence of the reactivity of NMs is that the atomic structure of the facet may not be perfect. Indeed in many cases, it was proved that surface crystal defects exist. For instance, the smallest maghemite NMs (γ-Fe_2O_3) which belong to the spinel group, exhibit a highest level of vacancy of the Fe tetrahedral surface sites than for the larger particles.[32,33] Surface defects in relation with surface relaxation and reconstruction are also supposed to control gold nanorods' growth.[34] It is suggested that the defects relieve stresses introduced during the nanorod synthesis. TiO_2 photocatalytic NMs also exhibit surface defects.[26] In the latter case, Kong et al.[26] proved that the decrease of the concentration ratio between bulk to surface defects enable to increase the photocatalytic efficiency.

2.7 Agglomeration as a Consequence of Surface Properties

Because of their surface charge associated with very small sizes, it is important to remember that NMs exhibit the aptitude to form stable colloidal suspensions over very long periods of time through homo-agglomeration. Two stabilization processes are known which account for both electrostatic and steric interactions.[35] Electrostatic stabilization of NMs, described by the DLVO theory (Derjaguin and Landau, Verwey and Overbeek theory[36,37]), is specific to objects with a surface charge and in aquatic media. The latter stabilization mechanism can be considered as being governed by kinetic energy resulting from attractive charge compensation that intrinsically depends on the chemical nature of the solid (van der Waals forces) and from electrostatic repulsive forces existing between two charged surfaces. Under these equilibrium conditions, particles spontaneously diffuse with the Brownian motion as a driving force. The electrostatic stabilization requires a strict control of the pH, and the nature and ionic strength of the solution. This stabilization process presents the significant advantage to maintain a pristine NM surface i.e. the atoms at the surface are in direct contact with the solution. Steric stabilization differs from electrostatic stabilization in the sense that the surfaces of NMs are chemically modified through adsorption or grafting of polymers from natural solutions (most of the time) forming a molecular corona. Such corona can serve as particle surface solvation and can prevent their agglomeration by steric repulsive forces. Chemical surface modification is at the origin of quasi-elastic interparticle collisions, which lead to the redispersion of all particles. The stability and efficiency of the dispersion depend on the type of solvent responsible for the size increase of the corona, the appropriateness between polymer chains at the surfaces and the surface coverage. Steric stabilization offers various interesting advantages. For instance, steric stabilization is a thermodynamic-based process that can be applied to a large number of systems including non-charged particles and to the dispersion of particles in nonpolar solvent.

3. SURFACE PROPERTIES CHARACTERIZATION IN COMPLEX MEDIA: (X-RAY, ELECTRON, ... BASED TECHNIQUES)

3.1 Samples Preparation: How to Preserve Surface Properties?

Up-to-date characterization techniques of the mineral surfaces have tremendously evolved in the past 10 years. However, the experimental conditions that are required to obtain high signal quality may be quite complex or difficult to apply to complex and heterogeneous bio-inorganic samples in which NMs are dispersed or embedded. Indeed a majority of electron (i.e. transmission electron microscopy (TEM), X-ray photoelectron spectroscopy (XPS), etc.) or ion beam (secondary ion mass spectrometry (nano-SIMS), etc.) techniques require the sample to be under vacuum, which can affect the surface properties of NMs including the dehydration of surfaces. Other X-ray-based techniques may require one to work at low temperature to avoid X-ray beam damages (X-ray absorption spectroscopy (XAS), X-ray diffraction (XRD), etc.). This is particularly obvious for synchrotron based techniques since the X-ray flux on samples becomes higher and higher with technological developments. More over the most powerful techniques to characterize the first atomic layers of the mineral surface can only work if the mineral is not embedded in other media (atomic force microscopy (AFM), scanning tunnelling microscopy (STM), etc.). All of these specifications are generally difficult to achieve in the case of samples which are NMs reacting with living organisms or incorporated in sediment or soils.

Therefore it is essential to develop protocols for sample preparations that avoid or minimize NM surface modifications. Vacuum is one of the most 'stressful' conditions for mineral surfaces. Indeed drying sample protocols to support vacuum need to be performed with special care. Evaporation requires crossing the interfaces between liquid to gaseous phases of the water phase diagram. Water possesses high surface tension to air which indicates that air drying will lead to tangential forces that may affect the nano-structure and therefore surface structure of samples.

Critical point drying might be a very good alternative. Indeed at the critical point it is not possible to distinguish between liquid and gas. Therefore liquid can be converted into gas without crossing the liquid–gaseous interface and therefore this avoids damage. However, the critical point of water implies an increase in T to 374 °C and the pressure to 229 bars that are too high to preserve surface properties. The replacement of water by CO_2 whose critical point is at 31 °C and 74 bar prevents damaging the samples.

Drying can also be performed not at the critical point, but at low temperature using freeze drying, decreasing, but not always, avoiding the effects of water surface tension. In the case of freeze drying, the dehydration occurs through water sublimation and like critical point drying avoids crossing the water liquid–gas interface. However, the critical stage that can alter the samples is therefore the freezing stage. Indeed, as a function of the freezing protocol (using liquid nitrogen, ethanol, etc.) water can convert to vitreous (fast freezing) or crystallized ice. In the case of water crystallization, mechanical constrain will occur and alter biological samples in which NMs can be incorporated. However the use of cryoprotectants that enable a lower water freezing point and increases viscosity in combination to slow the freezing process is an efficient protocol for cryopreservation of biological samples prior to dehydration using freeze drying.

In some cases it is difficult to use the sample preparations described above, and more classical techniques can be used. In this case it can be valuable to adopt protocols that are specifically developed for biological samples, i.e. highly hydrated and fragile samples. Even when a vacuum is required a solution can be to prepared and studied under cryogenic conditions. Many protocols can be found in the literature. In some cases the protocols require one to use specific resins prior to decreasing the temperature of the sample. Resins like Tissue-Tek® O.C.T. compound can be used to coat the samples before cryogenization. These types of resin that set at low temperature do not content alcohol or solvent and do not penetrate into biological tissues and small pores that minimize sample modification. In many cases samples cannot be analyzed like this. Generally thin sections need to be prepared out of larger samples. Cryomicrotome or ultra-cryomicrotome can be used to prepare thin or ultra-thin sections.

Of course if removing water can affect NM surfaces the other possibility is to use techniques that allow one to maintain a certain vapour pressure around the specimen. It is possible to perform electron microscope analysis with a specific chamber that allows for this. Those microscope are named 'environmental scanning electron microscopy (SEMs) or TEMs' to indicate that it is not necessary to apply a high vacuum around the sample.

In conclusion the surface analysis of NMs remains a very difficult task and even more difficult for specimens in which NMs are incorporated or included. In the next section we will detail the specificity of each tool and also mentioned how to best prepare samples.

3.2 Surface Atomic Structure and Atomic Arrangement Characterization

The characterization of surface properties most often requires the determination of the crystalline nature of atoms at the surface. In some cases the high-energy surface tension can lead to modifications of the atomic arrangement or even to amorphization of the surface of crystalline NMs (e.g. Ref. 32,38). Such surface modifications may strongly modify NMs reactivity and therefore their fate and behaviour in complex media. The following techniques can help one to decipher atomic arrangements in detail.

3.2.1 Non-Surface-Specific Techniques (XRD-PDF, XAS, Mössbauer Spectroscopy, Raman, NMR...)

Even if not specifically designed to study the first atomic layers of NM surfaces, many spectroscopies, scattering or microscopy techniques can turn out to be very surface sensitive as soon as the size of particles decrease and the surface to core atomic ratio becomes large.

XRD is dedicated to the study of long-range order of the atomic arrangement in crystals. Interatomic distances and unit cells parameters influence the position, intensity and shape of the diffraction peaks. However, XRD can only provide information for ordered NMs and amorphization as size decrease can strongly affect the intensity and width of the diffracted peaks. The size of the crystallized domain can be calculated based on the Sherrer formula, which states:

$$S = \frac{\lambda}{\omega} \cos\theta \qquad (5)$$

where S is the size of the coherent domain (crystallized domain) of the particle, λ is the wavelength of the X-ray beam, θ the diffraction angle and ω is the width of the peak at half-maximum.

If amorphization occurs at the surface of NM as size decreases, XRD may be useful to extract the size of the core coherent domain when the XRD size is compared to the TEM size determination. However since XRD provides average information and if amorphous phases co-exist with well-crystallized phases XRD would not be able to help decipher whether amorphization occurs at the surface of crystallized NMs or if two phases co-exist.

XRD is used to probe the periodic structure of minerals over distances above 100 Å. But the conversion of the signal obtained in reciprocal space to the real space is perfectly suited to study the non-periodic structure of matter in non-crystalline materials (glasses).[39] The technique called pair distribution function (PDF) requires scanning over a large Q range $Q = 4\pi \sin(\theta/\lambda)$ and at least greater than 20 $Å^{-1}$. Therefore to achieve high Q range values, and to obtain a high interatomic size resolution it is necessary to measure data using high X-ray energy (>100 keV), that is to say, low wavelength (<0.12 Å). Synchrotron are the best X-ray sources to achieve such high X-ray energy.

In addition to long-range order, techniques sensitive to local order can also provide valuable information of atomic structure at the surface of NMs. XAS is well suited to determine both the electronic structure (c.f. further) and the local order (geometry of the site, chemical nature of the neighbours, etc.) of a selected atom in a sample. As opposed to XRD, XAS is not limited to the study of well-ordered NMs but can also be used for either amorphous, poorly ordered or well-crystallized samples.[40] One of the powerful attributes of XAS is that it is element-specific. Therefore, one target atom can be selected in a very heterogeneous specimen. For instance, in the case of NMs composed of non-common element in nature like Ce, Ag or Pt, it is possible to be quite specific to NMs in natural media. XAS can determine the effect of external conditions on the structure of surface atoms. For instance, the pioneering studies of Chemseddine et al.[41] and Rockenberger et al.[42] have shown that the addition of acetate or thiolate can modify the symmetry of the cadmium surface site for CdS particles smaller than 20 nm. Whereas in the bulk of the particles, Cd is coordinated to sulphur to form tetrahedra, CdS_6 octahedra appeared at the surface. Another example from Auffan et al.[43] highlights the sensitivity of XAS to the surface structure of NMs. Extended X-ray absorption fine structures (EXAFS) spectra analysis of Fe_2O_3 coated nano-maghemite revealed that the surface Fe-sites were strongly nonsymmetric due to surface ligand exchange. Indeed the adsorption of di-mercaptosuccinic acid (DMSA) led to the exchange between OH surface Fe groups with SH from DMSA. Therefore the first atomic Fe layer was strongly distorted due to an increase of the Fe-surface ligand distance by at least 0.10/0.15 Å.

The latter examples demonstrate that XAS can probe the surface layer structure of particles, but only if objects are within the nanometric size range. Indeed, XAS gives average information concerning all crystallographic sites that exist in a mineral. Unfortunately, XAS is only sensitive to the largest fraction of sites when several sites exist in a compound. In the case of large particles, the surface sites represent a limited number of atoms that cannot be detected by XAS. As soon as the fraction of surface atoms becomes higher than 15–20%, XAS is sensitive enough to determine the modification of the surface site.

While XAS spectroscopy is based on the measurement of electronic transitions, the Mössbauer effect involves the interaction of γ radiation (i.e. the resonant absorption) with the nuclei of the atoms of a solid. To date, Mössbauer spectroscopy has been mainly used to study Fe NMs, but Au and Pt materials can also be studied by following nuclear transitions (see, for example, Ref. 44). Like XAS, the Mössbauer spectroscopy is element-specific and is sensitive to the atomic arrangement of the surface layer in a bulk solid. For metallic clusters of Au and Pt, Mulder et al.[44] showed that for NMs composed of 55, 147, 309 atoms (2, 3 or 4 atom shells) with cuboctahedral (fcc) arrangement and for a given n-shell compound, the 'surface' metal atoms in the outer nth-shell of the metal core do not show purely 'metallic' behaviour. Furthermore, experiments with two shell Au-55 NMs recovered by different types of ligands (triphenylphosphine, tri-*para*-tolylphosphine, meta-sulphonatophenyldiphenylphosphine, and trianisylphosphine) demonstrated that they do not exhibit a perfect metallic character due to the influence of the surface layer. But of course, like for any 'bulk' spectroscopy, the atomic arrangement can only be determined if particles are very small (below 10–15 nm).

Raman spectra are powerful markers of the structural and morphological properties of solids. Indeed Raman spectra can help to identify the nature of a crystal phase, and for some specific cases assess the crystalline or amorphous nature of NMs. Very similar to XRD, the position of Raman peaks can change as a function of the type of mineral. However, different minerals may exhibit overlapping peaks that can disrupt the exact determination of the NM mineral phase. In the specific case of NMs and even more specifically the surface atomic arrangement, Raman spectroscopy may be less helpful than the previous detailed tools. Raman peaks may be affected when the size of the particles decrease. For instance, in the case of TiO_2 a peak at $142\ cm^{-1}$ is measured for large particles. The peak shifts to 146 and $148\ cm^{-1}$ for particle sizes of 8 and 5 nm, respectively.[45] In the same time the full weight at half-maximum increases from 10 to $18\ cm^{-1}$.[45,46] However, it remains difficult with Raman spectroscopy to be specifically sensitive to surface atoms.

Nuclear magnetic resonance (NMR) can also provide valuable information on the structure of NMs. NMR is based on the interaction between electromagnetic fields and atom nuclei. When a magnetic field is applied, nuclei can be excited to a higher energy level that corresponds to a change in spin. When the spin on the nuclide returns to its base level, it emits energy at a wavelength that corresponds to the energy transfer that excited the nuclide. By measuring the resonance frequency of the nuclides, information on chemical structure can be obtained. All isotopes that contain an odd number of protons and/or of neutrons are suitable for NMR. A strong limitation to NMR is that the presence of paramagnetic elements strongly affects the signal. For example, the presence of iron in a system leads to a strong decrease of the peak intensity and therefore limits the application of NMR in these cases. In the specific case of NMs, elements of interest that can be used by solid-state NMR are as follows: ^{27}Al, ^{29}Si, ^{47}Ti, ^{49}Ti, ^{67}Zn, ^{109}Ag, etc.

The surface structure of NMs can be inferred from solid-state NMR. For example, naturally occurring NMs such as imogolite or allophane can be identified using solid-state NMR.[47,48] Since only Al atoms are at the surface ^{27}Al solid-state NMR can identify surface structural modifications.

Indirect technique can be used to address the different structural surface sites of NMs. For instance, it is possible by using chemical probes to obtain structural information of NM surfaces. Indeed as detailed previously, adsorption of molecules at the surface of minerals depends on many parameters, but most strongly on the reactivity of surface sites. The reactivity is based on the atomic arrangement at the surface. Therefore, the surface structure of the adsorbed molecule will change as a function of the modification of the surface structure of NMs. The following example will help to illustrate such effects. It is, in fact, possible to probe surface adsorption sites by using specific chemical probes.[32,49,50] For instance, As(III) was used to probe the surface structural heterogeneity of 6-nm nano-maghemite (Fe_2O_3). Even if the adsorption of As(III) at oxide surfaces was well studied (e.g. Refs 51–56), new surface sites were identified for iron oxides smaller than 10 nm. Those surface sites are generated by structural modification at the NM surface as their size decrease.[33] While micro-maghemites are composed by octahedral and tetrahedral surface iron sites, nano-maghemite surfaces are preferentially rich in octahedral sites with a high density of vacancies in the tetrahedral sites.[33] Such vacancies are very reactive potential adsorption sites for arsenic. By combining XRD and XAS, it was shown that As(III) was able to get adsorbed preferentially at the Fe(tetrahedral) vacancy site. Then As(III) adsorption enables to probe indirectly the surface sites defects.

3.2.2 Surface-Specific Techniques (TEM, XPS, AFM...)

TEM analyzes the transmitted or forward-scattered electron beam travelling through a sample. Thanks to extremely high spatial resolution modern TEMs can image atomic columns of thin samples and therefore can provide information of the structure and chemistry of surface atoms (Figure 6) of any particle down to the smallest sizes (few nanometres). So even if strictly speaking, TEM is not a 'true surface-specific technique', it is one of the most powerful techniques to provide details of the atomic arrangement at the surface of NMs. Different TEM techniques can be used to obtain structural images with subnanometre resolution and most often atomic level resolution. For instance, high-resolution TEM is based on the interference of coherent electron waves. The interaction between the sample and the (nearly) planar incident waves results in a modification of the electron wave at the exit plane. The modification of the waves is related to the arrangement of atoms within

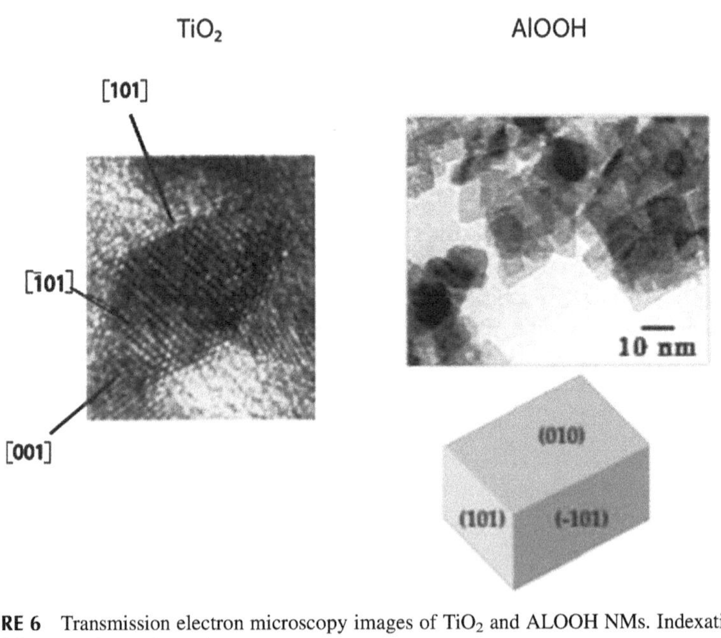

FIGURE 6 Transmission electron microscopy images of TiO_2 and ALOOH NMs. Indexation of crystal faces can be performed.

the object (e.g. Refs 57−59). High-angle annular dark-field TEM images are formed by electrons that scatter at high angles (a few degrees or even more) i.e. inelastic electron scattering. Because inelastic scattering depends on the number of electrons within the sample, the strength of scattering will increase with the atomic numbers. With an electron beam of less than 3 Å, imaging at the atomic level is possible.

The structural characterization of NMs can then be obtained from the measurement of the atomic interlayer distances. When the spatial resolution is high enough, it is possible to determine whether or not the surface atomic arrangement is modified. Figure 7 tends to show that the surface layer of the studied nano-CeO_2 might slightly differ from the core. However, Figure 7 also reveals that for NM surfaces, the determination of atomic structure remains an arduous task. TEM structural characterization can also be performed through electron diffraction. By doing so, information that can be obtained by diffraction (see XRD for instance) is possible on only one particle!

As detailed previously, particle shape is a consequence of surface energy minimization. TEM is a perfect tool to determine the nature of crystal faces of NM and their relative distribution.

SEM may not be considered as perfectly sensitive to the NM surface. Indeed for the smallest NM, the beam spot size may be as large as the NM itself. However, for the largest objects (few tens on nanometres) the secondary electron may provide details on the texture of the surface.

AFM and STM are not based on the interaction between electromagnetic beams and samples, but are based on a simple approach on how a probe 'feels'

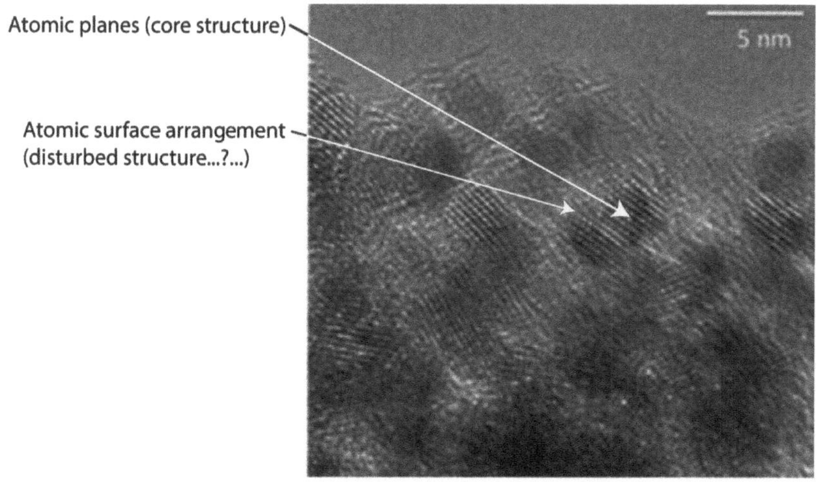

FIGURE 7 Transmission electron microscopy image of nano-CeO_2 particles exhibiting the atomic layers, and possible modification of the atomic arrangement at the surface. *Adapted from Ref. 60.*

its way along a surface. The location between the probe and the surface is determined by the change in cantilever deflection and tunnelling current in AFM and STM, respectively. Of course, the implementation of the principle was extremely complex. Both advanced microscopy techniques are developed to measure surface topography. In both cases, the operating principle consists of scanning a probe in close proximity to the surface. The principle is therefore perfectly sensitive to the surface of any object with no information from the bulk. In terms of resolution because of the sensitivity of the change in cantilever deflection, AFM is limited to a few nanometres and cannot determine the surface atomic structure. However, in the case of STM, the tunnelling current sensitivity enables the spatial resolution to be decreased, and therefore the STM can visualize surface atoms and defects. However, the detection system is based on the tunnelling current, STM is only able to image conducting and semiconducting samples. In the case of NMs in complex media STM requires that the NMs are directly at the surface of the sample.

XPS is one of the most well-known surface specific techniques. Indeed XPS as detailed in previous chapter is an electron spectroscopy technique for which the emitted electrons are ejected from the first 1–5 nm of a surface. However, XPS only provides information on the surface chemistry and not directly on the atomic arrangement and structure of the surface. XPS will be discussed below.

Surface-enhanced Raman scattering (SERS) is a specific case very sensitive to the surface of metallic NM. SERS will be detailed further since as XPS it provides information on the surface chemistry instead of surface structure of NMs.

3.3 Surface Chemistry

3.3.1 Non-Surface Specific Techniques (XAS, Raman, SEM-EDX...)

As detailed earlier, XAS is not specifically sensitive to the surface of NMs. But as the proportion of surface atoms of the smallest NMs represents more than 20% of all atoms, XAS turns out to be highly relevant. Indeed XAS not only provides interatomic distances between the target and neighbouring atoms but also redox state of the target atoms and also the nature of surrounding atoms. For example, Whitley et al.[61] using XAS (and more specifically EXAFS) were able to determine that Ag coated NMs aged with sewage sludge were sulphidized i.e. oxidized and complexed to sulphur. More interestingly, EXAFS revealed an accurate determination of all Ag forms after ageing with sewage sludge including initial Ag, AgS, AgCl and Ag-Acetate. Yin et al.[62] also found Ag-NP sulphidation with plants (*Lolium multiflorum*) by combining micro-XRF and micro-XAS X-ray absorption near-edge structure (XANES). XAS can also reveal the level of surface reduction in the case of cerium NMs. For instance, Auffan et al.,[1] Zeyon et al.[29] and Thill et al.[28] found evidence for surface Ce^{4+} reduction to Ce^{3+} when CeO_2 NP are in contact with bacteria and cells based on XANES spectra. The strong Ce^{3+} peak at 5729 eV as opposed to the doublet at 5733 and 5740 eV for Ce^{4+} are sufficiently distinguishable that it is possible to quantify the $Ce^{3+}/(Ce^{3+} + Ce^{4+})$ ratio with ±8% accuracy/sensitivity. The evolution of the surface reduction state of other metals can be probed by XAS, like Cu, Fe, Mn, etc. However as discussed previously, XAS cannot determine unambiguously the fact that a new chemical phase is formed at the surface or if a new type of particles is formed by dissolution of the original NMs. XAS needs to be associated with other tools like TEM to decipher between both hypotheses.

XAS can also be used to determine how molecules can adsorb at the NM surface. Adsorption can chemically modify NMs leading to variations in behaviour in aqueous and biological media. As detailed in Section 3, chemical or physical adsorption of molecules can occur. This also means that reversible or quasi-irreversible adsorption can occur. Irreversible adsorption of organic molecule can lead to surface charge modification. Indeed the chemical adsorption of DMSA at the nano-Fe_2O_3 surface not only alters the atomic arrangement but also the zeta potential and therefore the agglomeration state. XAS has proved that DMSA can replace OH on surface Fe with thiol groups leading to Fe–S bonds at the surface of nano-maghemite.[43,63] The Fe–S bond is easily detected since the Fe–S distance is 2.15–2.25 Å (like for pyrite (FeS_2[64]), greigite (Fe_3S_4[65]) or chalcopyrite ($CuFeS_2$[66])) whereas Fe–O distance never exceeds 2.1 Å for Fe^{3+} minerals. But the modification of the charge is related to the fact that carboxylic groups exhibit negative charge at pHs higher than 2.5/3.0. Then with the zeta potential for nano-maghemite at physiological pH (~7.4), the chemical adsorption of DMSA at their surface leads to negative zeta potential. Therefore, this negative charge brought by DMSA maintains particles in suspension instead of agglomeration and settling for pristine particles at the same pH.

Raman spectroscopy can also be very sensitive to the surface chemistry of metallic NMs. This phenomenon is so interesting that it has been developed to enhance the detection of specific molecules. Intrinsically Raman is a nonspecific surface technique, but with metallic NM this notion needs to be strongly reconsidered. The phenomenon is called SERS and was discovered 40 years ago.[67] Since the 1980s, it was shown that the signal of molecules adsorbed at the surface of silver NMs can be enhanced by 2–6 orders of magnitude.[68] However, the technique was not designed to characterize surface chemistry of NMs but to enhance the signal of molecule of interest to be detected and analyzed. Therefore SERS was implemented for the most advantageous NMs in support of understanding better adsorbed molecule.[69] Many metal NMs including silver, gold, copper, lithium, palladium, cadmium and nickel[69] can be used. Gold and silver are the usual choice since they combine good enhancement and reasonable chemical stability. Many adsorbed molecules can lead to enhance signal from adsorbed proteins, DNA bases, explosives, therapeutic agents, food additives and many others.

The mechanisms require that an analyte is adsorbed on a surface. Then a selected excitation frequency excites a plasmon and creates scattering. In a second step, the energy from the plasmon is transferred to the molecules previously adsorbed and the Raman phenomenon occurs on the molecule. Then the energy is back transferred to the plasmon. But as previously mentioned even if SERS was not designed to study the chemistry of NM surfaces, in the specific case of metallic NMs like gold and silver, SERS can be of great help to decipher what kind of molecules are adsorbed even when NMs are incorporated in complex media.

Other spectroscopic techniques like NMR can provide information on surface chemistry of NMs. Two of the most important elements to provide information on adsorbed molecules are ^{1}H and ^{13}C. For instance, Hens and Martins reviewed various NMR toolboxes to determine the mode of ligand adsorption at the surface of NMs[70] using ^{1}H solution NMR. In the specific case of quantum dots solution NMR is able to distinguish bound ligands from free ligands and tightly bound ligands from dynamic ligands in a rapid adsorption/desorption equilibrium.[70] Of course, many other examples can take advantage of element-specific NMR. For instance ^{31}P, ^{33}S solid-state NMR can be used to study adsorbed S- and P-rich molecules (sulphide, thiol, phosphate, phosphonate, etc.). As for many other tools, this technique is very efficient for a simple system i.e. NMs and adsorbed molecules, but encounters many limitation for complex systems. For instance, ^{33}S solid-state NMR of biological samples will not be able to discriminate easily free S-rich molecules from S-rich molecules adsorbed at the surface of NMs.

SEM is not surface-sensitive but can provide information on the chemistry of NMs. For instance, energy-dispersive X-ray spectroscopy (EDS, EDX or XEDS) coupled to SEM can be used for elemental analysis of samples. Therefore if molecules (inorganic) are adsorbed at the surface of NMs, they can be detected by EDS. Some limitations exist however. First of all even if the electron beam is small and can reveal details close to the nanometre scale, X-ray fluorescent beam analyzed by EDS detector derives from deeper in the sample than both secondary and backscatter electrons leading to a spatial resolution of a few cubic micrometres.

3.3.2 Surface-Specific Techniques (TEM-EDS/EELS, XPS, Nano-SIMS, STM...)

Coupled to EDS or to energy electron loss spectroscopy (EELS), TEM can provide information about the elemental distribution at a very low spatial scale (several nanometres). The energy resolution of EELS is 0.2 eV, while for EDX is >140 eV at 6 keV. In addition to elemental information, EELS can be used to determine the electronic structure, bonding and nearest neighbour distribution of the specimen atoms. The high-loss energy is related to electrons that have interacted with inner-shell or core electrons of the specimen atoms. Thus, the information obtained is similar to that given by XAS (see the above discussion) – that is, K-, L-, M-, and such ionization edges of the elements present in the sample appear in the EELS spectrum. Therefore, EELS can provide information on the speciation of elements. However, the very low intensity of the EELS signal at energies higher than 1500−2000 eV is a strong limitation in studying the K-edge of elements with an atomic number higher than silicon. Therefore EELS is well adapted to study elements in a 100−1500 eV energy range corresponding to the K-edge of low atomic number elements like carbon and oxygen or the L-edge of transition metals like Fe or Cu. Figure 8 reveals the efficiency of EELS to decipher the mechanisms of nano-zero valent iron oxidation. Indeed for particles with size from 50 to 100 nm, the spatial resolution of EELS coupled to TEM enables one to determine the thickness of the iron oxide layer as function of time; thanks to chemical maps with spatial resolution of 3 nm.

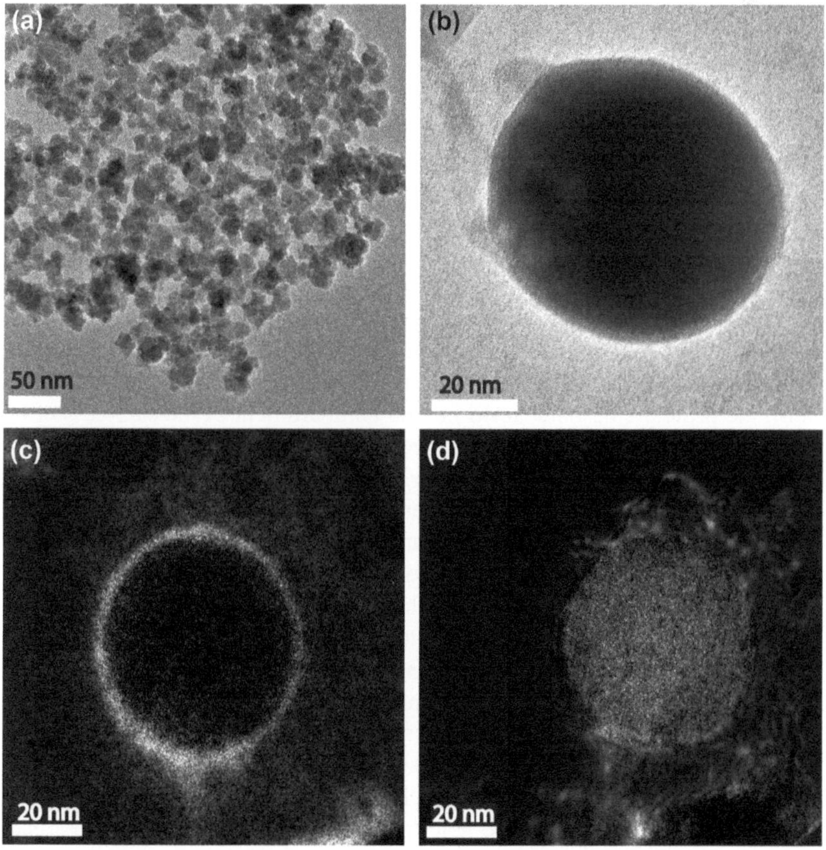

FIGURE 8 Bright-field transmission electron microscopy images of freshly prepared nano-Fe^0 (a and b). Oxygen (c) and Iron (d) elemental mappings of the single nano-Fe^0 particle shown in (b) acquired by EFTEM technique. EFTEM (Energy Filtered TEM) *Adapted from Ref. 71.*

As detailed previously XPS is based on the measurement of the energies of photoelectrons emitted as a result of the interaction between an incident X-ray beam with matter. As it measures the energies emitted from photoelectrons, XPS is classified as an electron spectroscopy technique. The main difference between electron and X-ray spectroscopies is the depth to which the surface is characterized. For example, electrons travel through an extremely short distance in a solid before losing their energy, while X-rays penetrate deeper into the solid matrix. Therefore, XPS is a surface or near-surface-sensitive technique, as it is sensitive to a depth between 1 and 5 nm. However, for NMs smaller than 20 nm, XPS can also be considered as a bulk sensitive technique! The ejected electrons correspond to core photoemission. The core level peaks for a given atom can exhibit different binding energies due to symmetry or oxidation state effects. As an example, for TiO_2 optical irradiation can lead to the formation of charge carriers by optical absorption across the band gap. These charge carriers can directly participate in redox processes on the TiO_2 surface. XPS can distinguish the different Ti oxidation states – that is, the $Ti_2p3/2$ photoemission varies from 455.3 eV for TiO (Ti^{2+}) to 456.7 eV for Ti_2O_3 (Ti^{3+}), 457.6 eV for Ti_3O_5 ($2xTi^{3+}$, Ti^{4+}) and 458.8 eV for TiO_2 (Ti^{4+}).[72] Since core level peak energies are different from one atom to another, and if the chemistry of adsorbed molecules is different from NP chemistry, XPS can be sensitive to NP surface chemistry even for particles smaller than 20 nm.

High-resolution SIMS can also be very sensitive to surface chemistry of NMs. Indeed, recent nano-SIMS work is able to map elemental distributions at high resolution, down to 50 nm, with a very high sensitivity to detect very low concentrations (milligrams per kilogram) for some elements.[73] However, even if the lateral resolution is not sufficient to determine surface chemistry of NMs (nm), the principle of nano-SIMS i.e. the bombardment of the sample section via energetic primary ions (e.g. Cs^+) results in the sputtering of only the top few atomic layers of the sample surface. This means that the vertical resolution is at the atomic level! Then secondary particles are emitted, which are a combination of electrons, uncharged particles and atomic or cluster ions. Some emitted particles from the surface are ionized, and can then be analyzed with a mass spectrometer to determine the chemical distribution of the sample. As mentioned, it is theoretically possible to obtain information from only one or very few atomic layers at each point and then to obtain information on the surface chemistry of particles even smaller than the lateral resolution. However, in practice such lateral resolution prevents from very accurate surface chemistry characterization. In the case of complex media nano-SIMS can provide valuable information on the location of NP. For instance, Georgantzopoulou et al.[74] were able to locate Ag NMs in *Daphnia magna*. This very innovative result comes from one of the most interesting advantages of nano-SIMS i.e. high sensitivity with low concentrations.

Although originally conceived as an imaging device, because the operating principle of the AFM is based on the measurement of force between a small tip and a surface with piconewton sensitivity, this method can also be used to characterize interactions between surfaces and NMs that are modified by modifications of surface chemistry.[75] The interaction force occurring between the AFM probe and the sample surface is then calculated according to Hooke's law

$$F = k\Delta z \tag{6}$$

where F is the force; k is the cantilever spring constant and Δz is the vertical deflection of the cantilever. AFM force measurements may be carried out using either a standard silicon nitride-tipped cantilever or a probe with attached material such as a colloid or NM. By attaching NMs such as a single-wall nanotube, information on interacting forces between the NM and an approaching surface may be measured directly. Alternatively, a colloid probe can be attached to the AFM tip and used to measure interactions with an area of deposited NMs.

4. CONCLUDING REMARKS

The characterization of NMs remains a difficult task because for most of the techniques technical limits are reached. But with recent development, it is now possible to obtain information on the atomic structure, chemical composition, reactivity, etc. of almost all NMs. It becomes less trivial when we focus on the surface of NMs, since it only concerns a small fraction of atoms. However, we have detailed most of the important techniques that can be used to obtain some information on the surface properties of NMs.

The major remaining challenge will be to determine surface properties of NMs when they are located in complex media for instance in most samples from nano-ecotoxicity/toxicity studies. It is almost impossible to address specifically the structure and chemistry of NM surfaces in soil, living organisms, sediments and natural waters. Only by combining various techniques, with all their advantages and limitations, it is sometimes possible to indirectly determine NM surface properties in pseudo natural and therefore complex systems. One has to be aware of all the advantages and limitations of the techniques described above, but one must also remember that sample preparation remains essential to preserving the surface makeup and properties of NMs.

REFERENCES

1. Auffan M, Rose J, Bottero JY, Lowry GV, Jolivet JP, Wiesner MR. Towards a definition of inorganic nanoparticles from an environmental, health and safety perspective. *Nat Nanotechnol* 2009a;**4**(10):634–41.

2. Daniel M, Astruc D. Gold nanoparticles: assembly, supramolecular chemistry, quantum-size-related properties, and applications toward biology, catalysis, and nanotechnology. *Chem Rev* 2004;**35**(16).
3. Haruta M. Size- and support-dependency in the catalysis of gold. *Catal Today* 1997;**36**(1): 153–66.
4. Sau TK, Pal A, Pal T. Size regime dependent catalysis by gold nanoparticles for the reduction of eosin. *J Phys Chem B* 2001;**105**(38):9266–72.
5. Alivisatos AP. Semiconductor clusters, nanocrystals, and quantum dots. *Science* 1996; **271**(5251):933–7.
6. Kelly KL, Coronado E, Zhao LL, Schatz GC. The optical properties of metal nanoparticles: the influence of size, shape, and dielectric environment. *J Phys Chem B* 2003;**107**(3):668.
7. Jolivet JP, Froidefond C, Pottier A, Chaneac C, Cassaignon S, Tronc E, et al. Size tailoring of oxide nanoparticles by precipitation in aqueous medium. A semi-quantitative modelling. *J Mater Chem* 2004;**14**:3281–8.
8. Dick K, Dhanasekaran T, Zhang Z, Meisel D. Size-dependent melting of silica-encapsulated gold nanoparticles. *J Am Chem Soc* 2002;**124**(10):2312.
9. Lai SL, Guo JY, Petrova V, Ramanath G, Allen LH. Size-dependent melting properties of small tin particles: nanocalorimetric measurements. *Phys Rev Lett* 1996;**77**(1):99.
10. Zhang M, Efremov MY, Schiettekatte F, Olson EA, Kwan AT, Lai SL, et al. Size-dependent melting point depression of nanostructures: nanocalorimetric measurements. *Phys Rev B* 2000;**62**(15):10548.
11. Pauling L. The principles determining the structure of complex ionic crystals. *J Am Chem Soc* 1929;**51**:1010–26.
12. Adamson AW, Gast AP. *Physical chemistry of surface*. 6th ed. New York: John Wiley & Sons; 1997.
13. Jolivet JP, Chaneac C, Tronc E. Iron oxide chemistry. From molecular clusters to extended solid networks. *Chem Commun* 2004;**5**:481–7.
14. Abbas Z, Labbez C, Nordholm S, Ahlberg E. Size-dependent surface charging of nanoparticles. *J Phys Chem C* 2008;**112**:5715–23.
15. Holmberg JP, Ahlberg E, Bergenholtz J, Hassellöv M, Abbas Z. Surface charge and interfacial potential of titanium dioxide nanoparticles: experimental and theoretical investigations. *J Colloid Interface Sci* 2013;**407**:168–76.
16. Jolivet J-P, Cassaignon S, Chanéac C, Chiche D, Durupthy O, Portehault D. Design of metal oxide nanoparticles: control of size, shape, crystalline structure and functionalization by aqueous chemistry. *C R Chim* 2010;**13**:40–51.
17. Vayssieres L. On the effect of nanoparticle size on water-oxide interfacial chemistry. *J Phys Chem C* 2009;**113**:4733–6.
18. Milani S, Bombelli FB, Pitek AS, Dawson KA, Radler J. Reversible versus irreversible binding of transferrin to polystyrene nanoparticles: soft and hard corona. *ACS Nano* 2012;**6**(3):2532–41.
19. Fujishima K, Hashimoto A, Watanabe T. *TiO_2 photocatalysis: fundamentals and applications*. Tokyo: BKC, Inc.; 1999.
20. Pelizzetti E, Serpone N. *Homogeneous and heterogeneous photocatalysis*. Dordrecht: D. 1Reidel Publishing Company; 1986.
21. Serpone N, Pelizzetti E. *Photocatalysis: fundamentals and applications*. New York: John Wiley & Sons; 1989.
22. Almquist CB, Biswas P. Role of synthesis method and particle size of nanostructured TiO_2 on its photoactivity. *J Catal* 2002;**212**:145–56.

23. Linsebigler AL, Lu G, Yates Jr JT. Photocatalysis on TiO_2 surfaces – principles, mechanisms, and selected results. *Chem Rev* 1995;**95**:735–58.
24. Nozik AJ. In: Ollis DF, Al-Ekabi H, editors. *Photocatalytic purification and treatment of water and air.* Amsterdam: Elsevier; 1993. p. 39.
25. Serpone N, Lawless D, Khairutdinov R. Size effects on the photophysical properties of colloidal anatase TiO_2 particles – size quantization or direct transitions in this indirect semiconductor. *J Phys Chem* 1995;**99**:16646.
26. Kong M, Li Y, Chen X, Tian T, Fang P, Zheng F, et al. Tuning the relative concentration ratio of bulk defects to surface defects in TiO_2 nanocrystals leads to high photocatalytic efficiency. *J Am Chem Soc* 2011;**133**:16414–7.
27. Auffan M, Rose J, Orsiere T, De Meo M, Thill A, Zeyons O, et al. CeO_2 nanoparticles induce DNA damage towards human dermal fibroblasts in vitro. *Nanotoxicology* 2009b;**3**:161–71.
28. Thill A, Zeyons O, Spalla O, Chauvat F, Rose J, Auffan M, et al. Cytotoxicity of CeO_2 nanoparticles for *Escherichia coli*. Physico-chemical insight of the cytotoxicity mechanism. *Environ Sci Technol* 2006;**40**:6151–6.
29. Zeyons O, Thill A, Chauvat F, Menguy N, Cassier-Chauvat C, Orear C, et al. Direct and indirect CeO_2 nanoparticles toxicity for *E. coli* and synechocystis. *Nanotoxicology* 2009;**3**:284–95.
30. Wulff G. On the question of speed of growth and dissolution of crystal surfaces. *Z Krystallogr Mineral* 1901:449–530.
31. Mutaftschiev B. In: Hull R, Osgood RM, Sakaki H, Zunger A, editors. *The atomistic nature of crystal growth.* Springer; 2001. p. 368.
32. Auffan M, Rose J, Proux O, Borschneck D, Masion A, Chaurand P, et al. Enhanced adsorption of arsenic onto maghemites nanoparticles: As(III) as a probe of the surface structure and heterogeneity. *Langmuir* 2008;**24**(7):3215–22.
33. Brice-Profeta S, Arrio MA, Tronc E, Menguy N, Letard I, Cartier dit Moulin C, et al. Magnetic order in [gamma]-Fe_2O_3 nanoparticles: a XMCD study. *J Magn Magn Mater* 2005;**288**:354.
34. Gai PL, Harmer MA. Surface atomic defect structures and growth of gold nanorods. *Nano Lett* 2002;**2**:771–4.
35. Cao G. *Nanostructures and nanomaterials: synthesis, properties and applications.* Imperial College Press; 2004.
36. Derjaguin B, Landau L. Theory of the stability of strongly charged lyophobic sols and of the adhesion of strongly charged particles in solutions of electrolytes. *Acta Phys Chem URSS* 1941;**14**:633.
37. Verwey EJW, Overbeek JThG. *Theory of the stability of lyophobic colloids.* Amsterdam: Elsevier; 1948.
38. Naicker PK, Cummings PT, Zhang HZ, Banfield JF. Characterization of titanium dioxide nanoparticles using molecular dynamics simulations. *J Phys Chem B* 2005;**109**(32):15243–9.
39. Warren BE. *X-ray diffraction.* Dover Publications; 1990.
40. Modrow H. Tuning nanoparticle properties – the X-ray absorption spectroscopic point of view. *Appl Spectrosc Rev* 2004;**39**:183–290.
41. Chemseddine A, Fieber-Erdmann M, Holub-Krappe E, Boulmaaz S. XAFS study of functionalized nanoclusters and nanocluster assemblies. *Z Phys D Atoms Mol Clust* 1997;**40**(1–4):566–9.
42. Rockenberger J, Tröger L, Kornowski A, Vossmeyer T, Eychmüller A, Feldhaus J, et al. EXAFS studies on the size dependence of structural and dynamic properties of CdS nanoparticles. *J Phys Chem B* 1997;**101**:2691–701.

43. Auffan M, Decome L, Rose J, Orsiere T, De Meo M, Briois V, et al. In vitro interactions between DMSA-coated maghemite nanoparticles and human fibroblasts: a physicochemical and cyto-genotoxical study. *Environ Sci Technol* 2006;**40**(14):4367–73.
44. Mulder E, Thiel RC, de Jongh L, Gubbens PCM. Size-evolution towards metallic behavior in nano-sized gold and platinum clusters as revealed by ^{197}Au Mössbauer spectroscopy. *Nanostruct Mater* 1996;**7**(3):269–92.
45. Kelly S, Pollak FH, Tomkiewicz M. Raman spectroscopy as a morphological, probe for TiO_2 aerogels. *J Phys Chem B* 1997;**101**:2730–4.
46. Choi HC, Jung YM, Kim SB. Size effects in the Raman spectra of TiO_2 nanoparticles. *Vib Spectrosc* 2005;**37**:33–8.
47. Denaix L, Lamy I, Bottero J-Y. Structure and affinity towards Cd^{2+}, Cu^{2+}, Pb^{2+} of synthetic colloidal amorphous aluminosilicates and their precursors. *Colloids Surf A Physicochem Eng Aspects* 1999;**158**:315–25.
48. Levard C, Rose J, Thill A, Masion A, Doelsch E, Maillet P, et al. Formation and growth mechanisms of imogolite-like aluminogermanate nanotubes. *Chem Mater* 2010;**22**(8): 2466–73.
49. Madden AS, Hochella MF, Luxton TP. Insights for size-dependent reactivity of hematite nanomineral surfaces through Cu^{2+} sorption. *Geochim Cosmochim Acta* 2006;**70**(16): 4095–104.
50. Villiéras F, Michot LJ, Bardot F, Chamerois M, Eypert-Blaison C, François M, Gérard G, Cases J-M. Surface heterogeneity of minerals. *C R Geosci* 2002;**334**(9):597–609.
51. Manceau A. The mechanism of anion adsorption on iron oxides: evidence for the bonding of arsenate tetrahedra on free $Fe(O, OH)_6$ edges. *Geochim Cosmochim Acta* 1995;**59**(17):3647.
52. Manning BA, Fendorf SE, Goldberg S. Surface structures and stability of arsenic(III) on goethite: spectroscopic evidence for inner-sphere complexes. *Environ Sci Technol* 1998;**32**(16):2383.
53. Sherman DM, Randall SR. Surface complexation of arsenic(V) to iron(III) (hydr)oxides: structural mechanism from ab initio molecular geometries and EXAFS spectroscopy. *Geochim Cosmochim Acta* 2003;**67**(22):4223.
54. Thoral S, Rose J, Garnier JM, Van Geen A, Refait P, Traverse A, et al. XAS study of iron and arsenic speciation during Fe(II) oxidation in the presence of As(III). *Environ Sci Technol* 2005;**39**(24):9478–85.
55. Waychunas GA, Davis JA, Fuller CC. Geometry of sorbed arsenate on ferrihydrite and crystalline FeOOH: re-evaluation of EXAFS results and topological factors in predicting sorbate geometry, and evidence for monodentate complexes. *Geochim Cosmochim Acta* 1995;**59**(17):3655.
56. Waychunas GA, Rea BA, Fuller CC, Davis JA. Surface chemistry of ferrihydrite: part 1. EXAFS studies of the geometry of coprecipitated and adsorbed arsenate. *Geochim Cosmochim Acta* 1993;**57**(10):2251.
57. Balmes O, Malm JO, Pettersson N, Karlsson G, Bovin JO. Imaging atomic structure in metal nanoparticles using high-resolution cryo-TEM. *Microsc Microanal* 2006;**12**(2):145–50.
58. Jose-Yacaman M, Marin-Almazo M, Ascencio JA. High resolution TEM studies on palladium nanoparticles. *J Mol Catal A Chem* 2001;**173**(1–2):61–74.
59. Yacaman MJ, Zorrilla C, Ascencio JA, Mondragon G, Reyes-Gasga J. Study of high resolution TEM images of nanoparticles either supported on amorphous films or embedded in a crystalline matrix. *Mater Trans JIM* 1999;**40**(2):141–5.
60. Auffan M, Masion A, Labille J, Diot MA, Liu W, Olivi L, et al. Long-term aging of a CeO_2 based nanocomposite used for wood protection. *Environ Pollut* 2014;**188**:1–7.

61. Whitley AR, Levard C, Oostveen E, Bertsch PM, Matocha CJ, von der Kammer F, et al. Behavior of Ag nanoparticles in soil: effects of particle surface coating, aging and sewage sludge amendment. *Environ Pollut* 2013;**182**:141–9.
62. Yin LY, Cheng YW, Espinasse B, Colman BP, Auffan M, Wiesner M, et al. More than the ions: the effects of silver nanoparticles on *Lolium multiflorum*. *Environ Sci Technol* 2011;**45**(6):2360–7.
63. Auffan M, Shipley HJ, Yean S, Kan AT, Tomson M, Rose J, et al. Nanomaterials as adsorbent. In: McGraw Hill, Wiesner, Bottero, editors. *Environmental nanotechnology: applications and impacts of nanomaterials*; 2007.
64. Bayliss P. Crystal structure refinement of a weakly anisotropic pyrite. *Am Mineral* 1977;**62**:1168–72.
65. Skinner BS, Erd RC, Grimaldi FS. Greigite, the thiospinel of iron: a new mineral. *Am Mineral* 1964;**49**:543–55.
66. Hall SR, Stewart YM. The crystal structure refinement of chalcopyrite, $CuFeS_2$. *Acta Crystallogr* 1973;**B29**:579–85.
67. Fleischmann M, Hendra PJ, McQuillan AJ. Raman spectra of pyridine adsorbed at a silver electrode. *Chem Phys Lett* 1974;**26**:163–6.
68. Kerker M. Electromagnetic model for surface-enhanced Raman-scattering (SERS) on metal colloids. *Acc Chem Res* 1984;**17**(8):271–7.
69. McNay G, Eustace D, Smith WE, Faulds K, Graham D. Surface-enhanced Raman scattering (SERS) and surface-enhanced resonance Raman scattering (SERRS): a review of applications. *Appl Spectrosc* 2011;**65**(8):825–37.
70. Hens Z, Martins JC. A solution NMR toolbox for characterizing the surface chemistry of colloidal nanocrystals. *Chem Mater* 2013;**25**(8):1211–21.
71. Kumar N, Auffan M, Gattacceca J, Rose J, Olivi L, Borschneck D. Molecular Insights of Oxidation Process of Iron Nanoparticles: Spectroscopic, Magnetic, and Microscopic Evidence. *Environ Sci Technol* 2014;**48**(23):S. 13888–13894.
72. Song Z, Hrbek J, Osgood R. Formation of TiO_2 nanoparticles by reactive-layer-assisted deposition and characterization by XPS and STM. *Nano Lett* 2005;**5**(7):1327–32.
73. Moore KL, Lombi E, Zhao FJ, Grovenor CRM. Elemental imaging at the nanoscale: NanoSIMS and complementary techniques for element localisation in plants. *Anal Bioanal Chem* 2012;**402**(10):3263–73.
74. Georgantzopoulou A, Balachandran YL, Rosenkranz P, Dusinska M, Lankoff A, Wojewodzka M, et al. Ag nanoparticles: size- and surface-dependent effects on model aquatic organisms and uptake evaluation with NanoSIMS. *Nanotoxicology* 2013;**7**(7): 1168–78.
75. Hodges CS. Measuring forces with the AFM: polymeric surfaces in liquids. *Adv Colloid Interface Sci* 2002;**99**(1):13–75.

Part III

Case Studies

Chapter 8

Control of Nanomaterials Used in Chemical Mechanical Polishing/Planarization Slurries during On-site Industrial and Municipal Biological Wastewater Treatment

Xiangyu Bi,[1] Robert Reed and Paul Westerhoff
School of Sustainable Engineering and the Built Environment, Arizona State University, Tempe, AZ, USA
[1]*Corresponding author: E-mail: xiangyu.bi@asu.edu*

Chapter Outline

1. INTRODUCTION TO NANOMATERIALS IN CHEMICAL MECHANICAL POLISHING/PLANARIZATION FLUIDS

Chemical mechanical polishing/planarization (CMP) is a process that removes materials by a combination of chemical and mechanical (or abrasive) actions to achieve highly smooth and planar material surfaces. CMP can be used to planarize a variety of materials including dielectrics, semiconductors, metals, polymers, and composites, and in semiconductor production processes it is crucial for achieving the performance goals of modern microprocessor and memory chips.[1,2] A typical CMP operation scheme is shown in Figure 1. Inorganic abrasive particles are an important component of CMP slurries. The three most commonly used abrasive particles in CMP slurries are the metal/metalloid oxide species alumina (Al_2O_3), silica (SiO_2), and ceria (CeO_2).[3] These three engineered nanomaterials are among the top 10 in production and use. Worldwide consumption amounts of 5500 ton/year for silica and 55 ton/year for alumina and ceria have been reported.[4] The CMP process plays a significant role in utilizing the considerable amount of the three nanomaterials.[5,6] CMP is estimated to be the second largest market for nanomaterials with a total annual usage of 9400 ton,[7] and CMP nanomaterials (NM) constituted nearly 60% of the total $1 billion worldwide market for nanopowders by 2005.[8,9] During the CMP process, a large volume of ultrapure water is used to clean the surface of polishing substrate (e.g., wafers), generating wastewater containing NM from the CMP slurry.[10] Thus CMP wastewater is potentially a major source of NM into sewer systems, and eventually into the environment. However, essentially no data exist on these materials because Si, Ce, and Al are not regulated metals in industrial discharges or at municipal wastewater treatment plants (WWTPs). Silica is the most commonly used abrasive, and Si concentrations of 1000–2000 mg/L have been reported in CMP tools effluent.[11,12] Because they are used in lesser quantities than silica slurries, alumina and ceria

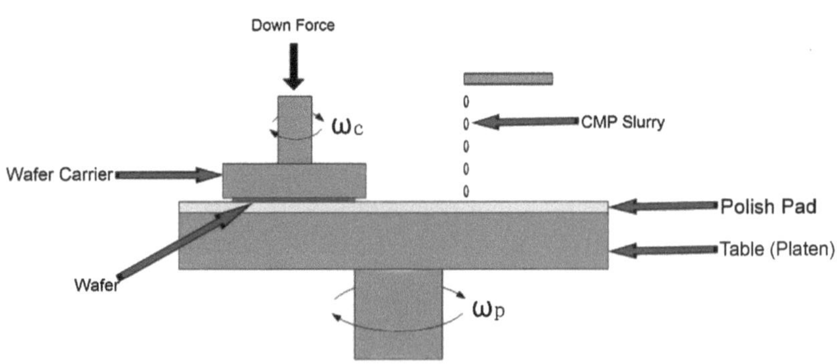

FIGURE 1 Schematic of representative chemical mechanical polishing/planarization (CMP) operation. *Modified from http://www.ntu.edu.sg/home/mdlbutler/Research/Research_CMP.htm [accessed 24.11.14].*

concentrations in CMP tools effluent (e.g., 1−100 mg/L) are expected to be lower than SiO_2. Current industrial discharge limits into sewers from semiconductor fabrication facilities generally include turbidity, biochemical oxygen demand, and specific chemicals for which municipal WWTPs have limits (e.g., arsenic, copper, fluoride, zinc, etc.). Not all fabrication facilities require on-site industrial treatment; this depends on their flow contribution to the local sewer system and local regulations.

The abrasive particles in CMP slurries are usually roughly spherical. Depending on different applications, particle size in CMP slurries can vary from 50 nm to 1 μm, and trends are toward CMP particles less than 100 nm in diameter to achieve highly polished surfaces.[3,13] The three most common oxide abrasives used in CMP slurries, SiO_2, Al_2O_3, and CeO_2 NM, usually have a relatively uniform shape and size. Silica can be distinguished as fumed silica or colloidal silica based on the synthesis method. Fumed silica is formed in a pyrogenic process by oxidizing chlorosilane ($SiCl_4$) at high temperature.[14] Colloidal silica is formed in liquid phase from a Si precursor (e.g., Na_2SiO_3).[15] A widely referenced method of synthesizing colloidal silica was presented by Stöber,[16] and a tetraalkylsilicate was used as the Si precursor in this method. Ceria NM used in CMP slurries typically have a crystalline structure, thus often yielding sharp edges, corners, and apexes.[17] Alumina NM used in CMP slurries can be α-alumina, ϒ-alumina, δ-alumina, and fumed alumina.[18] Alumina is softer than silica or ceria and is sometimes being coated with harder material such as silica.[3] NM below 100 nm can still be detected from CMP wastewater after the manufacturing lifetime.[12,19]

Considering the widespread use of nanomaterials in industrial CMP, it is important to balance their benefit to society with knowledge about the potential risks associated with release of these materials to the environment. Specifically, the ecological impacts and toxicity of CMP nanomaterials are of increasing interest to environmental researchers and agencies.[20,21] Figure 2 illustrates potential intervention points (i.e., on-site treatment, municipal WWTPs) to remove nanomaterials. The majority of on-site industrial treatments involve precipitation of metals, or sorption of pollutants (e.g., arsenic sorption) onto precipitates. For example, lime softening is commonly designed to remove dissolved and particulate copper from fabrication facilities' waste streams. Precipitated materials are gravitationally settled and disposed of in landfills. Materials not removed are generally discharged to municipal sewer systems and enter municipal WWTPs that often use biological treatment designed to remove nutrients (carbon, nitrogen, phosphorous) but are also capable of removing nanomaterials.[22−24]

In this chapter, we present lab-scale experiments designed to simulate the fate and removal of CMP NM through wastewater treatments including lime-softening and biomass adsorption processes. The objective is to investigate how well the CMP NM can be removed. We also present analytical techniques and limitations for quantifying engineered nanomaterials.

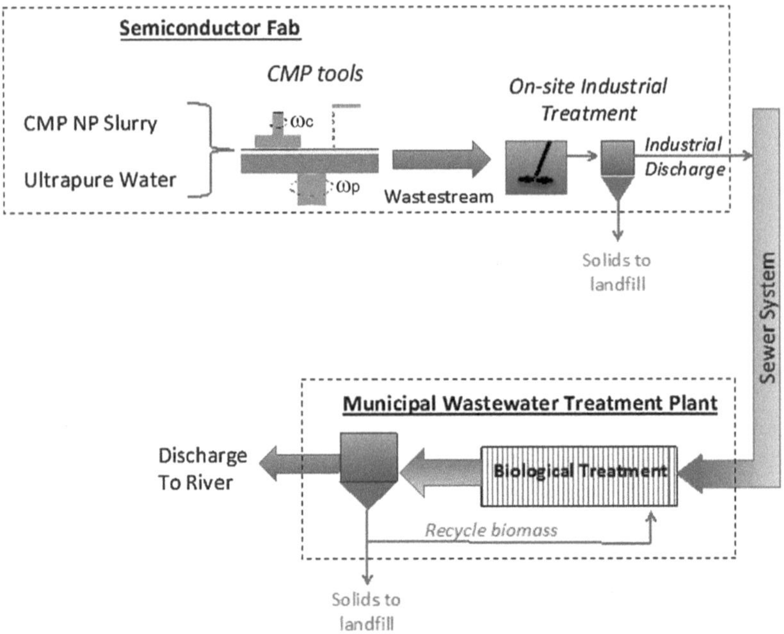

FIGURE 2 Source, treatment, and fate of chemical mechanical polishing/planarization (CMP) nanomaterials (NM) used in the semiconductor industry.

2. MATERIALS AND METHODS

2.1 Selection of CMP Slurries that Constitute Representative NM Species

We obtained four CMP slurry products from Cabot Microelectronics (Aurora, IL) as part of a collaboration with the Semiconductor Research Corporation. CMP slurries generally contain suspended NM, surfactants, oxidants, and other proprietary constituents. For use in this study, simplified slurries were provided absent of additives other than NM and acids or bases as pH-adjusting agents. The NM were dispersed and stabilized in suspension using mechanical processes, by Cabot, and the NM purportedly remain stable in suspension for months to years. These four CMP slurry products contain different metal/metalloid oxide NM species: colloidal silica, fumed silica, ceria, and alumina, generally covering the most commonly used NM in CMP slurries. The manufacturer-provided information about the four CMP slurries is shown in Table 1 and their complete characterization is the focus of another paper by the chapter authors. Here we present measurements for NM size distribution and characterization by dynamic light scattering (DLS), scanning electron microscopy (SEM), and single particle inductively coupled plasma—mass spectrometry (spICP-MS). Chemical compositions of the four CMP slurries were

TABLE 1 Characterization of Four CMP Slurries Provided by the Manufacturer

Slurry	Nanomaterial Concentration	Nanomaterial Size (nm)	pH	Additives
Colloidal silica	3 wt%	50–60	2.5–4.5	Acetic acid <1%
Fumed silica	5 wt%	120–140	10	KOH <1%
Ceria	1 wt%	60–100	3–4	–
Alumina	3 wt%	80–100	4.5–5.0	Nitric acid <1%

analyzed by ICP-MS after appropriate chemical and microwave digestion. For cases in which dissolved species are required to be distinguished from particulates, we used centrifugal ultrafilters to remove NM from solution.

2.2 Simulated Chemical Softening and Biological Treatment

To simulate the on-site lime-softening treatment process, jar tests were performed on diluted CMP slurries with $CaCl_2$ added under different pH conditions. The four types of CMP slurries were diluted 500–1000x to get around 20–30 mg/L metal concentrations of each NM species, which is fairly close to the corresponding concentrations in CMP wastewater. $CaCl_2$ was selected as the softening agent instead of lime or $Ca(OH)_2$ for the sake of better controlling pH conditions. All jar tests were operated in order of: (1) rapid mixing at 100 rpm for 2 min; (2) flocculation at 30 rpm for 30 min; and (3) sedimentation for 60 min. Samples were collected from the supernatant.

For the biomass adsorption test, we followed methods as previously published.[22] Fresh biomass suspension was cultured in lab-scale sequencing batch reactors (SBRs), which was inoculated with return activated sludge from a WWTP in the metro Phoenix area. The SBRs were operated in a 10-h cycle mode, which consisted 8 h of aeration, 90 min of sludge settling, and 30 min of feed solution replacement. The settled supernatant was discharged and 2-L fresh synthetic feed solution was added in each replacement. The cultured biomass was diluted to 1000 mg/L to be used for the NM adsorption experiment. All adsorption systems were buffered at pH 7 with 1 mM $NaHCO_3$. The four CMP slurries were tested with two concentration levels: (1) a concentration to match those used in the jar test, which is 26 mg/L as Si for colloidal silica, 24 mg/L as Si for fumed silica, 20 mg/L for ceria, and 31 mg/L for alumina; (2) a lower concentration level which was 4 mg/L for each NM. After addition of the CMP slurries into the biomass adsorption system, the solutions were shaken for 3 h and then allowed to gravitationally settle for 30 min. The supernatants were sampled and their metal concentrations were analyzed by

ICP-MS. In previous work, experiments with biomass from lab-scale SBRs and full-scale municipal WWTPs showed nearly identical NM removal.[25] The benefit of the lab-scale SBR biomass is the low concentrations of trace metals compared with those often present in full-scale WWTP biomass.

3. ANALYTICAL METHODS

3.1 DLS Analysis

A Brookhaven ZetaPals Analyzer (Brookhaven Instruments, Holtsville, USA) was used to perform DLS analysis with the particle-sizing function. The wavelength and the scattering angle were set at 660 nm and 90°, respectively. A multimodal size distribution (MSD) algorithm was used to characterize the particle size distribution. The MSD output format was set as intensity, the direct instrument measurement. Refractive indexes were used as: 1.765 for alumina, 2.200 for ceria, and 1.542 for silica.

3.2 spICP-MS Analysis for Sizing NM

spICP-MS is an emerging technique used to size and quantify NM in aqueous matrices. Details about the operation of spICP-MS can be found in recent publications.[26–28] Briefly, short signal integration (dwell) times are used to detect pulses of analyte above background, with the pulses representing particle detection events. A Thermo X series II ICP-MS was used for all analyses. The instrument was placed in time-resolved analysis mode in which the signal was recorded every dwell time (integration time of one reading by the detector) of 10 ms duration. A glass nebulizer and a conical spray chamber constitute the sample introduction system. The sample flow rate was set in the range of 0.6–0.7 $mL \cdot min^{-1}$. ^{140}Ce and ^{27}Al were used as the analyte isotopes for ceria and alumina NM. The analysis of silica was not performed because the detection limit for Si by spICP-MS is relatively high (>400 nm).[29]

3.3 CMP NM Chemical Digestion

Digestion is performed for the purpose of transforming NM into ions, allowing robust quantification of metal concentrations by ICP-MS. Conventional digestion methods[30] use acid (e.g., HNO_3, HCl, etc.) and heat the sample on hot plate or in a microwave digestion system. Herein we determined the digestion method for the four types of CMP NM as follows.

For colloidal silica and fumed silica, an alkaline solution, tetramethylammonium hydroxide (TMAH, 25%, J.T. Baker) was used to digest silica NM with microwave assistance. Specifically, 4 mL TMAH stock solution was added to each sample, with suitable amount of pure water to make the total sample volume at 15 mL. The microwave was operated by ramping the temperature up to 150 °C in the first 15 min; ramping the temperature up to 180 °C in the second 15 min; holding the temperature at 180 °C for 30 min. For ceria and alumina, 2 mL hydrofluoric acid (50%, J.T. Baker), 2 mL HCl (30%, J. T. Baker), and 6 mL HNO_3 (70%, J. T. Baker) were added into the sample and total volume was adjusted to 15–20 mL with pure water. The microwave procedure was identical to that used for silica.

3.4 Separation of NM from Dissolved Ions

Centrifugation is a traditional method to separate particulates from solution. Instead of relying solely upon mass density separation, which is ineffective for very small NM, a centrifugal ultrafiltration device with 30,000 Da nominal molecular weight limit (Millipore, Darmstadt, Germany) was used to separate NM and the ionic (dissolved) species for all four CMP slurries. Samples in centrifugal ultrafiltration devices were centrifuged at 5000 G for 30 min.

To demonstrate the effectiveness of the centrifugal ultrafiltration device to separate NM and dissolved species, we used a commercial SiO_2 NM (PM1040, Nissan Chemical, Houston) and ionic SiO_2 standard solution (HACH, Loveland). We made three solutions: (1) solution containing 1000 ppb (as SiO_2) ionic standard and 1000 ppb (as SiO_2) NM; (2) 1000 ppb NM; and (3) 1000 ppb ionic standard. Ten milliliter of each solution was pipetted into the centrifugal ultrafiltration devices. Thus the three solutions should give (1) 4.7 μg ionic Si and 4.7 μg NM Si; (2) only 4.7 μg NM Si; and (3) only 4.7 μg ionic Si. Figure 3 shows the recovery for the filtrate and concentrate for the three solutions. The recoveries of both filtrate (i.e., constituents that flux through the filter) and concentrate (i.e., constituents that are rejected by the filter) for all cases are close to the spiked amount. For the mixture case (solution (1)), both ionic Si and NM Si can be nearly completely recovered ($102 \pm 6\%$ for ionic Si and $106 \pm 8\%$ for NM Si, error bars are based on triplicate measurements by the instrument). For the solution containing only NM Si, 0.1 μg of Si (out of the total 4.7 μg) was detected from the filtrate (considered as ionic Si), which accounts for only ~2% of the Si mass. This indicates that some Si present in the silica NM solution passed through the ultrafiltration membrane. It is likely that the filterable Si was present as dissolved Si species resulting from dissolution of the original silica NM. For the solution containing only ionic Si, it gives recovery of 94% (4.4 μg out of 4.7 μg), which is the minimum recovery of the three experimental permutations.

4. RESULTS

4.1 Characterization of CMP NM

Table 2 summarizes different techniques that were used to characterize the NM size in the four CMP slurry samples. By comparing the results of each technique with the size provided by the manufacturer, we see the colloidal silica size measured by DLS or SEM is lower than the reported value. For the fumed

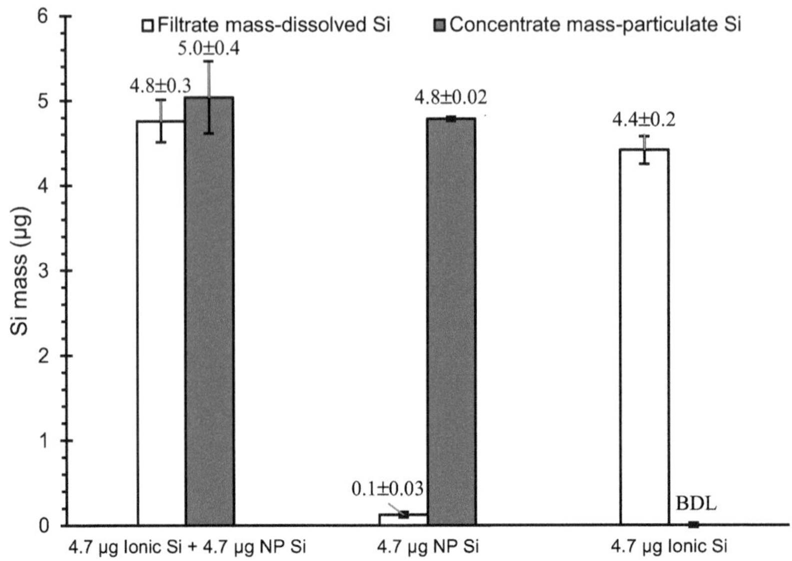

FIGURE 3 Test results of the centrifugal ultrafiltration devices for silica nanomaterials (NM) and ionic Si. Filtrate solution has passed through the ultrafiltration membrane, which represents dissolved Si species. Concentrate refers to solution rejected by the ultrafiltration membrane, which represents the particulate Si species. In the three cases, solution containing 4.7 μg ionic Si and 4.7 μg NM Si, solution containing only 4.7 μg ionic Si, and solution containing only 4.7 μg NM Si were tested with the centrifugal ultrafiltration devices. The filtrate recovery and concentrate recovery were measured by ICP-MS and shown as Si mass in the figure. Recoveries of ≥94% can be obtained for all cases for both filtrate (dissolved Si) and concentrate (silica NM).

TABLE 2 Nanomaterial Size Obtained by Different Analytical Techniques and the Manufacturer of the Chemical Mechanical Polishing/Planarization Slurries

	Technique	Colloidal Silica	Fumed Silica	Ceria	Alumina
Size (nm)	DLS	45.1 ± 13.4	183.5 ± 60.8	184.9 ± 76.2	156 ± 60.3
	SEM	37 ± 7	38 ± 14	42 ± 16	84 ± 21
	spICP-MS	NA	NA	40 ± 19	78 ± 20
	Manufacturer-provided	50–60	120–140	60–100	80–100

DLS, dynamic light scattering; SEM, scanning electron microscopy; spICP-MS, single particle inductively coupled plasma–mass spectrometry.

FIGURE 4 Scanning electron microscopy images of the four types of chemical mechanical polishing/planarization nanomaterials.

silica NM, there is a big difference between the measured results of DLS and SEM, with DLS closer to the reported value. To understand this we need to refer to the SEM images of the four types of CMP NM, which are shown in Figure 4. Colloidal silica gives discrete spherical images under SEM whereas fumed silica shows clusters of particles. This confirms the known character of fumed silica that aggregation unavoidably occurs during the combustion synthesis process.[15] In this case, size obtained by DLS is based on the Brownian motion of particles in water and gives the effective diameter closer to the cluster size. For the size obtained by SEM, we measured the diameters of discrete particles on the image. Similarly with ceria and alumina, DLS analysis reports larger diameter than SEM for both NM. From the SEM image, ceria particles appear more like crystals with clear edges and corners, which may cause the overestimation of DLS based on their motions in water. Some aggregation occurred for alumina NM based on the SEM images, which may cause the larger observed size by DLS since DLS is easily biased toward larger particles.[31] The aggregation states of fumed silica and alumina are also confirmed by transmission electron microscopy (data not shown) analysis.

Ceria and alumina NM were also characterized by spICP-MS. Figure 5 shows the time-resolved data and converted particle size distributions by spICP-MS for ceria and alumina NM in CMP slurries. The time-resolved data

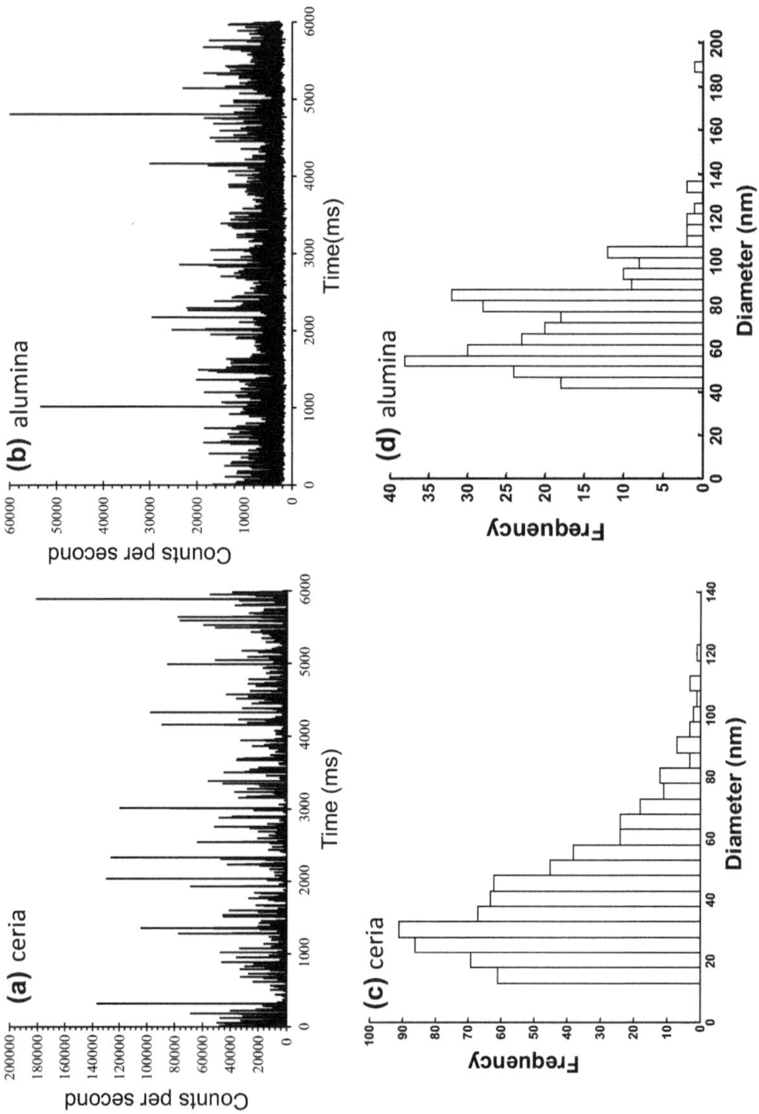

FIGURE 5 Single particle inductively coupled plasma–mass spectrometry (spICP-MS) analysis for ceria and alumina nanomaterials in CMP slurries: (a, b) time-resolved data (intensity versus time) of spICP-MS for ceria and alumina; (c, d) converted particle size distributions for ceria and alumina.

(Figure 5(a) and (b)) contains the instrument intensity versus time, in which each pulse represents a particle signal. By relating the pulse height to the particle size via a dissolved calibration curve,[26,27] particle size distributions (Figure 5(c) and (d)) can be converted from the time-resolved data. spICP-MS is a promising nanoanalysis technique for environmental samples with the advantage of low detection limit (ng/L levels) and sensitive size resolution ability. Herein it is demonstrated that spICP-MS can be used to detect the ceria and alumina NM in CMP slurries. Due to mass spectrum interferences in ICP-MS analysis for Si, the size detection limit by spICP-MS analysis of silica NM is around a few hundred nanometers.[32]

The surface charge of metal oxide NM influences their interaction with other surfaces, and hence their fate during different treatment processes.[33–35] The isoelectric point ranges from 1.7 to 3.5 for silica, 6.7 to 8.6 for ceria, and 8 to 9 for alumina.[36] Collaborators at Colorado School of Mines (J. Ranville) measured zeta potential for the CMP NM at pH 8.4 in 10 mM $NaHCO_3$ to be −16 mV for colloidal silica and more negative for fumed silica (>−35 mV).

4.2 Removal of NM in CMP Slurries through Calcium-Based Softening Process

During the on-site treatment of semiconductor wastewater, lime-softening (e.g., adding lime to adjust pH to 10−12) is often adopted to remove ionic metal chemicals (e.g., Cu). NM are potentially removed as well through this process. We used $CaCl_2$ as the calcium source to simulate the softening process. The Ca dosage was set at 1.9 mM and the removal of NM was examined under different pH conditions. The concentration of Ca was selected according to the fact that equivalent amount of lime gives rise to a pH about 11, which is a desired pH for typical lime-softening process.[37] Four types of NM in CMP slurries were investigated separately and their concentrations in the supernatant were determined by ICP-MS after digestion. All concentrations are reported as the elements of Si, Ce, and Al. Figure 6 shows the removal efficiencies for the four types of NM in CMP slurries under different pH conditions. Other than colloidal silica, all the tested pH conditions were in the basic range to mimic lime-softening conditions. For colloidal silica, a wide pH range was examined. This showed that the lowest removal for colloidal silica occurred around pH 5 to 6, which is about equal to the original pH of the diluted slurry. Only ~15% removal of Si was observed around pH 2 to 4. This peculiar stability of silica in vicinity of the zero point of charge (~pH 2) can be due to the lack of hydroxyl ions, which catalyzes the formation of siloxane bonds among particles and contributes to aggregation occurrence.[15] In the pH range of 6−7, NM removal increased from <10% to ~ 90%. Above pH 7, the removal of colloidal keeps increasing and >95% removal can be achieved at higher pH > 10. Similar trends are observed for fumed silica under basic pH conditions. Above pH 8, ~95% removal efficiency can be achieved for fumed

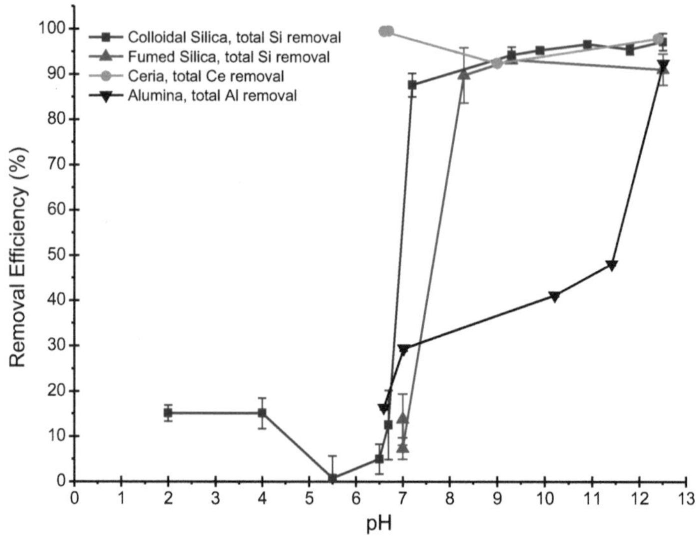

FIGURE 6 Removal efficiencies of Si, Ce, and Al for four types of nanomaterials in CMP slurries under different pH conditions. Ca dosage was 1.9 mM in all cases. The starting concentrations are 26 mg/L as Si for colloidal silica, 24 mg/L as Si for fumed silica, 20 mg/L as Ce for ceria, and 31 mg/L as Al for alumina. Concentrations are based on total metal measurement by inductively coupled plasma—mass spectrometry. pH values are obtained after jar test when reactions are considered in equilibrium.

silica NM. High removal efficiency, greater than 90%, was observed under all basic conditions tested for ceria NM. Alumina showed different behavior compared with the other three types of NM, which is reflected as a continuous and slow increase of the removal efficiency from pH ~6.5 to 12.5. The removal efficiency was raised from 16% (pH 6.6) to 92% (pH 12.5).

For the colloidal and fumed silica NM, we assume the dissolved Si (e.g., silicate) plays a role in the removal of total Si. Silica particles can have solubility of ~100 mg/L under neutral pH, and the solubility significantly increases when pH becomes higher.[15] Using the centrifugal ultrafiltration method, we found the dissolved Si concentrations in original colloidal silica and fumed silica slurries are 30.2 ± 0.9 mg/L and 37.0 ± 1.5 mg/L, respectively. Thus the dissolved Si concentrations in both slurries are lower than expected from silica solubility. After the original slurries were diluted to simulate the wastewater concentrations, we detected lower dissolved Si concentrations before performing the jar test. Figure 7 gives the dissolved Si concentrations in colloidal and fumed silica CMP slurries under different pH conditions before and after jar test. In Figure 7(a), a clear increase of the dissolved Si before jar test was observed for colloidal silica from pH 7 to 12.5. Dissolved Si also increased slightly when the pH was lower than 7. This trend is similar to the reported silica solubility versus pH.[15] After the jar test, the dissolved Si generally remained at concentrations below 1 mg/L. This result

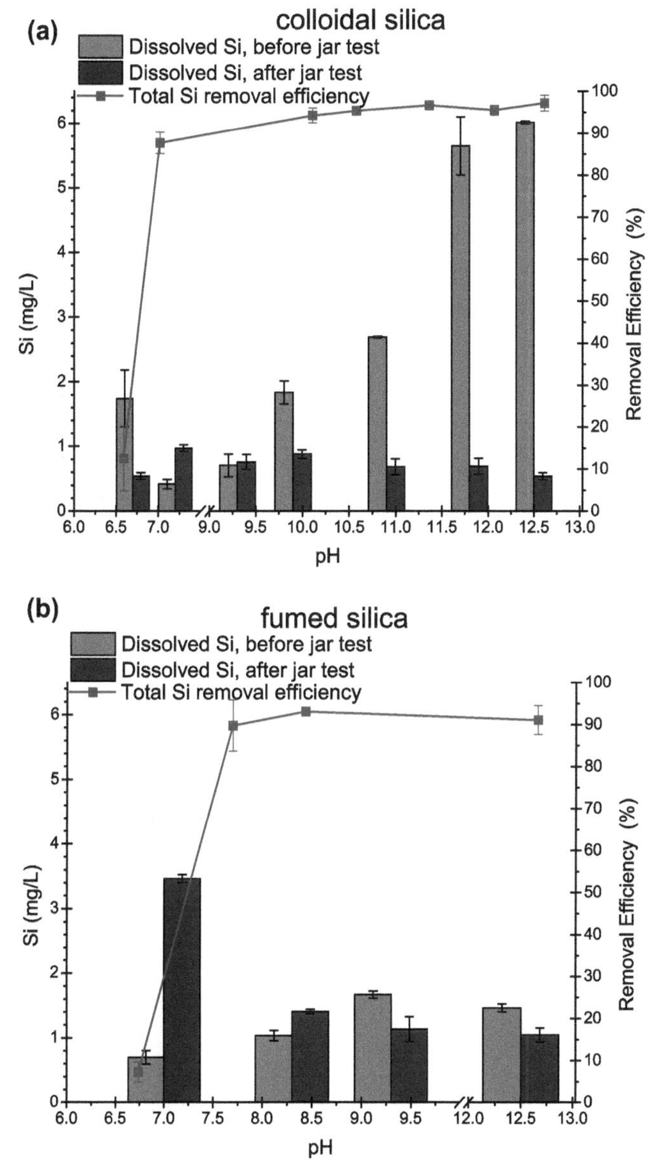

FIGURE 7 Dissolved Si concentrations before and after jar test for the CMP slurries containing: (a) colloidal silica and (b) fumed silica. Total Si removal efficiency (right axis) is shown as comparison.

indicates the dissolved Si after jar test may be dominated by, other than the dissolution of silica, other phenomena that are less influenced by the base pH conditions. Above pH 7, the removal efficiency increased with the increasing initial dissolved Si.

The observations lead to a possible mechanism for the removal of silica NM. The reaction of dissolved Si (e.g., silicate) and Ca^{2+} to form calcium silicate hydrate may bridge silica NM to increase agglomeration and therefore particle settling. Previous work has reported this reaction in cement chemistry.[38] In contrast, as seen in Figure 7(b), the fumed silica NM slurry did not show an apparent trend as did colloidal silica. The dissolved Si increased with pH but not as rapidly as with colloidal silica. After the jar test, dissolved Si remained at ~1 mg/L under most pH conditions except for pH 7, which gave ~3.5 mg/L dissolved Si. Overall, chemical precipitation showed an ability to remove a very large percentage of CMP NM when practiced at elevated pH levels. While some NM appear in the effluent, most of the residual metal content of the settled water appears to be dissolved metals (Si, Ce, Al) rather than NM. Thus, chemical softening using lime appears a good strategy for on-site industrial treatment of CMP NM. Future work needs to conduct similar work in the presence of surfactants, oxidants, and other additives that also present in mixed semiconductor waste streams—and jar tests similar to those employed here should be effective in assessing the impact of these additives.

Additional experiments were conducted by holding pH constant and varying the amount of added calcium. These experiments were conducted two different ways. First, colloidal silica was tested at pH 7.0 ± 0.2 in the presence of zero to 4 mM Ca^{2+} (Figure 8). Here, adding divalent calcium promoted aggregation and removal of the colloidal silica NM. At pH 7, colloidal silica is negatively charged and favorable interactions with calcium lead to aggregation. Second, colloidal silica was combined with ceria and alumina CMP NM

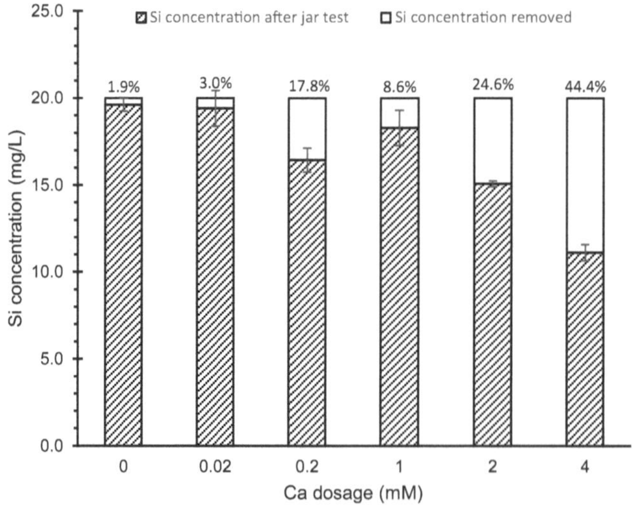

FIGURE 8 Effect of calcium addition on the removal of colloidal silica at pH 7.0 ± 0.2.

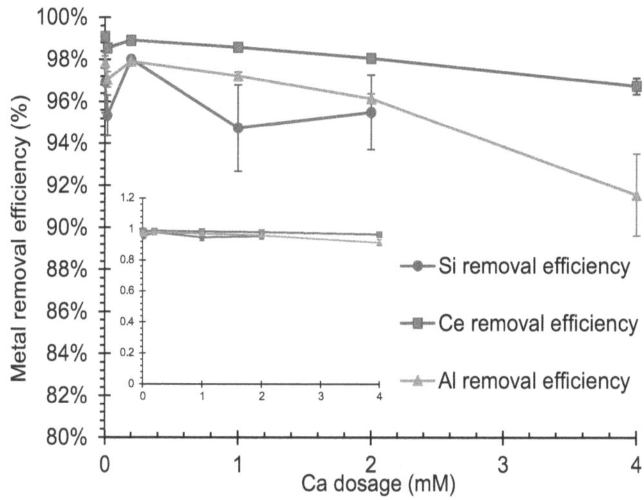

FIGURE 9 Effect of calcium addition on removal of colloidal silica, ceria, and alumina nanomaterials (NM) at pH 7.0 ± 0.5 from a mixed slurry (each NM at an initial concentration of 10 mg/L).

to form a mixed slurry that may be more representative of mixed waste streams at a Fab (Figure 9). The pH was 7.0 ± 0.5 and again calcium addition varied from zero to 4 mM Ca. Without any calcium, near complete removal of the Si occurred—which is in contrast to experiments with colloidal silica alone (Figure 8). At pH 7, the alumina NM are likely positively charged and favorably aggregate with the negatively charged colloidal silica; the ceria NM will be near neutral, near pH 7. Adding calcium to the mixed slurry only slightly decreased the NM removal efficiency, potentially by sorbing to the colloidal silica, increasing its surface charge (more positive) and decreasing its aggregation potential with the other NM.

4.3 Removal of NM in CMP Slurries by Biomass Adsorption

In the biomass adsorption test, two concentration levels of NM were examined. The first concentration level was set the same as the jar test, i.e., ~20–30 mg/L as the corresponding metal for each type of NM. The second concentration level was set at 4 mg/L as the corresponding metal, thus taking into account that the NM concentrations will be lower when they enter a municipal WWTP.

Figure 10 shows the removal of four types of NM in CMP slurries. At the higher concentration level (Figure 10(a)), greater than 60% NM in all CMP slurries can be removed by biomass. Colloidal silica shows the highest removal efficiency of 88.8%, followed by ceria NM with 82.7% removal efficiency. It is interesting that fumed silica behaves quite differently than

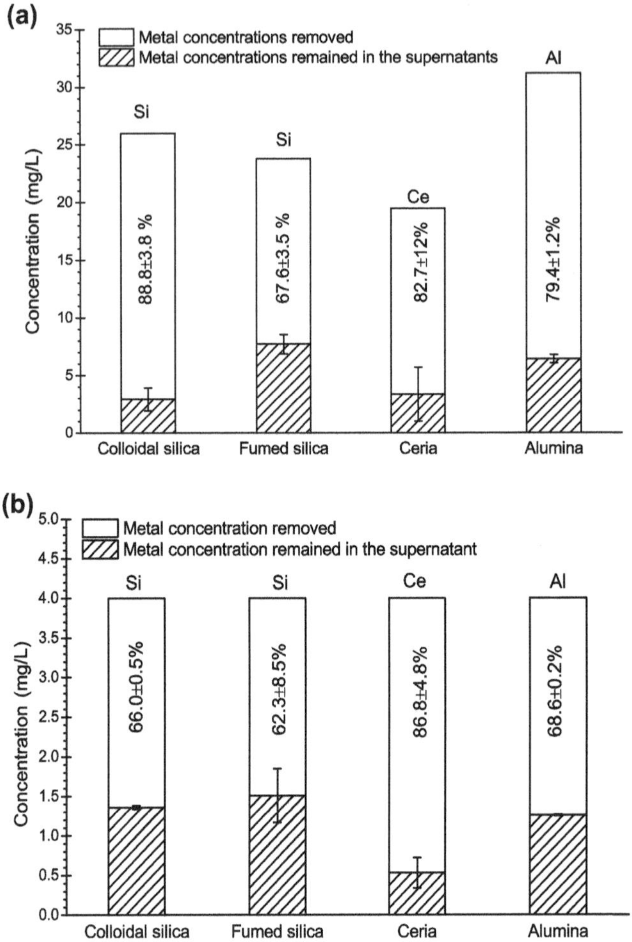

FIGURE 10 Removal of four types of nanomaterials (NM) in CMP slurries by biomass adsorption (1000 mg/L dry weight of biomass): (a) starting concentration at 20–30 mg/L as each element; (b) starting concentration at 4 mg/L as each element. Labeled percentage values on the white bar represent the removal efficiency of each type of NM by biomass.

colloidal silica, and at 67.6% shows the minimum removal efficiency of the four slurries. With a lower starting concentration of 4 mg/L, which is shown in Figure 10(b), slightly different behaviors were observed. In this case, ceria NM showed the highest removal efficiency of 86.8%. Colloidal silica, fumed silica, and alumina NM gave comparable removal efficiencies in the range of 60–70%. By cross-comparing the two concentration levels, lower removal efficiencies were obtained with the 4 mg/L concentration level than the higher concentration level for colloidal silica, fumed silica, and alumina. For ceria, similar removal efficiencies were obtained for the two concentrations levels. Overall, NM removals of 60–80% were observed at the screening biomass dose of 1000 mg/L. However, compared to similar experiments conducted with silver, gold, or titanium dioxide NM performed elsewhere,[22] the removals of CMP NM are much lower. It is likely that the mechanical stabilization process employed in preparing CMP slurries created NM that are more difficult to remove than commercially available, stock NM.

5. CONCLUSIONS

As engineered nanomaterials are tested for their potential toxicity and adverse ecological effects, it is also important to understand the ability of existing industrial and municipal treatment processes to remove NM used in widely used applications such as CMP slurries. By simulating the lime-softening and biomass treatment processes on a laboratory scale, the removal effectiveness of these two treatment processes on NM in CMP slurries were demonstrated. Testing with four representative types of NM, including colloidal silica, fumed silica, ceria, and alumina, it shows that, in the calcium-based softening process, greater than 90% removal efficiencies can be achieved for all types of NM by manipulating the pH conditions. Generally higher pH values (e.g., pH 11–12.5) give rise to higher removal efficiencies. For colloidal silica and fumed silica, dissolved Si concentration increases with pH and may contribute to the removal of total Si. After the softening treatment, dissolved Si generally remains at ~1 mg/L for both colloidal silica and fumed silica. In the biomass adsorption process, about 60–80% NM in CMP slurries can be removed. When the NM concentrations change from 20–30 mg/L to 4 mg/L (corresponding to the potential concentrations in fresh CMP wastewater and potential concentrations in municipal wastewater, respectively), removal efficiencies by the biomass adsorption process decrease for colloidal silica, fumed silica, and alumina, whereas for ceria a high removal efficiency (>80%) can still be obtained.

Throughout this study, analysis techniques such as DLS, spICP-MS, microwave digestion, and centrifugal ultrafiltration demonstrated their viability for different data objectives. Perhaps the most novel analysis shown here was demonstrating the applicability of spICP-MS for detection and sizing of ceria and alumina NM contained in CMP slurries. The application of spICP-MS for silica NM is currently very limited by background noise, but is an area of active research. Despite this analytical challenge, we demonstrate quantification of dissolved Si and SiO_2 NM by centrifugal ultrafiltration, and recommend this method as an effective approach to separate ionic and particulate species.

ACKNOWLEDGMENT

This research was funded by the Semiconductor Research Corporation (SRC, Task 425.040).

REFERENCES

1. Brown KH, Grose DA, Lange RC, Ning TH, Totta PA. Advancing the state of the art in high-performance logic and array technology. *IBM J Res Dev* 1992;**36**(5):821–8.
2. Krishnan M, Nalaskowski JW, Cook LM. Chemical mechanical planarization: slurry chemistry, materials, and mechanisms. *Chem Rev* 2009;**110**(1):178–204.
3. Zantye PB, Kumar A, Sikder A. Chemical mechanical planarization for microelectronics applications. *Mater Sci Eng R* 2004;**45**(3):89–220.
4. Piccinno F, Gottschalk F, Seeger S, Nowack B. Industrial production quantities and uses of ten engineered nanomaterials in Europe and the world. *J Nanopart Res* 2012;**14**(9):1–11.
5. Hermann A, Diesner M-O, Abel J, Hawthorne C, Greßmann A. *Assessment of impacts of a European register of products containing nanomaterials*. Dessau-Roßlau, Germany: Federal Environment Agency (Umweltbundesamt); 2013.
6. Willems & van den Wildenberg. *Roadmap report on nanoparticles*. Barcelona, Spain: W&W España s.I.; 2005.
7. Pitkethly MJ. Nanomaterials—the driving force. *Mater Today* 2004;**7**(12):20–9.
8. Singh RK, Bajaj R. Advances in chemical-mechanical planarization. *MRS Bull* 2002;**27**(10):743–51.
9. Feng X, Sayle DC, Wang ZL, Paras MS, Santora B, Sutorik AC, et al. Converting ceria polyhedral nanoparticles into single-crystal nanospheres. *Science* 2006;**312**(5779):1504–8.
10. Makino H, Emi H, Yamaguchi A, Iritani E, Namiki N, Myojo T, et al. In: *Chapter 7: Environmental and safety issues with nanoparticles*. Elsevier Science; 2008. p. 385–417.
11. Kin K-T, Tang H-S, Chan S-F, Raghavan S, Martinez S. Treatment of chemical–mechanical planarization wastes by electrocoagulation/electro-Fenton method. *Semicond Manuf IEEE Trans* 2006;**19**(2):208–15.
12. Kuan W-H, Hu C-Y. Chemical evidences for the optimal coagulant dosage and pH adjustment of silica removal from chemical mechanical polishing (CMP) wastewater. *Colloids Surf A* 2009;**342**(1–3):1–7.
13. Liu Y, Zhang K, Wang F, Di W. Investigation on the final polishing slurry and technique of silicon substrate in ULSI. *Microelectron Eng* 2003;**66**(1):438–44.
14. Willardson RK, Weber ER, Li SMH, Miller RM. *Chemical mechanical polishing in silicon processing*, vol. 63. Academic Press; 1999.
15. Iler RK. *The chemistry of silica: solubility, polymerization, colloid and surface properties, and biochemistry*. 1979.
16. Stöber W, Fink A, Bohn E. Controlled growth of monodisperse silica spheres in the micron size range. *J colloid Interface Sci* 1968;**26**(1):62–9.
17. Wang ZL, Feng X. Polyhedral shapes of CeO_2 nanoparticles. *J Phys Chem B* 2003;**107**(49):13563–6.
18. Schroeder DJ, Moeggenborg KJ, Chou H, Chamberlain JP, Hawkins JD, Carter P. *CMP method utilizing amphiphilic nonionic surfactants*. 2005. In Google Patents.
19. Lai CL, Lin SH. Electrocoagulation of chemical mechanical polishing (CMP) wastewater from semiconductor fabrication. *Chem Eng J* 2003;**95**(1–3):205–11.
20. Klaine SJ, Alvarez PJ, Batley GE, Fernandes TF, Handy RD, Lyon DY, et al. Nanomaterials in the environment: behavior, fate, bioavailability, and effects. *Environ Toxicol Chem* 2008;**27**(9):1825–51.
21. Dunphy Guzman KA, Taylor MR, Banfield JF. Environmental risks of nanotechnology: National Nanotechnology Initiative Funding, 2000–2004. *Environ Sci Technol* 2006;**40**(5):1401–7.

22. Kiser MA, Ladner DA, Hristovski KD, Westerhoff PK. Nanomaterial transformation and association with fresh and freeze-dried wastewater activated sludge: implications for testing protocol and environmental fate. *Environ Sci Technol* 2012;**46**(13):7046–53.
23. Kiser MA, Ryu H, Jang H, Hristovski K, Westerhoff P. Biosorption of nanoparticles to heterotrophic wastewater biomass. *Water Res* 2010;**44**(14):4105–14.
24. Westerhoff P, Kiser A, Hristovski K. Nanomaterial removal and transformation during biological wastewater treatment. *Environ Eng Sci* 2013;**30**(3):109–17.
25. Wang Y, Westerhoff P, Hristovski KD. Fate and biological effects of silver, titanium dioxide, and C60 (fullerene) nanomaterials during simulated wastewater treatment processes. *J Hazardous Mater* 2012;**201–202**:16–22.
26. Degueldre C, Favarger PY, Wold S. Gold colloid analysis by inductively coupled plasma-mass spectrometry in a single particle mode. *Anal Chim Acta* 2006;**555**(2):263–8.
27. Pace HE, Rogers NJ, Jarolimek C, Coleman VA, Higgins CP, Ranville JF. Determining transport efficiency for the purpose of counting and sizing nanoparticles via single particle inductively coupled plasma mass spectrometry. *Anal Chem* 2011;**83**(24):9361–9.
28. Pace HE, Rogers NJ, Jarolimek C, Coleman VA, Gray EP, Higgins CP, et al. Single particle inductively coupled plasma-mass spectrometry: a performance evaluation and method comparison in the determination of nanoparticle size. *Environ Sci Technol* 2012;**46**.
29. Lee S, Bi X, Reed RB, Ranville JF, Herckes P, Westerhoff P. Nanoparticle size detection limits by single particle ICP-MS for 40 elements. *Environ Sci Technol* 2014;**48**(11):6391–400.
30. Eaton AD, Franson MAH. *Standard methods for the examination of water & wastewater.* 2005.
31. Filella M, Zhang JW, Newman ME, Buffle J. Analytical applications of photon correlation spectroscopy for size distribution measurements of natural colloidal suspensions: capabilities and limitations. *Colloid Surf A Physicochem Eng Asp* 1997;**120**(1–3):27–46.
32. Bi XY, Lee S, Ranville JF, Sattigeri P, Spanias A, Herckes P, et al. Quantitative resolution of nanoparticle sizes using single particle inductively coupled plasma mass spectrometry with the K-means clustering algorithm. *J Anal At Spectrom* 2014;**29**(9):1630–9.
33. Zhang Y, Chen Y, Westerhoff P, Crittenden JC. Impact of natural organic matter and divalent cations on the stability of aqueous nanoparticles. *Water Res* 2009;**43**(17):4249–57.
34. Zhang Y, Chen YS, Westerhoff P, Crittenden JC. Stability and removal of water soluble CdTe quantum dots in water. *Environ Sci Technol* 2008;**42**(1):321–5.
35. Zhang Y, Chen YS, Westerhoff P, Hristovski K, Crittenden JC. Stability of commercial metal oxide nanoparticles in water. *Water Res* 2008;**42**(8–9):2204–12.
36. Kosmulski M. pH-dependent surface charging and points of zero charge. IV. Update and new approach. *J Colloid Interface Sci* 2009;**337**(2):439–48.
37. USEPA. *Enhanced coagulation and enhanced precipitative softening guidance manual.* Office of Water (4607); 1999. EPA 815-R-99–012.
38. Greenberg S. Reaction between silica and calcium hydroxide solutions. I. Kinetics in the temperature range 30 to 85°. *J Phys Chem* 1961;**65**(1):12–6.

Chapter 9

Case Study – Characterization of Nanomaterials in Food Products

Karen Tiede,*[,1] Agnieszka Dudkiewicz,* Alistair Boxall[§] and John Lewis*
* *The Food and Environment Research Agency, Sand Hutton, York, UK;*
[§] *Environment Department, University of York, Heslington, York, UK*
[1]*Corresponding author: E-mail: Karen.Tiede@fera.gsi.gov.uk*

Chapter Outline

1. INTRODUCTION

Nanotechnology offers promising opportunities for the food sector. The numerous applications of nanomaterials (NMs) in the food industry have been summarized in several reviews,[1–4] and include usage for improvement of shelf life; sensory quality; and nutritional value of foodstuffs. A constantly updated inventory of consumer products containing NMs including food-related items

Frontiers of Nanoscience, Vol. 8. http://dx.doi.org/10.1016/B978-0-08-099948-7.00009-9

is available at www.nanotechproject.org. Examples of NMs used in the food sector are nano-silver – adding antimicrobial properties to food packaging[5] or other surfaces potentially in contact with food (e.g. fridges); titanium dioxide NMs – applied in food containers for UV protection[3]; nano-silica – acting as a taste intensifier in dietetic beverage[6]; and nano-encapsulates – rendering lipophilic vitamins suspendable in water,[7] or for improving vitamin thermal stability.[8] The development of applications for nanotechnology in the food industry raises a need for risk assessment and regulation of nano-enabled food products before they can be introduced into the market. In addition to this, these same methods will be required in order to monitor and assess the positive aspects of the performance of NM-enhanced products e.g. nutrient delivery.

An additional complication faced when attempting to detect and characterize intentionally added NMs is the possible presence of material which fits the NM definition, but is (1) a minor component of an ingredient whose properties are based on the 'bulk' material (unintentionally resulting from the manufacturing process); (2) a contaminant which just happens to be 100 nm or less or; (3) derived from a natural source. The literature contains many illustrations of where such unintentionally added NMs have been reported, for example: (1) E171 (titanium dioxide – used as a colourant), and E551 (synthetic amorphous silica – used as an anticaking agent in powdered foods, or as a filler in nutraceutical pills)[9,10]; (2) Contamination of bread and biscuits with metal debris (Cd, W, etc.), arising from the machinery used in the flour milling and bread-production process[11]; (3) Casein micelles in milk[12] are a classic example of where a food contains naturally occurring 'nano'.

The complexity of a food matrix presents a potential problem to all areas of food analysis, but is a particular issue when attempting to identify and characterize the presence of NMs. This is because many of the matrix components, e.g. surfactants, proteins, salts, are likely to affect the physical characteristics and behaviour of NMs, thereby altering the measured nanometric parameter. For example, changes in pH and/or ionic strength can cause NMs to undergo transformations such as agglomeration, aggregation and/or dissolution; whereas interactions with surfactants and proteins (or the pH and/or ionic strength of the food matrix generally) can alter the surface chemistry of an NM. Not only do these transformations make it more difficult to accurately characterize the NM, but they are also expected to influence factors such as the NM fate, potential toxicity, and the intended functionality[13] of the resulting NM entity. For example, agglomeration/aggregation and dissolution will alter the size of the NM, which in turn may affect its toxicological impact (because it is thought that smaller NMs are more easily taken up by cells and small organisms in comparison to larger ones).[14,15] Similarly, association of NMs with naturally occurring NMs may have similar toxicological implications as described for agglomeration, aggregation and surface transformations. Additionally, the production of ions from NMs undergoing dissolution can add to a potential toxic effect, as demonstrated for silver NMs, where dissolution resulted in increased toxicity to human mesenchymal stem cells.[16] Surface transformation, such as protein adsorption may be a decisive factor in NM uptake and translocation within the body.[17–19] Although there are no data on direct uptake of NMs by humans from foods, in vitro studies have shown that NMs can transform multiple times under conditions similar to that in the human digestive tract.[20,21] The NM transformations may have toxicological significance if, for example, they affect NM size/morphology/charge in a GI tract where NMs could potentially be taken up. Therefore, characterization of NMs in the food products themselves is crucial, rather than relying on data from the characterization of the 'as manufactured' NM.

Another key issue, when attempting to characterize NMs in a foodstuff, is that the integrity of the NM should be maintained throughout all stages of the analytical procedure, e.g. homogenization, extraction, cleanup, concentration/dilution. Due to the limited number of studies on NMs in food matrices, the adopted NM extraction and characterization methods have been taken from studies investigating NMs in other complex matrices, such as environmental samples or cosmetics, and will be discussed further.

To conclude, in this chapter we will discuss the application of analytical methods for the detection and characterization of intentionally added NMs in food, beverages, nutraceuticals and food contact materials, as well as recent advances in this field. There will be a focus on the characterization of unique NM properties, including size and size distribution (affected by dissolution, aggregation/agglomeration and association with the food components). The challenges relating to the preservation of the original state of NMs in food during sample preparation and analysis; the distinction between naturally occurring, incidental/contaminating and intentionally added NMs; as well as understanding analytical outputs, will also be addressed, using 'real-world' examples where appropriate.

2. ANALYTICAL TECHNIQUES USED FOR THE DETECTION AND CHARACTERIZATION OF NMs IN FOODS

Many analytical techniques that have been used to characterize NMs in other complex matrices, such as environmental and cosmetics, are also applicable to the analysis of NMs in foods. Extensive reviews on analytical techniques potentially suitable to detect and characterize NMs in complex media, including food and food contact materials, have been published.[13,22–24] An overview of analytical methods generally suitable for the characterization of NMs is also presented in Chapter 2 of this book. Here we discuss some of the most common analytical approaches specifically adopted, or potentially useful, for the analysis of NMs in foodstuffs. Special attention is given to electron microscopy (EM) as it is currently the most widely applied technique in NM characterization,[25] and is important in risk assessment studies of NMs in food and feed.[26]

Given the complex and variable nature of food matrices, plus the potential for NM/matrix interference and interactions, sample preparation will be the first aspect discussed.

2.1 Sample Preparation and Pretreatment

The main aims of sample preparation are to (1) make the food sample homogeneous so that a representative portion can be subsampled; (2) separate/extract NMs away from components in the matrix which would otherwise interfere with the analysis; (3) concentrate or dilute the sample/extract to optimize its detectability; (4) prepare the sample/extract in such a way as to make it compatible with the analytical technique being used. It is highly unlikely that any single sample preparation method is sufficient to achieve all of the above aims. Therefore, Table 1 describes general scenarios which may be met when undertaking the analysis of NMs in foods, and provides suggested approaches to deal with various analyte/matrix combinations.

A basic, but sometimes very effective approach to sample homogenization is to simply liquidize the food matrices in an aqueous solvent, and has been successfully applied to the extraction of titanium dioxide (E171) from chewing gum and personal care products.[9,27]

For more complex food matrices, some form of matrix dissolution may be required (as long as the NM is not affected by the dissolving reagent). For example, silica NMs were successfully isolated from rat lung tissue and human endothelial cell lysate[28] and tomato soup[29] by digestion of the matrix using nitric acid, leaving the silica NM intact. The NM was then retrieved and prepared for further analysis.

In situations where the NM may be affected by the dissolving reagent, the use of appropriate enzymes could be considered as a means to break down the matrix, thereby releasing the NMs. This approach has been successfully used to retrieve silver NMs from a chicken paste,[30] and is one that is regularly undertaken in the wider field of trace element speciation, where maintaining the integrity of the metal/organic moiety is paramount.

Liquid/liquid partitioning can be a useful approach for samples where the matrix is more liquid, e.g. homogenate of embryonic zebrafish and water, in particular, if the NM is sufficiently hydrophobic, e.g. a fullerene. For example, Fullerene C60 was extracted from biological media with toluene[31] and from water using solid phase extraction.[32]

A novel approach for the extraction of NMs from liquid matrices is the use of cloud point extraction. This was successfully applied as a pre-concentration method for silver NMs from environmental waters, reportedly preserving the size and shape of the NMs.[33] The use of phase-transfer agents might also be another nonstandard possibility.[34] It should be noted that the novelty of these last two approaches will mean that it will be quite a while before they are applicable to many of our regular food matrices.

TABLE 1 Advantages and Disadvantages of Some Sample Preparation Techniques Applied for Nanomaterial (NM) Analysis in FoodStuff

Sample Preparation Technique	Aim	Advantages	Disadvantages	Examples of the Samples to Use with
Dissolution of sample matrix in a solvent	To dissolve the matrix, leaving the NMs intact	Simple, low cost and fast method to use	Applicable to a narrow range of foods, not suitable for use with NMs that dissolve in a given solvent. Further matrix removal steps, e.g. filtration, centrifugation, dialysis, may be needed	Food product of simple composition easily soluble in the solvent e.g. chewing gums, sweets, method might be applicable for inorganic and organic NMs that do not dissolve in the solvent
Enzymatic digestion	To digest the matrix leaving intact NMs	Usable with most food matrices, it can 'liquidize' solid food samples and may help to remove naturally occurring NMs which interfere with the analysis	Slow compared to other methods. The digest may require further cleanup, e.g. filtration, centrifugation. Enzymes may also damage organic NMs	All types of food matrices are applicable, but mainly suitable for inorganic NMs, although may be also used for organic NMs with a different composition than the matrix
Chemical digestion	To digest the matrix leaving intact NMs	Can be used with all types of food matrices. This approach may be useful if needing to remove naturally occurring NMs which are interfering with the analysis	Applicable only to narrow range of NMs that do not dissolve in aggressive digestion conditions. This approach is not suitable for use with organic NMs	All types of food matrices, suitable only for some inorganic NMs, e.g. silica

Continued

TABLE 1 Advantages and Disadvantages of Some Sample Preparation Techniques Applied for Nanomaterial (NM) Analysis in FoodStuff—cont'd

Sample Preparation Technique	Aim	Advantages	Disadvantages	Examples of the Samples to Use with
Filtration	To remove matrix particles that are larger than NMs and to retain NMs in filtrate or to retain NMs on the filter separating from dissolved substances	Simple, inexpensive and common in use method	May cause NM losses in filtrate if electrostatic interactions between NM and filter are present, does not remove naturally occurring NMs from the sample, solid food samples need pretreatment allowing 'liquidization'	Liquid food matrices, all types of NMs
Centrifugation	To remove matrix particles that are larger than NMs and retain NMs in the supernatant or to collect the NMs in sediment	Simple, inexpensive and common in use method	May cause NM losses if electrostatic interactions between NM and centrifuge tube walls are present, does not remove naturally occurring NMs from the sample, may cause aggregation of NMs, solid food samples need pretreatment allowing 'liquidization'	Liquid food matrices, all types of NMs
Dialysis	To remove dissolved substances from the sample	Effective method for purifying NM suspensions from matrix-dissolved substances	Limited applicability for preparation of food samples containing NMs, most of the methods can tolerate the presence of dissolved substances but not suspended solids	Liquid food samples, any type of NMs
Extraction	To extract NMs from the sample	Effective in separation of NMs from the sample, may be selective to a certain type of NM	Need to be developed for each type of NM, limited range of methods so far published	Liquid food samples, all types of NMs

Many of the techniques used in NM detection are compromised by the presence of dissolved or co-suspended materials in the sample/extract solution. In these situations, it is reasonable to consider the use of physical approaches, e.g. filtration, centrifugation, dialysis, to remove the interfering components. However, applying these approaches may result in the removal (loss) of NMs as well, so would need to be assessed carefully before reporting data. Similarly, the fate of any analyte ions (present due to NM dissolution) should also be properly understood in order to correctly interpret any resulting toxicological data. A means by which these areas of potential analytical error can be properly assessed and understood is by performing a mass-balance on the overall procedure. As the science of nanometrics matures, it is encouraging to see this parameter becoming increasingly common in good quality publications.

Removal of dissolved substances is performed in order to limit the interference from ions in some measuring instruments (e.g. gas-phase electrophoretic mobility molecular analyser (GEMMA) will create readings for dissolved nonvolatile substances,[35] salts will create crystals when dry specimens for EM are prepared).

Physical approaches, such as ultracentrifugation and ultrafiltration, are often used as part of the cleanup/extraction procedure. Ultracentrifugation has been used for the separation of gold ions from gold NM,[36] and sedimentation of titanium dioxide NMs from a dissolved bubble gum suspension.[27] Whereas, ultrafiltration was used to separate ionic silver from Ag-NM prior to spiking chicken paste with the NMs. This was done in order to evaluate possible interferences from ions during inductively coupled plasma–mass spectrometry (ICP-MS) analysis of the Ag-NM concentration.

Dialysis is commonly used for separation of dissolved substances from NM dispersions (e.g. Ref. 37) and was applied for the assessment of NM dissolution rates.[38]

Chromatographic and some field-flow fractionation (FFF) techniques can be used to separate NMs from food matrices containing suspended solids of limited size (the authors used coarse separation – e.g. through filters with pore size of 450 nm[10,30] prior to sample introduction, with no deleterious effects observed). Hydrodynamic chromatography (HDC) was used to separate synthetic amorphous silica NMs (E551) from coffee with added powdered coffee creamer[10] as well as silver NMs from activated sludge.[39]

In addition to the procedures described above, a special type of sample preparation is required when preparing samples for analysis by EM. When performing EM, a sample introduced to the vacuum chamber of the microscope needs to be dry. Different sample preparation procedures to achieve this are described, for example, in Ref. 40. In short, depending on the type of sample preparation methods such as drying, electrostatic trapping of NMs on the EM grid, sedimentation of NMs onto an EM grid, resin embedding or freezing can be applied. The first three methods are generally useful for preparation of NMs in liquid samples and the latter two for solid samples or samples where spatial organization of NMs is of interest.

2.2 Imaging of NMs in Samples and Sample Extracts

Imaging methods are widely used for NM detection and characterization of particle size and shape. Some studies have also applied these techniques to the quantification of NMs[41] and other objects in the nano range such as viruses[42] and natural aquatic colloids.[43]

Optical microscopy is limited to a resolution of only 200 nm due to the large wavelengths it operates at. Thus, with the exception of near-field scanning optical microscopy (NSOM) and confocal laser scanning microscopy, optical microscopy is generally not suited for the analysis of NMs. Atomic force microscopy and EM provide sufficient resolution to image NMs, with EM being the one more often used for the characterization of NMs in foods.

The two main challenges when applying imaging methods to the characterization of NMs in foods is the selection of appropriate sample preparation methods, and the sampling of a representative subsample. Microscopes are only able to analyze a very limited sample volume (order of picolitres), incomparably smaller to other analytical techniques. This means that samples need to be highly homogeneous, and contain NMs at sufficiently detectable numbers. Sample preparation usually involves pre-concentration of NMs, but this step often suffers from the formation of artefacts such as agglomerates,[22,40] or a less efficient separation of larger NMs compared to smaller NMs, causing the reported size distribution to be inaccurately focused around smaller size values.[44,45] Data extraction and analysis are also major issues associated with imaging techniques because of the high number of representative images (showing a certain number of objects of interest) which are required to produce reliable data. Manual interpretation of micrographs is almost impossible due to the vast number of images, but automated image analysis of NMs in biological samples, using the Definiens Developer™ Software, has recently been reported.[39]

2.2.1 Electron Microscopy

EM is widely applied for characterizing NMs, and is often used as validation for other analytical methods e.g. Refs 46,47. In addition to providing data regarding the size and shape of NMs, EM allows investigation of crystallinity, primary particle size (if aggregates/agglomerates have been formed),[48] associations with matrix components,[49] and chemical composition, by means of coupling the EM to spectroscopic methods. The European Food Safety Authority (EFSA) recommended EM as the primary method (with a second method being optional) for the detection and size characterization of NMs in food and feed products,[26] due to its versatile applicability. Please refer to Table 2 for a comparison of EM against other NM-characterization techniques.

2.2.1.1 High Vacuum Electron Microscopy

To date, EM operating under high-vacuum has been applied in the majority of studies looking at NMs in foods. Examples include measurement of silver

TABLE 2 Expression of Size Depending on the Chosen Analytical Technique

Measurement	Techniques Providing Measurement	Definition
Projected area diameter	Microscopy	Diameter of a circle corresponding to the projected two-dimensional image area of the particle
Electric mobility diameter	Capillary electrophoresis, GEMMA	Diameter of a sphere that shows the same electric mobility as the particle
Sedimentation diameter	CLS	Diameter of a sphere corresponding to the same sedimentation velocity as the particle
Hydrodynamic diameter	HDC, SEC, FFF, LC, DLS, NTA	Diameter of a sphere corresponding to the same diffusional behaviour as the particle
Mass equivalent diameter	SP-ICP-MS	Diameter of a sphere corresponding to the same mass as the particle and the density of the chemical component

CLS, centrifugal liquid sedimentation; HDC, hydrodynamic chromatography; SEC, size exclusion chromatography; FFF, field-flow fractionation; LC, liquid chromatography; DLS, dynamic light scattering; NTA, nanomaterial tracking analysis; SP-ICP-MS, single particle inductively coupled plasma–mass spectrometry.

NMs in chicken meat homogenate,[50] synthetic amorphous silica in tomato soup,[29] characterization of primary particle size in titanium dioxide (E 171) separated from chewing gum,[27] and chemical detection of zinc oxide and titanium dioxide NMs in ash from starches and flour.[51] We also relied on high-vacuum EM methods in the in-house study presented below as a case study, where the aim was to detect and characterize NMs in a commercially-available food concentrate and a nutraceutical.

Case Study: Application of High Vacuum Electron Microscopy for Detection and Characterization of NMs in a Food Concentrate and a Nutraceutical

In this study, we applied EM for detection, as well as the chemical and morphological characterization of NMs in a commercially-available food product and a nutraceutical. We also focused on the application of EM to identify chemically different particles. Identification and characterization of NMs in mixtures is an area of potential future interest because this is how NMs often appear in food products (and environmental samples), but the literature is, thus far, lacking such research. This lack is likely to be due to the analytical challenges associated with analyzing and distinguishing NMs in mixtures. Sample preparation and instrumental settings were established so as to allow the repeatability of the study to be assessed. NM detection was based on classification of majority of particles smaller than 100 nm being present in the images. This was done in order to follow recommendation for the European Commission (EC) definition of NM,[52] where an NM is termed as a substance containing at least 50% of particles smaller than 100 nm. This study was conducted within the EU FP7 funded NanoLyse project (www.nanolyse.eu).

Samples and sample preparation. Two commercially-available products, with declared content of food additives potentially containing a fraction of particles with size below 100 nm, were purchased from a local supermarket: A food concentrate (powdered tomato soup) with declared content of E551 (silica, used as anticaking agent) and a multivitamin nutraceutical containing E551, E171 (titanium dioxide) and E530 (magnesium oxide). Although only detection of E551 NM in food concentrate was required by the NanoLyse project, the nutraceutical was included because of the interesting mixture of potentially 'nano' additives, which provided us with an opportunity to establish the possibility of distinguishing chemically different particles by morphological and chemical analyses.

Two sample preparation protocols were applied: (1) sedimentation of NMs on the EM grid, and (2) capturing NMs on the EM grid by electrostatic attraction. Both of these protocols limit particle agglomeration, as well as reducing salt crystal formation.[44,45] Electrostatic attraction is also a semi-selective method, retaining only charged particles on the EM grid (in this case, negatively charged silica), thus potentially limiting interferences from suspended solids that may be present in the food matrix. This method was previously applied to the analysis of synthetic amorphous silica NMs in tomato soup.[29] As sedimentation of NMs on the EM grid is not reliant on NM charge, it has potential for use in the preparation of samples containing mixtures of NMs.

The food supplement was prepared for transmission and scanning electron microscopy (TEM and SEM) analyses by sedimentation on a formvar—carbon-coated TEM grid (Agar Scientific, Stansted, UK) using a Beckman XL-100 (Brea, US) ultracentrifuge operated at RCF = 100.000 g at 20 °C for 60 min. Prior to centrifugation, the pills were crushed in a mortar, and the powder then diluted with demineralized water in a 1:9 ratio (by weight). The sample was subsequently mixed on a magnetic stirrer for 6 h. The homogenate (0.5 mL) was added to a polyallomer centrifuge tube (SW40Ti swing bucket rotor), and topped up with borate buffer (pH 8.0). Each of the tubes contained two TEM grids and a flat platform made of Agar 100 resin for grid support (Agar Scientific, Stansted, UK). The grids were then analyzed by TEM and SEM. For SEM analysis grids were attached to aluminium SEM stubs using carbon tape.

The powdered soup was added to boiling tap water (following instructions given on the product packaging), allowed to cool, and the pH adjusted to 8.0

using 0.8M NaOH. This solution was then left on a magnetic stirrer for 6 h. Negatively charged particles from the soup sample (such as silica) were extracted onto the surface of formvar–carbon-coated TEM grids using electrostatic attraction: The grids were placed 'floating' on a drop of 0.1% solution of gelatin for ~5 min and then rinsed three times using a drop of demineralized water. Gelatin-coated grids were put in contact with the soup sample for ~2 min and subsequently rinsed with two drops of demineralized water. Any excess moisture remaining between rinses and sample application was removed using filter paper. Prepared grids were attached to aluminium SEM stubs using carbon tape. Stubs were then coated with 10 nm Pt/Pd using 2300HR high resolution fine coater with a JEOL FC-TM20 thickness controller (Tokyo, Japan). The coating was applied to improve the contrast of silica during imaging, and to avoid imaging artefacts such as charging or contamination.

Imaging. SEM images were acquired using an FEI Sirion S field emission gun SEM (Hillsboro, US) equipped with a through-the-lens detector (TLD) and electron energy dispersive X-ray spectrometer (EDS) (Thermo Fisher NS7 system S/N 0409235) with NSS112E NORAN operating software. The images of silica extracted from soup were taken at an operating voltage of 5 kV and spot size of 3 nm, whereas samples of the food supplement containing different types of food additives were imaged at 10 kV and spot size of 4 nm. The increased SEM voltage allows the possibility that NMs comprised of higher atomic number elements, such as Ti, can be distinguished from NMs containing lower atomic number elements, such as Si. However, increasing the voltage also increases the penetration depth of the electron beam into the sample, and is therefore not recommended for imaging of small NMs comprised of low atomic number elements (low contrast). The increase of current (spot size) allows the signal, and thus the contrast, to be increased. However, this also results in an increase in noise and a decrease in resolution. The EDS spectra of point analyses were recorded at spot size 4 and 5 kV for 30 s.

TEM imaging was performed using a JEOL JEM 2011 TEM (Tokyo, Japan) operating at 200 kV. The microscope was equipped with a digital camera (Gatan 794, Pleasanton, Canada) and EDS system (Thermo Fisher NS7 system S/N 0409237) operated using the NSS112E NORAN software.

Results and conclusions. Figures 1–3 show representative EM images of the food additives that could be potentially classed as NMs in a food concentrate and a nutraceutical product. Additionally EDS spectra of the nutraceutical sample were included in Figures 2 and 3.

As shown in Figure 1, the identified silica in the food concentrate was composed of aggregates of small primary particles, some of which were also present in a non-aggregated form. The appearance of the E551 food additive as aggregates is very typical for this substance, and is a result of the manufacturing process.[48,53] It can also be noticed that the majority of the

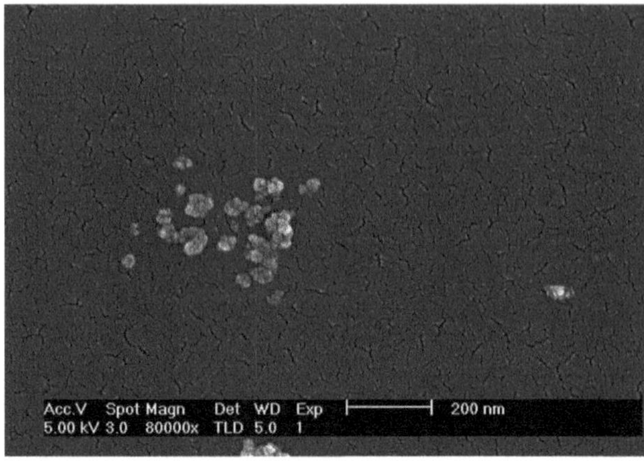

FIGURE 1 Silica food additive in a commercially available soup concentrate.

particles, and even aggregates (as shown in Figure 1), were smaller than 100 nm. Thus the food additive used in this product would be classed as an NM.

The food additives found and identified by EDS in the nutraceutical were composed of structures larger than 100 nm, as it appeared that all three additives formed mixed aggregates. However, the size of primary particles of titanium dioxide and silica was found to be slightly smaller than 100 nm (as shown on Figure 2). Therefore these two additives of the nutraceutical product could also potentially be classified as NMs, whereas the observed primary particles of magnesium oxide were clearly larger than 100 nm. Both SEM and TEM images of the food additives (silicon dioxide, titanium dioxide, magnesium oxide) in the nutraceutical showed different morphology and contrast traits. Contrast was clearly weakest for synthetic amorphous silica, whereas titanium dioxide and magnesium oxide particles were both characterized by good contrast, and could be distinguished from each other by morphology and primary particle size using TEM. Additionally, SEM/EDS allowed us to obtain elemental maps of the sample as shown in Figure 4. All the aggregates visible were identified as containing Mg, Ti or Si, although it is also noticeable that Ti and Si maps are characterized by high background interference. During elemental map acquisition, the detection of particles smaller than 100 nm is not easy due to this interference and sample drift. However, overall, the sample preparation methods chosen here for NM-containing samples were able to validate the presence of targeted food additives by EM imaging and analysis.

2.2.1.2 EM of Partly and Fully Hydrated Samples

Advances in the development of EM have led to techniques which are capable of imaging partly (or even fully) hydrated samples, and include alternative preparation and imaging techniques such as environmental SEM or TEM, liquid EM and cryo-EM. These techniques have been applied to the analysis of

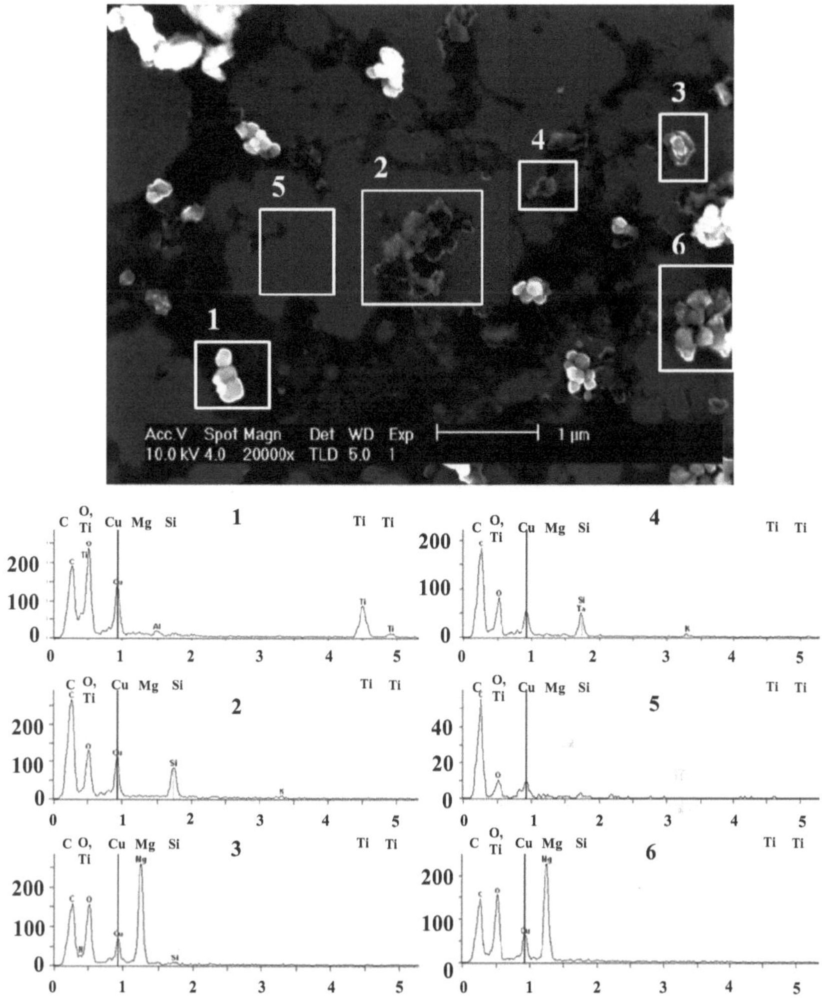

FIGURE 2 Scanning electron microscopy image of silica, titanium dioxide and magnesium oxide food additives in a commercially available nutraceutical with energy dispersive X-ray spectrometer spectra of selected areas (y-axis: number of spectral counts and x-axis: energy in keV).

NMs in cosmetics, food and environmental media,[54–56] as well as for food structures in general, e.g. cheese,[57] bread[58] or chocolate.[59]

These methods have the advantage of visualizing virtually unaltered food samples containing NMs, preserving or partly preserving the spatial distribution and original state of the NMs within the sample. This, however, comes at the cost of loss of other information, for example, environmental and liquid EM suffer a loss of resolution compared to high-vacuum methods. Further, these methods are often limited to samples with a high NM content. However,

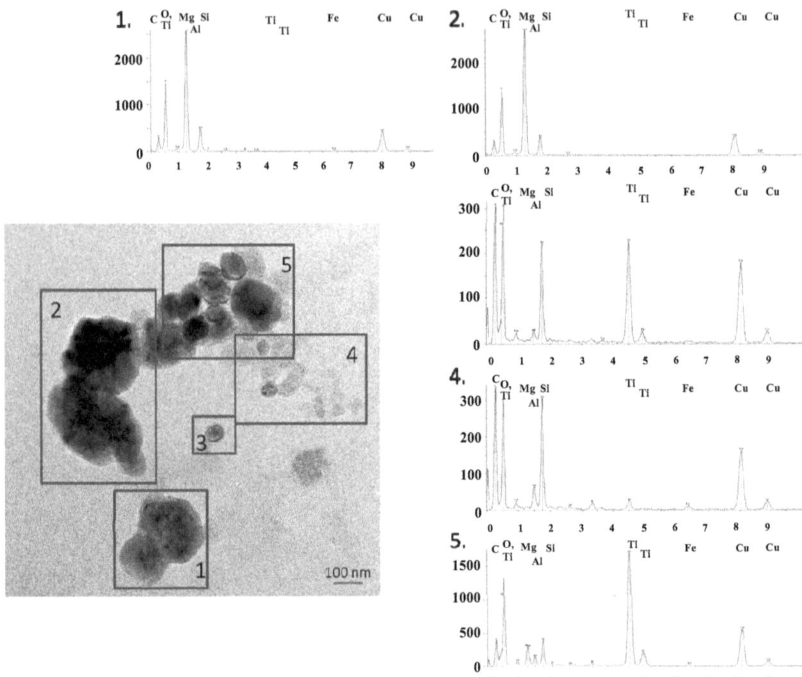

FIGURE 3 Transmission electron microscopy image of silica, titanium dioxide and magnesium oxide in a commercially available nutraceutical with EDS spectra of selected areas.

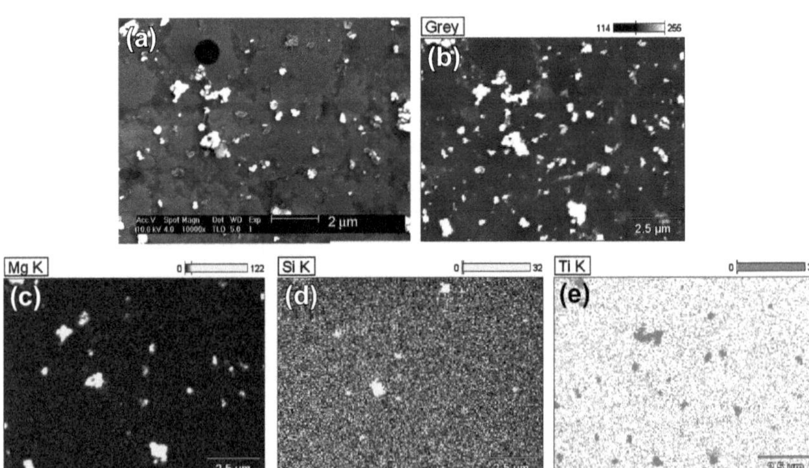

FIGURE 4 Scanning electron microscopy (a) and energy dispersive X-ray spectrometer (b) images of food additives in a commercially available nutraceutical with elemental maps of Mg (c), Si (d) and Ti (e).

sample preparation approaches allowing to pre-concentrate the NMs on liquid EM images have already been described, e.g. partial dehydration of the fully liquid sample controlled by modification of vapour pressure inside of environmental SEM was applied for imaging of aquatic colloids.[60] Also the electron transparent windows of capsules containing liquid samples for imaging in a high-vacuum microscope have been charged to attract NMs from the sample and allow imaging of a higher number of NMs, e.g. for imaging silica NMs in tomato soup[59] and metal and metal oxide NMs in environmental aquatic samples.[55] These developments represent an exciting area of research, offering much to the NM scientific community.

2.3 Mass Quantification and Chemical Analysis

Until relatively recently, the only way to get quantitative data regarding the presence of inorganic NMs, especially metal-based NMs, in a foodstuff was to analyze the sample using two different techniques: one to determine the size and morphology of the NM, e.g. EM, and the second to determine the total amount of the element present, e.g. atomic absorption spectrometry, optical emission spectrometry, inductively coupled plasma–mass spectrometry (ICP-MS), inductively coupled plasma–optical emission spectrometry or particle induced X-ray emission.

There are several downsides to using this two-technique approach, and include the cost and time requirements; increased potential variability due to measuring two different aliquots of sample (issues with homogenous distribution of the NM throughout the sample matrix); an assumption that the measured element has originated solely from the NM, and did not have a contribution from the foodstuff or from ions dissolving from the NM itself. Hence, great effort has been invested in the development of techniques capable of accessing both sets of data in a single analytical run. Coupling of ICP-MS online to separation techniques such as FFF or HDC is now commonly practised for the detection and quantification of NMs in foods and other complex matrices,[10,29,30,39] as is the use of ICP-MS in single-particle mode (SP-ICP-MS).[61] This subject is discussed in greater depth in **Chapter 3** of this book, but it should be noted that this is an area of analytical chemistry that is still in its early stages of maturity.

Regarding organic NMs, there are methods available for their chemical analysis, with quantification by mass using MS.[62] Organic NMs can be differentiated by MS in combination with ion mobility spectrometry.[63] Matrix-assisted laser desorption/ionization–time of flight – MS (MALDI-TOF-MS) was used for the determination of surfactant NM fingerprints[64] and for the detection and characterization of small NMs such as proteins, peptides, oligosaccharides and oligonucleotides.[65] The number of studies reported using MS-based techniques for the quantification of organic NMs in complex matrices is limited, but this will need to be a priority area for future research.[66]

2.4 Screening Methods

Detection of NMs in complex matrices can be achieved using methods such as enzyme linked immunosorbent assay (ELISA) or surface plasmon resonance (SPR). Thus far, ELISA has been successfully applied for the detection of cross-linked gelatin NMs in soft drinks, milk and fruit juices,[67] whilst SPR sensors have successfully detected silver NMs in tomatoes, cucumbers and river water.[68] Although these methods do not provide detailed characteristics of the NMs, they may be useful in the routine detection of NMs in foods in the future.[69]

2.5 Size Separation and Measurement

There are a range of methods capable of generating data regarding the size and polydispersity of NMs (specifically spherical NMs, i.e. nanoparticles – NPs) which are routinely used for screening purposes. However, of those, there is a much smaller subset (in terms of numbers used in laboratories) that can also provide data regarding the chemical composition of the analyte materials.

Methods which are able to provide chemical identification of NMs tend to involve the use of mass spectrometry, often hyphenated to separation techniques (with the exception of single particle ICP-MS (SP-ICP-MS) into which the sample is directly nebulized).

The separation techniques that are routinely coupled online to MS, and provide information regarding the size of the NM include size exclusion chromatography, liquid chromatography (LC), HDC, FFF and electrophoresis. For example, HDC coupled to MALDI-TOF-MS has been applied for the size separation and chemical analysis of liposome-type NMs dispersed in a beverage matrix,[69] and LC coupled to electrospray ionization (ESI)-MS has been used to separate and identify proteins and peptides in complex samples.[23,70,71]

Inorganic NMs in complex matrices were characterized in multiple studies by separation methods coupled to ICP-MS. For example, HDC-ICP-MS was used for detection of silica NMs in coffee creamer[10] as well as detection and size estimation of silver NMs in activated sludge.[39] Asymmetric flow FFF coupled to ICP-MS was successfully applied to the characterization of synthetic amorphous silica in tomato soup[29] and silver NMs in chicken paste.[30]

Recently there is an increasing interest in application of SP-ICP-MS for the characterization of NMs in foods. SP-ICP-MS was used to characterize silver NMs in chicken meat[30,72] and titanium dioxide E171 in food and consumer products.[73]

Regarding those size characterization techniques which do not have the ability to chemically classify NMs, because of their relatively low cost and ease of use, they are probably the most regularly used instruments for NM screening activities. However, due to the underlying measurement or detection principle, natural NMs occurring in food cannot be distinguished from intentionally added NMs using these nonspecific approaches, nor can mixtures of NMs. Nevertheless, where these methods are able to be usefully used is in the measurement of NMs in complex matrices where no other particles of similar size are expected. For example, Fede et al.[74] used dynamic light scattering (DLS) for the measurement of silica NMs in microbiological media composed of dissolved substances.[74] Where food samples spiked with NMs have been studied, the background can be monitored in the non-spiked sample and subsequently subtracted from the target NM, as measured in the spiked sample. This approach was used in a number of studies e.g. measurement of gold NMs in orange juice by nanoparticle tracking analysis (NTA),[75] and measurement of silver and silica NMs in coffee and milk by centrifugal liquid sedimentation (CLS).[76] From some of the nonselective separation methods, such as chromatography and FFF coupled to light scattering/absorption detectors, it is possible to collect sample fractions which then can be further analyzed offline e.g. by using EM- or MS-based techniques.[46,77]

Another size separation method potentially suitable for NMs in food matrices is GEMMA, although the collection of liquid samples for subsequent analysis with, e.g., MS is not possible. This is due to the fact that the liquid sample is transformed into an aerosol by electrospray ionization prior to size separation. However it is possible to equip the instrument with an electrostatic particle sampler that allows capturing of size-separated NMs straight onto an EM grid.[78] This way the captured NMs should allow EM imaging without particle overlap and agglomeration, a typical outcome of other EM sample preparation protocols, which is not desirable.

2.6 Quality Assurance and Control

In order to obtain reliable analytical data for the analysis of NMs in any matrix (including foodstuffs), the use of validated techniques is of utmost importance. A recent publication by Linsinger et al.[79] proposed a validation protocol for methods detecting and quantifying NMs in food (referring to concentration and size).[79] This protocol is based on standard procedures used for general chemical analysis (for example see Ref. 80) and suggests the establishment of following parameters: (1) limit of detection in terms of quantification (minimum amount of particles that need to be present) and size (smallest and biggest size that can be determined); (2) limit of quantification; (3) precision; (4) selectivity; (5) robustness; (6) trueness/recovery and (7) linearity/working range (e.g. upper and lower size limit). To date, the numbers of published studies on validation of analytical methods for characterization of NMs, especially in complex matrices, is limited.[81] An example is the validation of CLS and DLS for characterization of spherical NMs in pristine dispersions.[82] Method uncertainty regarding the size characterization of NMs has however been discussed, mainly relating to the reference materials of NMs in pristine dispersions,[36,83–85] although recently a study on reference material of synthetic amorphous silica in tomato soup has also been published.[29]

3. CHALLENGES ASSOCIATED WITH THE CHARACTERIZATION OF NMs IN FOOD

3.1 Measurement of NMs and Comparability of Analytical Methods

For a full characterization of an NM-containing sample, it is important to characterize not only the mean or median size of the NMs, but also the particle size distribution (PSD) weighted by number[25,86–88] rather than mass. The reason being that a decrease in particle diameter leads to a calculated increase in particle number concentration. Therefore, a number-weighted PSD is shifted towards smaller sizes compared to a mass-weighted PSD, and is therefore thought of as a more relevant approach for characterization of NMs (e.g. toxicologically).[89] Additionally, a recommendation on the definition of NMs, issued by the European Commission, relates to the measurement of the number-weighted PSD.[52] However, only few techniques, such as EM, NTA, SP-ICP-MS or GEMMA, are capable of directly measuring PSD based on particle number. Other techniques like FFF or HDC with ICP-MS detection measure PSD based on particle mass, whereas CLS and DLS measure an intensity-weighted PSD (based on detection of light scattering/absorption) which is then converted into a mass-based PSD. However, there is ongoing discussion regarding the conversion of mass- or intensity-based PSD into a number-based PSD.[90]

Another challenge in the measurement of the weighted-number PSD is the availability of reference materials. Certified reference materials for different sizes of NMs are commercially available. However, thus far, no materials certified for particle number, as well as size, can be found. These are obviously required for the determination of accuracy of any number-based PSD measurement. In view of nonavailability of these materials, many authors use confirmatory techniques for result validation, e.g. Refs 10,20,30,45,46,91. However, this approach is only valid if the NMs are composed of perfectly spherical particles[45,92–96] that are not aggregated/agglomerated or associated with food components, and have not undergone swelling or shrinkage whilst being measured under different conditions. (Note: the authors suspect that they observed swelling when comparing measurements made of a latex particle system by CLS and EM. It is suggested that the latex NMs must have swollen whilst in the electron beam, because the PSD measured by CLS appeared narrower in range compared to that measured by EM.[47]) Another area of potential confusion arises from the fact that analytical techniques assume that all NMs are spherical (equivalent spherical diameter), and base their measurement procedures (and the calculation parameters) upon this assumption (see Table 1 for different size definitions depending on analytical technique). This means that different PSDs may be generated from different instrumental techniques due to their underlying measurement principles, and can lead to a lack of comparability of results from two seemingly similar experiments.

Approaches that allow comparability of measurements are urgently required, especially for NMs that occur in an agglomerated/aggregated form, as some of these are already found in foods as a fraction of approved food additives, and so should therefore also be included in toxicity assays.[97] Recommendations and guidance should also refer to specific equivalent measurements, and perhaps even to specific instruments and sample preparation methods (in cases where definition of the NM size is critical). For example, current EC recommendations for the definition of NM may pose a challenge of classifying some materials because no specific equivalent spherical unit is referred to in the document.[30]

Another reason for the incomparability of sizes measured by different analytical methods is the limit of detection, in terms of NM size and also concentration. Most of the analytical methods cannot measure NMs smaller than a given size; however, the exact limits of detection for specific NM size are very rarely quoted in the literature. Nevertheless, without prior knowledge of this size, the interpretation of analytical output may lead to inaccurate result. Similarly a biased result will be obtained if a method has too high limit of detection for NMs composing element concentration and a part of PSD (especially risk for small-sized particles of very small weight) is cut out from the size distribution output. These limits of detection need to be known (and quoted) before commencing any analysis of NMs in foods or other complex matrices. Greater discussion on lower size detection limit of the available analytical techniques is provided in **Chapter 3**.

In order to obtain accurate values for detection limits (in terms of NM size) reference materials featuring a wide number-based PSD are required. To date, commercially available reference materials are solely certified for their mean size rather than a full number-based PSD. For this reason the International Organization for Standardization (Geneva, Switzerland) is developing a guidance document on how to produce a reference material featuring a broad PSD.[98] Such a reference material would be of great aid to analytical method validation.

3.2 Sample Preparation and Potentially Occurring Artefacts

With entities as labile as NMs, there is a strong possibility that either (or both) of the sample preparation stages will introduce analytical artefacts, and thus affect the accuracy of resulting data.[13] It has been widely shown that sample preparation, and for that matter also sampling, sample handling, transport and storage (e.g. time until sample is processed, exposure to light, temperature and pressure changes) can have an effect on the state of particles. For example, even a simple preparation step such as dilution, will affect the sample by altering the suspended solids content, dissolved substances, pH, ionic strength, aggregate size, etc. Changes in pH and ionic strength are likely to affect the state of NM agglomeration,[20,91,99] while limiting the amount of suspended solids will also result in the removal of associated NMs. Losses of NMs can also occur during preparation due to their sorption onto to container walls or filters.[49,100]

As different analytical methodologies will typically require the use of differing sample preparation protocols, it is unlikely that potential artefacts can be easily predicted prior to starting an investigation. It is expected that, as the research area matures and/or analytical data on the behaviour of NM certified reference materials in foods becomes more widely available, our understanding of how best to control artefact production will improve.

Therefore, when planning a study, it is recommended that sample preparation should be kept to minimum,[13,22,40] with every effort made regarding how to keep artefact production to an absolute minimum (given each analyte/matrix/detection technique combination). We would also recommend confirmation of the results by use of different sample preparation protocols or analytical methods (preferably one that expresses measurement in the same way as the first method of choice (see Table 1)).

4. CONCLUSIONS AND FUTURE TRENDS

The development of nanotechnology in food and food-related products is steadily increasing. However, this growth is hindered by a lack of understanding regarding the behaviour and fate (and potential risks) associated with the use of NMs in food and other consumer products. During their life cycle, NMs can change fundamentally, e.g. agglomerates or aggregates may be formed or broken up; particles may bind to other materials; coatings may degrade; or particle surface areas may change due to chemical reactions such as oxidation. Therefore there is a crucial time-and-life-stage aspect to NMs and their analysis, which needs much greater consideration.

To date, as no single analytical method can provide all of the necessary information regarding NMs in food, a combination of several methods and approaches is still strongly recommended.

Future effort should not just focus on the development of advanced detection and characterization techniques, but should also address the many aspects of sampling, sample handling and preparation.[34] As the science of nanometrics matures, areas such as quality assurance and quality control will inevitably become a priority in order to improve comparability and meaningfulness of data (a factor which is currently missing). Once all of the above have been achieved, it should be possible to provide legislators with accurate assessments of the risks posed by NMs, so that relevant and enforceable regulation of NMs in foods can be put into place, thereby reassuring both the food industry and the general public.[97]

REFERENCES

1. Chaudhry Q, Scotter M, Blackburn J, Byrony R, Boxall ABA, Castle L, et al. Applications and implications of nanotechnologies for the food sector. *Food Addit Contam A* 2008;**25**:241–58.
2. Chaudhry Q, Castle L. Food applications of nanotechnologies: an overview of opportunities and challenges for developing countries. *Trends Food Sci Technol* 2011;**22**:595–603.

3. Blasco C, Picó Y. Determining nanomaterials in food. *TrAC Trends Anal Chem* 2011;**30**:84–99.
4. Cushen M, Kerry J, Morris M, Cruz-Romero M, Cummins E. Nanotechnologies in the food industry – recent developments, risks and regulation. *Trends Food Sci Technol* 2012;**24**:30–46.
5. Mahdi SS, Vadood R, Nourdahr R. Study on the antimicrobial effect of nanosilver tray packaging of minced beef at refrigerator temperature. *Global Vet* 2012;**9**:284–9.
6. MRC. *A response from the MRC Human Nutrition Research Unit to the House of Lords Science and Technology Select Committee call for evidence on nanotechnologies and food*. 2009.
7. Haham M, Ish-Shalom S, Nodelman M, Duek I, Segal E, Kustanovich M, et al. Stability and bioavailability of vitamin D nanoencapsulated in casein micelles. *Food Funct* 2012;**3**:737–44.
8. Chen L, Bai G, Yang R, Zang J, Zhou T, Zhao G. Encapsulation of β-carotene within ferritin nanocages greatly increases its water-solubility and thermal stability. *Food Chem* 2014;**149**:307–12.
9. Weir A, Westerhoff P, Fabricius L, Hristovski K, von Goetz N. Titanium dioxide nanoparticles in food and personal care products. *Environ Sci Technol* 2012;**46**:2242–50.
10. Dekkers S, Krystek P, Peters RJB, Lankveld DP, Bokkers BG, van Hoeven-Arentzen PH, et al. Presence and risks of nanosilica in food products. *Nanotoxicology* 2010;**5**:393–405.
11. Gatti AM, Tossini D, Gambarelli A, Montanari S, Capitani F. Investigation of the presence of inorganic micro- and nanosized contaminants in bread and biscuits by environmental scanning electron microscopy. *Crit Rev Food Sci Nutr* 2009;**49**:275–82.
12. Bijl E, de Vries R, van Valenberg H, Huppertz T, van Hooijdonk A. Factors influencing casein micelle size in milk of individual cows: genetic variants and glycosylation of κ-casein. *Int Dairy J* 2014;**34**:135–41.
13. Dudkiewicz A, Luo P, Tiede K, Boxall ABA. Detecting and characterizing nanoparticles in food, beverages and nutraceuticals. *Nanotechnol Food Beverage Nutraceutical Ind* 2012: 53–80.
14. Hillyer JF, Albrecht RM. Gastrointestinal persorption and tissue distribution of differently sized colloidal gold nanoparticles. *J Pharm Sci* 2001;**90**:1927–36.
15. Liu W, Wu Y, Wang C, Li HC, Wang T, Liao CY, et al. Impact of silver nanoparticles on human cells: effect of particle size. *Nanotoxicology* 2010;**4**:319–30.
16. Kittler S, Greulich C, Diendorf J, Köller M, Epple M. Toxicity of silver nanoparticles increases during storage because of slow dissolution under release of silver ions. *Chem Mater* 2010;**22**:4548–54.
17. Lynch I, Dawson KA. Protein-nanoparticle interactions. *Nano Today* 2008;**3**:40–7.
18. Lynch I, Salvati A, Dawson KA. Protein-nanoparticle interactions: what does the cell see? *Nat Nanotechnol* 2009;**4**:546–7.
19. Mahmoudi M, Lynch I, Ejtehadi MR, Monopoli MP, Bombelli FB, Laurent S. Protein–nanoparticle interactions: opportunities and challenges. *Chem Rev* 2011;**111**:5610–37.
20. Peters R, Kramer E, Oomen AG, Rivera ZE, Oegema G, Tromp PC, et al. Presence of nano-sized silica during in vitro digestion of foods containing silica as a food additive. *ACS Nano* 2012;**6**:2441–51.
21. Walczak AP, Fokkink R, Peters R, Tromp P, Herrera Rivera ZE, Rietjens IM, et al. Behaviour of silver nanoparticles and silver ions in an in vitro human gastrointestinal digestion model. *Nanotoxicology* 2012;**7**:1198–210.

22. Tiede K, Boxall A, Tear S, Lewis J, David H, Hassellov M. Detection and characterization of engineered nanoparticles in food and the environment. *TFAC* 2008;**25**:795–821.
23. Luykx DMAM, Peters RJB, van Ruth SM, Bouwmeester H. A review of analytical methods for the identification and characterization of nano delivery systems in food. *J Agric Food Chem* 2008;**56**:8231–47.
24. Alger H, Momcilovic D, Carlander D, Duncan TV. Methods to evaluate uptake of engineered nanomaterials by the alimentary tract. *Compr Rev Food Sci Food Saf* 2014;**13**:705–29.
25. Calzolai L, Gilliland D, Rossi F. Measuring nanoparticles size distribution in food and consumer products: a review. *Food Addit Contam A* 2012;**29**:1183–93.
26. EFSA Scientific Committee. Guidance on the risk assessment of the application of nanoscience and nanotechnologies in the food and feed chain. *EFSA J* 2011;**9**(5). 2140:1–36.
27. Chen X-X, Cheng B, Yang Y-X, Cao A, Liu JH, Du LJ, et al. Characterization and preliminary toxicity assay of nano-titanium dioxide additive in sugar-coated chewing gum. *Small* 2013;**9**:1765–74.
28. Tadjiki S, Assemi S, Deering CE, Veranth JM, Miller JD. Detection, separation, and quantification of unlabeled silica nanoparticles in biological media using sedimentation field-flow fractionation. *J Nanopart Res* 2008;**11**:981–8.
29. Grombe R, Charoud-Got J, Emteborg H, Linsinger TPJ, Seghers J, Wagner S, et al. Production of reference materials for the detection and size determination of silica nanoparticles in tomato soup. *Anal Bioanal Chem* 2014;**406**:3895–907.
30. Loeschner K, Navratilova J, Købler C, Mølhave K, Wagner S, von der Kammer F, et al. Detection and characterization of silver nanoparticles in chicken meat by asymmetric flow field flow fractionation with detection by conventional or single particle ICP-MS. *Anal Bioanal Chem* 2013;**405**:8185–95.
31. Isaacson CW, Usenko CY, Tanguay RL, Field J. Quantification of fullerenes by LC/ESI-MS and its application to in vivo toxicity assays. *Anal Chem* 2007;**79**:9091–7.
32. Chen Z, Westerhoff P, Herckes P. Quantification of C60 fullerene concentrations in water. *Environ Toxicol Chem* 2008;**27**:1852–9.
33. Liu J, Chao J, Liu R, Tan ZQ, Yin YG, Wu Y, et al. Cloud point extraction as an advantageous preconcentration approach for analysis of trace silver nanoparticles in environmental waters. *Anal Chem* 2009;**81**:6496–502.
34. Peters RJB, ten Dam G, Bouwmeester H, Helspera H, Allmaier G, von der Kammer F, et al. Identification and characterization of organic nanoparticles in food. *TrAC* 2011;**30**:100–12.
35. Hinterwirth H, Wiedmer SK, Moilanen M, Lehner A, Allmaier G, Waitz T, et al. Comparative method evaluation for size and size-distribution analysis of gold nanoparticles. *J Sep Sci* 2013;**36**:2952–61.
36. Kaiser DL, Waters RL. *Report of investigation, reference material 8012 gold nanoparticles, 30 nm diameter.* 2007.
37. Uboldi C, Bonacchi D, Lorenzi G, Hermanns MI, Pohl C, Baldi G, et al. Gold nanoparticles induce cytotoxicity in the alveolar type-II cell lines A549 and NCIH441. *Part Fibre Toxicol* 2009;**6**:6–18.
38. Franklin NM, Rogers NJ, Apte SC, Batley GE, Gadd GE, Casey PS. Comparative toxicity of nanoparticulate ZnO, bulk ZnO, and $ZnCl_2$ to a freshwater microalga (*Pseudokirchneriella subcapitata*): the importance of particle solubility. *Environ Sci Technol* 2007;**41**:8484–90.
39. Tiede K, Boxall ABA, Wang X, Tiede D, Baxter M, David H, et al. Application of hydrodynamic chromatography-ICP-MS to investigate the fate of silver nanoparticles in activated sludge. *J Anal At Spectrom* 2010;**25**:1149.

40. Dudkiewicz A, Tiede K, Loeschner K, Soegaard Jensend LH, Jensen E, Wierzbicki R, et al. Characterization of nanomaterials in food by electron microscopy. *TrAC Trends Anal Chem* 2011;**30**:28–43.
41. Baalousha M, Prasad A, Lead JR. Quantitative measurement of the nanoparticle size and number concentration from liquid suspensions by atomic force microscopy. *Environ Sci Process Impacts* 2014;**16**:1338–47.
42. Zheng YZ, Webb R, Greenfield PF, Reid S. Improved method for counting virus and virus like particles. *J Virol Methods* 1996;**62**:153–9.
43. Lienemann C-P, Heissenberger A, Leppard GG, Perret D. Optimal preparation of water samples for the examination of colloidal material by transmission electron microscopy. *Aquat Microb Ecol* 1998;**14**:205–13.
44. Balnois E, Wilkinson KJ. Sample preparation techniques for the observation of environmental biopolymers by atomic force microscopy. *Colloids Surf A Physicochem Eng Aspects* 2002;**207**:229–42.
45. Baalousha M, Lead JR. Rationalizing nanomaterial sizes measured by atomic force microscopy, flow field-flow fractionation, and dynamic light scattering: sample preparation, polydispersity, and particle structure. *Environ Sci Technol* 2012;**46**:6134–42.
46. Tiede K, Boxall ABA, Tiede D, Tear S, David H, Lewis J. A robust size-characterization methodology for studying nanoparticle behaviour in 'real' environmental samples, using hydrodynamic chromatography coupled to ICP-MS. *J Anal At Spectrom* 2009;**24**:964.
47. Anderson W, Kozak D, Coleman VA, Jämtingb A, Traua M. A comparative study of submicron particle sizing platforms: accuracy, precision and resolution analysis of polydisperse particle size distributions. *J Colloid Interface Sci* 2013;**405**:322–30.
48. Boldridge D. Morphological characterization of fumed silica aggregates. *Aerosol Sci Technol* 2010;**44**:182–6.
49. Tiede K, Hassellov M, Breitbarth E, Chaudhry Q, Boxall AB. Considerations for environmental fate and ecotoxicity testing to support environmental risk assessments for engineered nanoparticles. *J Chromatogr A* 2009;**1216**:503–9.
50. Lari L, Dudkiewicz A. Sample preparation and EFTEM of meat samples for nanoparticle analysis in food. *J Phys Conf Ser* 2014;**522**:012057.
51. Song X, Li R, Li H, Hu Z, Mustapha A, Lin M. Characterization and quantification of zinc oxide and titanium dioxide nanoparticles in foods. *Food Bioprocess Technol* 2014;**7**:456–62.
52. The European Commission. Commission recommendation of 18 October 2011 on the definition of nanomaterial (text with EEA relevance). *Off J Eur Union* 2011;**275**:38–40.
53. Barthel H, Heinemann M, Stintz M, Wessely B. Particle sizes of fumed silica. *Part Part Syst Charact* 1999;**16**:169–76.
54. Lorenz C, Tiede K, Tear S, Boxall A, Von Goetz N, Hungerbühler K. Imaging and characterization of engineered nanoparticles in sunscreens by electron microscopy, under wet and dry conditions. *Int J Occup Env Heal* 2010;**16**:406–28.
55. Tiede K, Tear SP, David H, Boxall AB. Imaging of engineered nanoparticles and their aggregates under fully liquid conditions in environmental matrices. *Water Res* 2009;**43**:3335–43.
56. Luo P, Morrison I, Dudkiewicz A, Tiede K, Boyes E, O'Toole P, et al. Visualization and characterization of engineered nanoparticles in complex environmental and food matrices using atmospheric scanning electron microscopy. *J Microsc* 2013;**250**:32–41.
57. Noronha N, Duggan E, Ziegler G, Stapletonc JJ, O'Riordana ED, O'Sullivana M, et al. Comparison of microscopy techniques for the examination of the microstructure of starch-containing imitation cheeses. *Food Res Int* 2008;**41**:472.

58. Barcenas M, Rosell C. Effect of frozen storage time on the bread crumb and aging of par-baked bread. *Food Chem* 2006;**95**:438.
59. James BJ, Smith BG. Surface structure and composition of fresh and bloomed chocolate analysed using X-ray photoelectron spectroscopy, cryo-scanning electron microscopy and environmental scanning electron microscopy. *LWT Food Sci Technol* 2009;**42**:929–37.
60. Doucet FJ, Lead JR, Maguire L, Achterbergb EP, Millwardb GE. Visualisation of natural aquatic colloids and particles? A comparison of conventional high vacuum and environmental scanning electron microscopy. *J Environ Monitor* 2005;**7**:115.
61. Tuoriniemi J, Cornelsi G, Hassellov M. Improving the accuracy of single particle ICPMS for measurement of size distributions and number concentrations of nanoparticles by determining analyte partitioning during nebulisation. *J Anal At Spectrom* 2014;**29**: 743–52.
62. Qian H, Eckenhoff WT, Zhu Y, Pintauer T, Jin R. Total structure determination of thiolate-protected Au38 nanoparticles. *J Am Chem Soc* 2010;**132**:8280–1.
63. Jackson SN, Wang H-YJ, Woods AS, Ugarov M, Egan T, Schultz JA. Direct tissue analysis of phospholipids in rat brain using MALDI-TOFMS and MALDI-ion mobility-TOFMS. *J Am Soc Mass Spectrom* 2005;**16**:133–8.
64. Buse J, Badea I, Verrall RE, El-Aneed A. Tandem mass spectrometric analysis of the novel gemini surfactant nanoparticle families G12-s and G18:1-s. *Spectrosc Lett Int J Rapid Commun* 2010;**43**:447.
65. Müller R, Allmaier G. Molecular weight determination of ultra-high mass compounds on a standard matrix-assisted laser desorption/ionization time-of-flight mass spectrometer: PAMAM dendrimer generation 10 and immunoglobulin M. *Rapid Commun Mass Spectrom* 2006;**20**:3803–6.
66. Singh G, Stephan C, Westerhoff P, Carlander D, Duncan TV. Measurement methods to detect, characterize, and quantify engineered nanomaterials in foods. *Compr Rev Food Sci Food Saf* 2014;**13**:693–704.
67. Dehalu V, Weigel S, Rebe S, Grombe R, Löbenberg R, Delahaut P. Production and characterization of antibodies against crosslinked gelatin nanoparticles and first steps toward developing an ELISA screening kit. *Anal Bioanal Chem* 2012;**403**:2851–7.
68. Rebe Raz S, Leontaridou M, Bremer MGEG, Peters R, Weigel S. Development of surface plasmon resonance-based sensor for detection of silver nanoparticles in food and the environment. *Anal Bioanal Chem* 2012;**403**:2843–50.
69. Helsper JPFG, Peters RJB, Brouwer L, Simpson JT. Characterization and quantification of liposome-type nanoparticles in a beverage matrix using hydrodynamic chromatography and MALDI-TOF mass spectrometry. *Anal Bioanal Chem* 2013;**405**:1181–9.
70. Hufnagel P, Rabus R. Mass spectrometric identification of proteins in complex post-genomic projects. Soluble proteins of the metabolically versatile, denitrifying '*Aromatoleum*' sp. strain EbN1. *J Mol Microbiol Biotechnol* 2006;**11**:53–81.
71. Carstens MG, van Nostrum CF, Verrijk R, de Leede LG, Crommelin DJ, Hennink WE. A mechanistic study on the chemical and enzymatic degradation of PEG-Oligo(epsilon-caprolactone) micelles. *J Pharm Sci* 2008;**97**:506–18.
72. Peters RJB, Rivera ZH, van Bemmel G, Marvin HJ, Weigel S, Bouwmeester H. Development and validation of single particle ICP-MS for sizing and quantitative determination of nano-silver in chicken meat. *Anal Bioanal Chem* 2014;**406**:3875–85.
73. Peters RJB, van Bemmel G, Herrera-Rivera Z, Helsper H, Marvin H, Weigel S, et al. Characterization of titanium dioxide nanoparticles in food products: analytical methods to define nanoparticles. *J Agric Food Chem* 2014;**62**:6285–93.

74. Fede C, Selvestrel F, Compagnin C, Mognato M, Mancin F, Reddi E, Celotti L. The toxicity outcome of silica nanoparticles (Ludox®) is influenced by testing techniques and treatment modalities. *Anal Bioanal Chem* 2012;**404**:1789–802.
75. Gallego-Urrea JA, Tuoriniemi J, Hassellöv M. Applications of particle-tracking analysis to the determination of size distributions and concentrations of nanoparticles in environmental, biological and food samples. *TrAC Trends Anal Chem* 2011;**30**:473–83.
76. Lozano O, Mejia J, Tabarrant T, Masereel B, Dogné JM, Toussaint O, et al. Quantification of nanoparticles in aqueous food matrices using particle-induced X-ray emission. *Anal Bioanal Chem* 2012;**403**:2835–41.
77. Von Der Kammer F, Legros S, Larsen EH. Separation and characterization of nanoparticles in complex samples (food/environment) by field flow fractionation. *TrAC* 2011;**30**:425–36.
78. Allmaier G, Laschober C, Szymanski WW. Nano ES GEMMA and PDMA, new tools for the analysis of nanobioparticles-protein complexes, lipoparticles, and viruses. *J Am Soc Mass Spectrom* 2008;**19**:1062–8.
79. Linsinger TPJ, Chaudhry Q, Dehalu V, Delahaut P, Dudkiewicz A, Grombe R, et al. Validation of methods for the detection and quantification of engineered nanoparticles in food. *Food Chem* 2013;**138**:1959–66.
80. Boque R, Maroto A, Riu J, Rius X. Validation of analytical methods. *Grasas Aceites* 2002;**53**:128–43.
81. Szakal C, Roberts SM, Westerhoff P, Bartholomaeus B, Buck N, Illuminato I, et al. Measurement of nanomaterials in foods: integrative consideration of challenges and future prospects. *ACS Nano* 2014;**8**:3128–35.
82. Braun A, Couteau O, Franks K, Kestens V, Roebben G, Lamberty A. Validation of dynamic light scattering and centrifugal liquid sedimentation methods for nanoparticle characterization. *Adv Powder Technol* 2011;**22**:766–70.
83. Braun A, Kestens V, Franks K, Roebben G, Lamberty A, Linsinger TP. A new certified reference material for size analysis of nanoparticles. *J Nanopart Res* 2012;**14**:1–12.
84. Small JA, Waters RL. *Report of investigation, reference material 8011 gold nanoparticles, 10 nm diameter.* 2012.
85. Kaiser DL, Waters RL. *Report of investigation, reference material 8013 gold nanoparticles, 60 nm diameter.* 2007.
86. Hagendorfer H, Kaegi R, Parlinska M, Sinnet B, Ludwig C, Ulrich A. Characterization of silver nanoparticle products using asymmetric flow field flow fractionation with a multidetector approach – a comparison to transmission electron microscopy and batch dynamic light scattering. *Anal Chem* 2012;**84**:2678–85.
87. Laborda F, Jiménez-Lamana J, Bolea E, Castillo JR. Critical considerations for the determination of nanoparticle number concentrations, size and number size distributions by single particle ICP-MS. *J Anal At Spectrom* 2013;**28**:1220–32.
88. Baalousha M, Lead JR. Nanoparticle dispersity in toxicology. *Nat Nanotechnol* 2013;**8**:308–9.
89. Yada RY, Buck N, Canady R, DeMerlis C, Duncan T, Janer G, et al. Engineered nanoscale food ingredients: evaluation of current knowledge on material characteristics relevant to uptake from the gastrointestinal tract. *Compr Rev Food Sci Food Saf* 2014;**13**:730–44.
90. Linsinger TPJ, Roebben G, Gilliland D, Calzolai L, Francois R, Gibson P, et al. *Requirements on measurements for the implementation of the European Commission definition of the term 'nanomaterial'*. JRC Refernce Reports. 2012. EUR 25404 EN:1–56.
91. Park S, Woodhall J, Ma G, Veinot JG, Cresser MS, Boxall AB. Regulatory ecotoxicity testing of engineered nanoparticles: are the results relevant to the natural environment? *Nanotoxicology* 2013:1–10.

92. Kasper G. Dynamics and measurement of smokes. Size characterization of nonspherical particles. *Aerosol Sci Technol* 1982;**1**:187–99.
93. Naito M, Hayakawa O, Nakahira K, Mori H, Tsubaki J. Effect of particle shape on the particle size distribution measured with commercial equipment. *Powder Technol* 1998;**100**:52–60.
94. Bowen P. Particle size distribution measurement from millimeters to nanometers and from rods to platelets. *J Dispersion Sci Technol* 2002;**23**:631–62.
95. DeCarlo PF, Slowik JG, Worsnop DR, Davidovits P, Jimenez JL. Particle morphology and density characterization by combined mobility and aerodynamic diameter measurements. Part 1: theory. *Aerosol Sci Technol* 2004;**38**:1185–205.
96. Domingos RF, Baalousha MA, Ju-Nam Y, Reid MM, Tufenkji N, Lead JR, et al. Characterizing manufactured nanoparticles in the environment: multimethod determination of particle sizes. *Environ Sci Technol* 2009;**43**:7277–84.
97. EFSA. *Input from EFSA for consideration during the revision of the commission recommendation.* 2012.
98. Draft ISO/WD 14411-2. Preparation of particulate reference materials – part 2: polydisperse spherical particles, in preparation.
99. Jiang J, Oberdörster G, Biswas P. Characterization of size, surface charge, and agglomeration state of nanoparticle dispersions for toxicological studies. *J Nanopart Res* 2008;**11**:77–89.
100. Bonard J-M, Stora T, Salvetat J-P, Maier F, Stöckli T, Duschl C, et al. Purification and size-selection of carbon nanotubes. *Adv Mater* 1997;**9**:827–31.

Index

Note: Page numbers followed by "f" and "t" indicate figures and tables respectively.

E

N

O

P

T

U

V

W

X

Z

FSC
www.fsc.org
MIX
Papier | Fördert
gute Waldnutzung
FSC® C083411

Kolibri 360 GmbH
Ferdinand-Jühlke-Straße 7
99095 Erfurt, Deutschland
produktsicherheit@kolibri360.de